PROBABILITY
AND STATISTICS
FOR SCIENCE AND ENGINEERING
WITH EXAMPLES IN R

BY HONGSHIK AHN

cognella® ACADEMIC PUBLISHING

Bassim Hamadeh, CEO and Publisher
Carrie Montoya, Manager, Revisions and Author Care
Kaela Martin, Project Editor
Christian Berk, Associate Production Editor
Miguel Macias, Senior Graphic Designer
Alexa Lucido, Licensing Associate
Natalie Piccotti, Senior Marketing Manager
Kassie Graves, Vice President of Editorial
Jamie Giganti, Director of Academic Publishing

Cover image copyright© by Depositphotos / ikatod.

Printed in the United States of America.

ISBN: 978-1-5165-3110-3 (pbk) / 978-1-5165-3111-0 (br)

Contents

Preface

This book is designed for a one-semester course in probability and statistics, specifically for students in the natural science or engineering. It is also suitable for business and economics students with a calculus background. The text is based on my past teachings over the course of many years at Stony Brook University. Most existing textbooks contain topics to fulfill at least one whole year of instruction, and therefore, they are excessive for a one semester course in probability and statistics. The purpose of this book is to cover just the necessary topics for a one-semester course, thus reducing the typical volume of a textbook and lowering the financial burden on students.

This book provides examples of how to use the R software to obtain summary statistics, calculate probabilities and quantiles, find confidence intervals, and conduct statistical testing. In addition to using distribution tables, students can calculate probabilities of various distributions using their smartphones or computers. Since R is available under an open-source license, everyone can download it and use it free of charge.

In the second edition, the new Section 1.4 on relationship between two variables has been added to the first addition. In Chapter 8 and Chapter 9, flowcharts have been added for identifying the appropriate test methods for population means. Also, 27% more exercise problems have been added, and some problems in the first edition have been modified.

This book is organized as follows: Chapter 1 covers descriptive statistics, Chapter 2 through Chapter 5 cover probability and distributions, Chapter 6 provides concepts about sampling, and Chapter 7 through 9 cover estimations and hypothesis testing. Hypothesis testing can be overwhelming for students, due to the inundating formulas needed for numerous cases. To help students understand and identify the correct formula to use, a comprehensive table for each type of test is provided. As mentioned earlier, flowcharts have been added in the second edition. Students can easily follow these tables to choose the appropriate confidence intervals and statistical tests. A summary is given at the end of each chapter. Distribution tables for various distributions are provided in the appendix.

I wish to thank Hyojeong Son for carefully reviewing the first edition, checking answers to the exercise problems, and creating the online tools; David Saltz for contributing some examples and exercise problems; Chelsea Kennedy for proofreading the preliminary edition and checking the solutions to the exercise problems; Yan Yu for proofreading the preliminary edition; Jerson Cochancela for proofreading the manuscript, contributing an exercise problem, and providing helpful suggestions; and Mingshen Chen for checking the solutions to the exercise problems for the preliminary edition. I also thank Cognella for inviting me to write this book.

Hongshik Ahn
Department of Applied Mathematics and Statistics
Stony Brook University

Describing Data

1. Display of Data by Graphs and Tables

There are various ways to describe data. In this section, we study how to organize and describe a set of data using graphs and tables.

A. FREQUENCY DISTRIBUTIONS

A *frequency distribution* is a table that displays the frequency of observations in each interval in a sample. To build a frequency distribution, we need the following steps.

Basic Steps
1. Find the minimum and the maximum values in the data set.
2. Determine class intervals: intervals or cells of equal length that cover the range between the minimum and the maximum without overlapping
   e.g., minimum 0, maximum 100: [0, 10), [10, 20), ..., [90, 100]
3. Find frequency: the number of observations in the data that belong to each class interval. Let's denote the frequencies as $f_1, f_2, \cdots$.
4. Find relative frequency:

$$\frac{\text{Class frequency}}{\text{Total number of observations}}$$

The relative frequencies are denoted as $f_1/n, \ f_2/n, \cdots$ if the total sample size is n.

EXAMPLE 1.1 Midterm scores of an introductory statistics class of 20 students are given below.

69 84 52 93 81 74 89 85 88 63 87 64 67 72 74 55 82 91 68 77

We can construct a frequency table as shown in Table 1.1.

TABLE 1.1   Frequency table for Example 1.1

Class interval	Tally	Frequency	Relative frequency
50–59	||	2	2/20 = 0.10
60–69	┼┼┼	5	5/20 = 0.25
70–79	||||	4	4/20 = 0.20
80–89	┼┼┼ ||	7	7/20 = 0.35
90–99	||	2	2/20 = 0.10
Total		20	1.00

There is no gold standard in selecting class intervals, but a rule of thumb is an integer near $\sqrt{n}$ for the number of classes.

B. HISTOGRAM

A *histogram* is a pictorial representation of a frequency distribution. Figure 1.1 is a histogram obtained from the frequency distribution in Example 1.1.

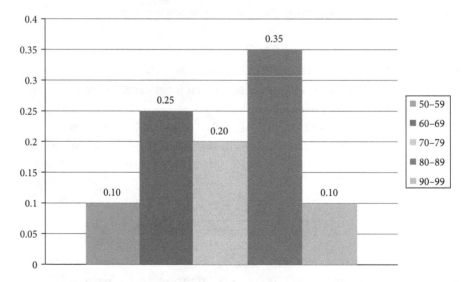

FIGURE 1.1   Histogram of the data in Example 1.1.

Figure 1.1 used the relative frequency as the height of the bar in each class. The histogram adequately visualizes the frequency distribution of the data. We expect that the class with a longer bar would have a higher count. However, the histogram may not appropriately display the frequency distribution and thus may mislead the data interpretation when we use the height as the relative frequency if the interval lengths are not equal. To avoid this, we can divide the relative frequency by the interval length for each class. Then the area of each bar becomes the relative frequency of the class, and thus the total area of the histogram becomes 1. This is necessary when the interval lengths are different. The height of this histogram is obtained as follows:

$$\text{Height} = \frac{\text{Relative frequency}}{\text{Width of the interval}}$$

For Example 1.1, the height of each bar in the histogram is:

$$
\begin{aligned}
\text{Height} &= \frac{\text{Relative frequency}}{\text{Width of the interval}} \\
&= \frac{0.10}{10} = 0.010 \text{ for } [50, 60) \\
&= \frac{0.25}{10} = 0.025 \text{ for } [60, 70) \\
&= \frac{0.20}{10} = 0.020 \text{ for } [70, 80) \\
&= \frac{0.35}{10} = 0.035 \text{ for } [80, 90) \\
&= \frac{0.10}{10} = 0.010 \text{ for } [90, 100)
\end{aligned}
$$

A histogram shows the shape of a distribution. Depending on the number of peaks, a distribution can be called *unimodal* (one peak), *bimodal* (two peaks) or *multimodal* (multiple peaks). A distribution can be *symmetric* or *skewed*. A skewed distribution is asymmetrical with a longer tail on one side. A distribution with a longer right tail is called skewed to the right (right skewed or positively skewed), and a distribution with a longer left tail is called skewed to the left (left skewed or negatively skewed). Figure 1.2 displays some typical shapes of distributions.

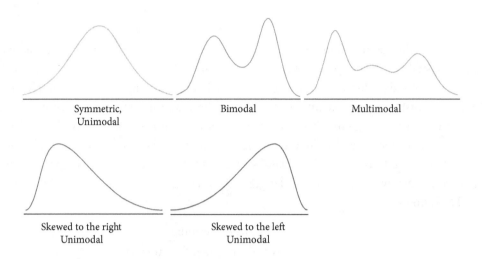

FIGURE 1.2   Shapes of distributions.

EXAMPLE 1.2   The following are the midterm exam scores of a probability and statistics course in a past semester at Stony Brook University.

30, 34, 38, 44, 45, 46, 47, 48, 50, 50, 51, 52, 53, 53, 53, 54, 55, 55, 55, 56, 56, 57, 57, 58, 58, 59, 59, 60, 60, 60, 60, 61, 61, 62, 62, 62, 62, 63, 63, 63, 63, 63, 63, 63, 64, 64, 65, 65, 65, 65, 65, 66, 66, 67, 67, 67, 68, 68, 68, 68, 68, 69, 69, 69, 69, 69, 69, 69, 70, 70, 70, 70, 70, 70, 71, 71, 71, 72, 72, 73, 73, 73, 73, 73, 73, 73, 73, 73, 74, 74, 74, 75, 75, 75, 76, 76, 76, 76, 76, 76, 77, 77, 77, 77, 77, 78, 78, 78, 78, 78, 79, 79, 79, 80, 80, 80, 80, 81, 81, 81, 81, 82, 82, 82, 82, 82, 83, 83, 83, 83, 84, 84, 84, 84, 84, 84, 84, 84, 85, 85, 86, 86, 87, 87, 87, 87, 88, 88, 88, 88, 88, 88, 89, 89, 89, 89, 89, 90, 90, 90, 91, 92, 93, 93, 94, 94, 94, 94, 95, 95, 95, 95, 96, 96, 96, 97, 97, 98, 98, 99, 100

A histogram of the above data can be obtained using R statistical software. R is a programming language and software for statistical analysis. It is freely available and can be downloaded from the Internet. You can read the data file as:

>midterm=read.csv("filename.csv")

or enter the data on R as

>midterm=c(30,34,38,...,100)

Here, > is the cursor in R. To read a data file from your computer, it must be a *comma separated values* file with extension .csv. You may need to list the directories containing the file, such as:

>midterm=read.csv("c:\\Users***\\filename.csv")

Here, *** is the name(s) of subdirectory (or subdirectories). Using the command "hist" as below,

>hist(midterm)

we obtain the histogram given in Figure 1.3.

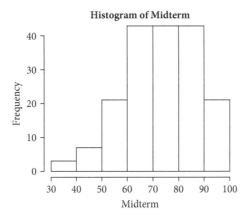

FIGURE 1.3   Histogram of the data in Example 1.2 generated by R.

C. STEM-AND-LEAF PLOT

A *stem-and-leaf plot* displays data in a graphical format, similar to a histogram. Unlike a histogram, a stem-and-leaf plot retains the original data and puts the data in order. Thus, a stem-and-leaf plot provides more details about the data than a histogram. A stem-and-leaf plot consists of two columns separated by a vertical line. The left column containing the leading digit(s) is called the *stem*, and the right column containing the trailing digit(s) is called the *leaf*. Figure 1.4 shows the shape of a stem-and-leaf plot.

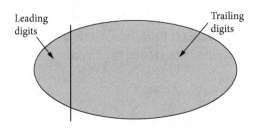

FIGURE 1.4   Shape of a stem-and-leaf plot.

To construct a stem-and-leaf plot, one or two leading digits are listed for the stem values. The trailing digit(s) become the leaf. The trailing digits in each row of the leaf are arranged in ascending order. Steps for constructing a stem-and-leaf plot are given below.

Basic Steps
1. List one or more leading digits for the stem values.
2. The trailing digit(s) become the leaves.
3. Arrange the trailing digits in each row so they are in increasing order.

EXAMPLE 1.3   Final examination scores of 26 students in an introductory statistics course are given below.

55 61 94 94 69 77 68 54 85 77 92 92 81 73 69 81 75 84 70 81 81 89 59 72 82 62

The following is a stem-and-leaf plot for the above data.

5	549
6	19892
7	773502
8	51141192
9	4422

$\rightarrow$

5	459
6	12899
7	023577
8	11112459
9	2244

Using R, a stem-and-leaf plot for the above data can be obtained by

```
>a=c(55,61,94,94,69,77,68,54,85,77,92,92,81,73,69,81,75,84,70,81,81,89,59,72,82,62)
```

```
>stem(a)
```

D. DOT DIAGRAM

The data in Example 1.3 can be displayed using a dot diagram, as shown in Figure 1.5.

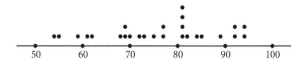

FIGURE 1.5   Dot diagram for the data in Example 1.3.

EXAMPLE 1.4   Heights of students (in inches) in a varsity wrestling team are given below.

67.2 65.0 72.5 71.1 69.1 69.0 70.2 68.2 68.5 71.3

67.5 68.6 73.1 71.3 69.4 65.5 69.5 70.8 70.0 69.2

A stem-and-leaf plot can have the tens digit or the first two digits in stem, but the latter will display the distribution more efficiently, as shown below.

65	05				65	05
66					66	
67	25				67	25
68	256				68	256
69	10452		→		69	01245
70	280				70	028
71	133				71	133
72	5				72	5
73	1				73	1

Figure 1.6 is a dot diagram of the above data.

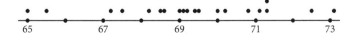

FIGURE 1.6   Dot diagram for the data in Example 1.4.

EXAMPLE 1.5   We can compare distributions of two sets of data using side-by-side stem-and-leaf plots. The same examination is given to two classes. The scores of the two classes are given below.

Class A: 78 80 60 74 85 100 51 60 40 67 100 90 58 40 89 100

Class B: 42 76 37 57 93 60 55 47 51 95 81 53 52 65 95

The following side-by-side stem-and-leaf plots compare the score distributions of the two classes.

Class A		Class B
	3	7
00	4	27
18	5	12357
007	6	05
48	7	6
059	8	1
0	9	355
000	10	

The above plots show that Class A performed better than Class B in general.

A histogram can be built for qualitative (categorical) data.

EXAMPLE 1.6   A frequency distribution of the enrollment of four classes in a high school is given in the following table.

Class	Frequency	Relative frequency
Algebra	26	0.26
English	30	0.30
Physics	19	0.19
Biology	24	0.24
Total	99	0.99

Note that the total of the relative frequencies is 0.99. This is due to a rounding error. The above table can be visualized using the bar graph in Figure 1.7.

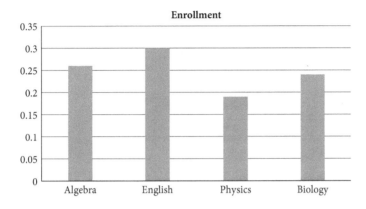

FIGURE 1.7   Bar graph for the enrollment data in Example 1.6.

The data can also be displayed using the pie chart given in Figure 1.8. In a pie chart, a circle is divided into slices according to the proportion of each group. The angle of each slice is obtained by (class frequency / sample size) $\times 360°$. It is equivalent to the relative frequency multiplied by $360°$. For the above data, the central angle of each slide is obtained as follows:

Algebra: $(26/99) \times 360° = 94.5°$

English: $(30/99) \times 360° = 109.1°$

Physics: $(19/99) \times 360° = 69.1°$

Biology: $(24/99) \times 360° = 87.3°$

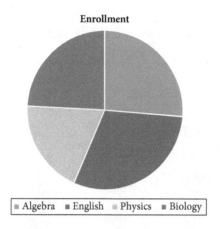

Enrollment

▪ Algebra ▪ English ▪ Physics ▪ Biology

FIGURE 1.8   Pie chart for the enrollment data in Example 1.6.

Using R, a pie chart for the data in Example 1.6 can be obtained by
> slices=c(26,30,19,24)
> lbls=c("Algebra","English","Physics","Biology")
> pie(slices,labels=lbls,main="Enrollment")

E. TIME PLOTS

A time plot is a plot of observations against time or the order in which they are observed. Time plots can show the following:

- Trends: increasing or decreasing, changes in location of the center, or changes in variation
- Seasonal variation or cycles: fairly regular increasing or decreasing movements

The following table shows the number of workers in a company who arrived late to work in the morning during a four-week period. Figure 1.9 is a time plot for the data. You can see a clear weekly pattern from the plot. More workers were late on Mondays than any other day of the week in general. On Fridays, the number of workers who were late decreased throughout the four-week period.

Week	Monday	Tuesday	Wednesday	Thursday	Friday
1	6	3	2	4	7
2	8	0	5	3	2
3	7	2	1	0	2
4	5	0	1	0	1

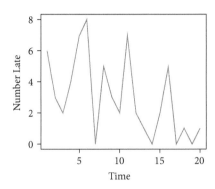

FIGURE 1.9   Time plot for the number of workers who were late for work.

Using R, the time plot in Figure 1.9 can be obtained by
>Number_late=c(6,3,2,4,7,8,0,5,3,2,7,2,1,0,2,5,0,1,0,1)
>ts.plot(Number_late)

2. *Measures of Central Tendency*

We often describe a set of data using a single value that identifies the central position within the range of the data. This is called a measure of central tendency. Three widely used measures of central tendency are mode, mean, and median. These three measures are explained below.

A. MODE

The *mode* is the number that occurs most often. For example, the mode is 81 in the examination score data in Example 1.3, 71.3 in the height data in Example 1.4, and English in the enrollment data in Example 1.6.

B. MEAN

The *mean* is the average of the numbers in the sample. It can be obtained by the sum of all the values in the data divided by the sample size. Let a data set consists of n observations: $x_1, x_2, \cdots, x_n$. Then the sample mean is:

$$\bar{x} = \frac{x_1 + x_2 + \cdots + x_n}{n} = \frac{1}{n} \sum_{i=1}^{n} x_i$$

EXAMPLE 1.7   The weights of five 7th grade girls (in pounds) are given below.

122   94   135   111   108

The mean weight of 114 pounds is obtained as follows:

$$\bar{x} = \frac{122 + 94 + 135 + 111 + 108}{5} = \frac{570}{5} = 114$$

A dot diagram for the data showing the sample mean is given in Figure 1.10.

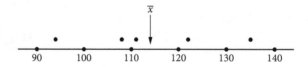

FIGURE 1.10   Dot diagram for the data in Example 1.7

The sample mean $\bar{x}$ represents the average value of the observations in a sample. We estimate the population mean (average of all the values in the population) using $\bar{x}$. The population mean is denoted by μ. The value of $\bar{x}$ using one digit of decimal accuracy beyond what is used in the individual x_i is acceptable.

C. MEDIAN

A sample *median* is the middle value of the ordered data set. Let's denote the sample median as $\tilde{x}$. An algorithm for obtaining the median is given below.

Algorithm for obtaining the median:

a.   Order data

Let $x_{(i)}$ denote the i-th smallest observation.

$$x_1, x_2, \cdots, x_n \rightarrow x_{(1)} \le x_{(2)} \le \cdots \le x_{(n)}$$

Here, $x_{(i)}$ denotes the i-th smallest observation. For example, the weight data in Example 1.7 is ordered as follows:

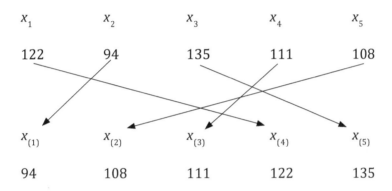

b.   i) If n is odd, take the middle value. It can be obtained as follows:

$$\tilde{x} = x_{\left(\frac{n+1}{2}\right)}$$

For the data in Example 1.7, $n = 5$ is odd. The median is the middle value:

94   108 $\boxed{111}$ 122   135

The sample median is: $\tilde{x} = x_{\left(\frac{n+1}{2}\right)} = x_{\left(\frac{5+1}{2}\right)} = x_{(3)} = 111$.

ii) If n is even, take the average of the two middle values. It can be obtained as follows:

$$\tilde{x} = \frac{x_{\left(\frac{n}{2}\right)} + x_{\left(\frac{n}{2}+1\right)}}{2}$$

EXAMPLE 1.8 The number of days that the first six heart transplant patients at Stanford University Hospital survived after their operations are given below.

15   3   46   623   126   64

Since $n = 6$ is even, we sort the data and find the middle two values as follows:

3   15 | 46   64 | 126   623

The sample median is the average of the middle two values:

$$\tilde{x} = \frac{46 + 64}{2} = 55$$

It can also be obtained as follows:

$$\tilde{x} = \frac{x_{\left(\frac{n}{2}\right)} + x_{\left(\frac{n}{2}+1\right)}}{2} = \frac{x_{\left(\frac{6}{2}\right)} + x_{\left(\frac{6}{2}+1\right)}}{2} = \frac{x_{(3)} + x_{(4)}}{2} = \frac{46 + 64}{2} = 55$$

The sample mean is:

$$\bar{x} = \frac{15 + 3 + 46 + 623 + 126 + 64}{6} = 146.2$$

Only one patient lived longer than 146.2 days. Thus, the sample median is a better indicator in this case.

The sample mean is greatly influenced by outlying values, while the sample median is not influenced by outlying values. If the distribution is skewed, then the mean will be closer to the longer tail of the distribution than the median. This is because the extreme values influence the mean. The mean is larger than the median if the distribution is skewed to the right. Typically, income or salary data are skewed to the right. The mean is smaller than the median if the distribution is skewed to the left. The mean is the same as the median if the distribution is symmetric. Therefore, the sample mean is appropriate as a central tendency for a bell-shaped population, and the median is appropriate for skewed data. Figure 1.11 illustrates these relationships.

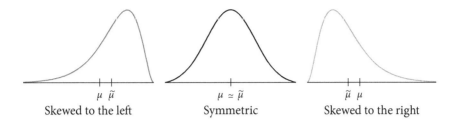

$$\mu \ \tilde{\mu} \qquad\qquad \mu \simeq \tilde{\mu} \qquad\qquad \tilde{\mu} \ \mu$$

| Skewed to the left | Symmetric | Skewed to the right |

FIGURE 1.11   The relationship between the mean and the median. Here, the population median is denoted as $\tilde{\mu}$.

The sample mean of grouped data can be obtained the same way as above. In a data set with k classes, let x_i be the class mark of the i-th class, f_i the corresponding class frequency. Then the sample mean is obtained as follows.

$$\bar{x} = \frac{\sum_{i=1}^{k} f_i x_i}{n}$$

EXAMPLE 1.9   A sample of professional baseball players' annual income is given below.

Income (in $1,000)	Number of players
30	180
50	100
100	50
500	20
1,000	10
2,000	15
3,000	15
5,000	10
Total	400

The sample mean is:

$$\bar{x} = \frac{\sum_{i=1}^{k} x_i f_i}{n} = \frac{180 \times 30,000 + 100 \times 50,000 + \cdots + 10 \times 5,000,000}{400} = \frac{160,400,000}{400}$$
$$= 401,000$$

the sample median is:

$$\tilde{x} = \frac{x_{\left(\frac{400}{2}\right)} + x_{\left(\frac{400}{2}+1\right)}}{2} = \frac{x_{(200)} + x_{(201)}}{2} = \frac{50{,}000 + 50{,}000}{2} = 50{,}000$$

The mode is 30,000.

Only 70 out of 400 baseball players in this data set earned more than the mean of $401,000 per year. As shown in Figure 1.12, the distribution is skewed to the right. For these data, the sample median or mode are better representative values of professional baseball players' annual income than the sample mean.

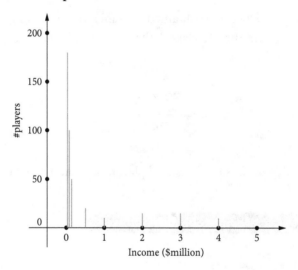

FIGURE 1.12   Professional baseball players' annual income.

In R, you can find the mean and median of data "a" as

```
>mean(a)
```

```
>median(a)
```

respectively.

3. *Measures of Variation*

A. VARIANCE AND STANDARD DEVIATION

A measure of location (or measure of central tendency) cannot give a complete summary of data. We also want to know how all the observations spread out. The measure that provides information about how much the data are dispersed is called the measure of variation. An intuitive choice of the measure of variation is the sample deviation. A deviation is defined as observation – sample mean, which is $x - \bar{x}$. For a sample containing observations $x_1, x_2, \cdots, x_n$, the i-th deviation is $x_i - \bar{x}$. However, the sample deviation is:

$$\sum_{i=1}^{n}(x_i - \bar{x}) = \sum_{i=1}^{n} x_i - \sum_{i=1}^{n} \bar{x} = n\bar{x} - n\bar{x} = 0$$

Because the sample deviation is always zero, this measure is useless. This problem can be resolved if we use the absolute deviations: $|x_1 - \bar{x}|, \cdots, |x_n - \bar{x}|$. The sample absolute deviation defined as $\sum_{i=1}^{n} |x_i - \bar{x}|$ is not zero. However, this measure leads to theoretical difficulties. Another choice of the measure of variation is the squared deviation, defined as $(x_1 - \bar{x})^2, \cdots, (x_n - \bar{x})^2$. The squared deviation is widely used for the measure of variation because it has nice statistical properties. The *sample variance* is obtained using the squared deviations. The sample variance is defined as follows:

$$s^2 = \frac{\sum_{i=1}^{n}(x_i - \bar{x})^2}{n - 1} \tag{1-1}$$

The sample standard deviation is the square root of the sample variance:

$$s = \sqrt{\frac{\sum_{i=1}^{n}(x_i - \bar{x})^2}{n - 1}}$$

Note that the divisor in the sample variance is $n - 1$ instead of n. This is because the x_i tend to be closer to $\bar{x}$ than μ. Thus, $n - 1$ is used as the divisor to compensate for this. Using n tends to underestimate σ^2.

THEOREM 1.1   The sample variance can also be written as the following:

$$s^2 = \frac{\sum_{i=1}^{n} x_i^2 - \left(\sum_{i=1}^{n} x_i\right)^2 \Big/ n}{n-1} \qquad (1\text{-}2)$$

PROOF

$$s^2 = \frac{\sum_{i=1}^{n}(x_i - \bar{x})^2}{n-1} = \frac{\sum_{i=1}^{n}(x_i^2 - 2x_i\bar{x} + \bar{x}^2)}{n-1} = \frac{\sum_{i=1}^{n} x_i^2 - 2\bar{x}\sum_{i=1}^{n} x_i + n\bar{x}^2}{n-1}$$

$$= \frac{\sum_{i=1}^{n} x_i^2 - 2n\bar{x}^2 + n\bar{x}^2}{n-1} = \frac{\sum_{i=1}^{n} x_i^2 - n\bar{x}^2}{n-1} = \frac{\sum_{i=1}^{n} x_i^2 - \left(\sum_{i=1}^{n} x_i\right)^2 \Big/ n}{n-1}$$

Calculation of the sample variance is usually simpler with the above formula than the formula in (1-1). As shown in the proof, (1-2) is equivalent to:

$$s^2 = \frac{\sum_{i=1}^{n} x_i^2 - n(\bar{x})^2}{n-1}$$

EXAMPLE 1.10   Find the sample variance of the following data:

6 12 6 6 4 8

Observation	x_i	x_i^2
1	6	36
2	12	144
3	6	36
4	6	36
5	4	16
6	8	64
Total	$\sum x_i = 42$	$\sum x_i^2 = 332$

The sample variance is:

$$s^2 = \frac{\sum_{i=1}^{n} x_i^2 - \left(\sum_{i=1}^{n} x_i\right)^2 / n}{n-1} = \frac{332 - 42^2/6}{6-1} = 7.6$$

The sample standard deviation is:

$$s = \sqrt{7.6} = 2.76$$

The sample variance of grouped data can be obtained the same way as above. In a data set with k classes, let x_i be the class mark of the i-th class, f_i the corresponding class frequency. Then the sample variance is obtained as follows.

$$s^2 = \frac{\sum_{i=1}^{k} f_i(x_i - \bar{x})^2}{n-1} = \frac{\sum_{i=1}^{k} f_i x_i^2 - \left(\sum_{i=1}^{k} f_i x_i\right)^2 / n}{n-1}$$

In Example 1.9, the sample variance is

$$s^2 = \frac{\sum_{i=1}^{k} f_i x_i^2 - \frac{\left(\sum_{i=1}^{k} f_i x_i\right)^2}{n}}{n-1}$$

$$= \frac{(180 \times 30,000^2 + 100 \times 50,000^2 + \cdots + 10 \times 5,000,000^2) - \frac{(160,400,000)^2}{400}}{400-1}$$

$$= 993,963,909,774$$

and the standard deviation is

$$s = \sqrt{993,963,909,774} = 996,977.$$

In R, the variance and standard deviation of data "a" can be obtained as

>var(a)

>sd(a)

respectively. The standard deviation can also be obtained by

>sqrt(var(a))

B. PERCENTILES

Another measure of variation is the sample range, which is the difference between the maximum and the minimum observations. Detailed information about the measure of location and measure of variation can be obtained using ordered data. Percent ranks of a data set are called percentiles. The $100p$-th *percentile* is defined as the value such that at least $100p\%$ of the observations are at or below this value, and at least $100(1 - p)\%$ are at or above this value.

> Sample $100p$-th percentile: value such that at least $100p\%$ of the observations are at or below this value, and at least $100(1 - p)\%$ are at or above this value.

An algorithm for finding the sample $100p$-th percentile is given below.
1. Order the n observations from smallest to largest.
2. Find np.
3. If np is <u>not</u> an integer, round it up to the next integer and find the corresponding ordered value.

For example, if $np = 3.4$, then round it up to 4 and take the fourth-smallest observation $x_{(4)}$.

If np is an integer, say k, then calculate:

$$\frac{x_{(k)} + x_{(k+1)}}{2}$$

Quartiles are frequently-used percentiles. Quartiles are the points of division into quarters. The first quartile, or the lower quartile, is the 25th percentile, which is denoted as Q_1. The second quartile, or the 50th percentile, is the median, which is denoted as Q_2. The third quartile, or the upper quartile, is the 75th percentile, which is denoted as Q_3.

First quartile (Q_1): 25th percentile
Second quartile (Q_2): 50th percentile, median
Third quartile (Q_3): 75th percentile

C. BOXPLOTS

A *boxplot*, or box-and-whisker plot, is a graphical display of data through their quartiles, as shown in Figure 1.13.

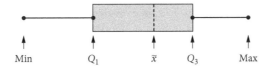

FIGURE 1.13   A boxplot.

The boxplot is based on the five number summary: minimum observation, maximum observation, first quartile, third quartile, and median. The sample *range* ($x_{(n)} - x_{(1)}$) is denoted by connecting the minimum and maximum values by a line. We define the *interquartile range* (IQR) as the difference between the third quartile and the first quartile ($Q_3 - Q_1$). The interquartile range is shown by a box starting with the first quartile and ending with the third quartile in a boxplot. This box shows the length of the middle half in the distribution. The median is identified inside this box using dashed lines. There are methods to identify outliers (or extreme values) using the quartiles. An observation is called an *outlier* if it is either less than $Q_1 - 1.5(\text{IQR})$ or greater than $Q_3 + 1.5(\text{IQR})$.

Sample range: $x_{(n)} - x_{(1)} = \text{max} - \text{min}$
Interquartile range: $Q_3 - Q_1$
Outlier: an observation that is less than $Q_1 - 1.5(\text{IQR})$ or greater than $Q_3 + 1.5(\text{IQR})$

EXAMPLE 1.11 The final examination scores of a statistics course are given below.

81 57 85 84 99 90 69 76 76 83

To find the quartiles, we order the data as follows:

57 69 76 76 81 83 84 85 90 99

Since the sample size of $n = 10$ is even, the median is the average of the two middle values. It can be obtained as follows:

$$\tilde{x} = Q_2 = \frac{x_{\left(\frac{n}{2}\right)} + x_{\left(\frac{n}{2}+1\right)}}{2} = \frac{x_{(5)} + x_{(6)}}{2} = \frac{81 + 83}{2} = 82$$

Because the median is the 50th percentile, it can also be calculated using $np = 10(0.5) = 5$. Since 5 is an integer, the median is the average of the fifth and sixth ordered observations. This is the value we obtained above. The first quartile and the third quartile are obtained as shown below.

First quartile: $np = 10(0.25) = 2.5 \uparrow 3, \qquad Q_1 = x_{(3)} = 76$

Third quartile: $np = 10(0.75) = 7.5 \uparrow 8, \qquad Q_3 = x_{(8)} = 85$

Alternatively, the first quartile can also be obtained as the median of the first half: 57, 69, 76, 76, 81. Likewise, the third quartile can be obtained as the median of the second half: 83, 84, 85, 90, 99. The same results are obtained from both approaches.

The minimum is 57, and the maximum is 99. The sample range is $99 - 57 = 42$, and the interquartile range is $Q_3 - Q_1 = 85 - 76 = 9$. Figure 1.14 shows the boxplot for these data. The plot shows that it is skewed to the left. The first quarter of the data is sparse, and the third quarter (between the median and the third quartile) is dense.

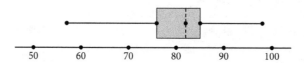

FIGURE 1.14   Boxplot for the data in Example 1.11.

To find outliers, we calculate:

$$Q_1 - 1.5(IQR) = 76 - 1.5(9) = 62.5$$

$$Q_3 + 1.5(\text{IQR}) = 85 + 1.5(9) = 98.5$$

Because $57 < 62.5$ and $99 > 98.5,$ 57 and 99 are outliers.

EXAMPLE 1.12 The final examination scores from a year ago for the same course in Example 1.11 are given below.

78 99 47 53 71 69 60 57 45 88 59

The scores are ordered as follows:

45 47 53 57 59 60 69 71 78 88 99

Because the sample size of $n = 11$ is odd, the median is the middle value. It can be obtained as follows:

$$\tilde{x} = Q_2 = x_{\left(\frac{n+1}{2}\right)} = x_{\left(\frac{12}{2}\right)} = x_{(6)} = 60$$

Alternatively, it can also be calculated using $np = 11(0.5) = 5.5,$ which is rounded up to 6. Hence, the median is the sixth ordered observation. The first quartile and the third quartile are obtained as shown below.

First quartile: $np = 11(0.25) = 2.75 \uparrow 3, \qquad Q_1 = x_{(3)} = 53$

Third quartile: $np = 11(0.75) = 8.25 \uparrow 9, \qquad Q_3 = x_{(9)} = 78$

The first quartile can also be obtained as the median of the first half: 45, 47, 53, 57, 59. Likewise, the third quartile can be obtained as the median of the second half: 69, 71, 78, 88, 99. The same results are obtained from both approaches. The boxplot for these data is given in Figure 1.15. This boxplot shows that the data are skewed to the right. The last quarter of the data is sparse and the second quarter (between the first quartile and the median) is dense.

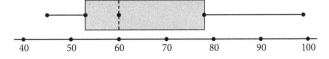

FIGURE 1.15   Boxplot for the data in Example 1.12.

Let's find the 90th percentile. We have $np = 11(0.9) = 9.9$, which is rounded up to 10. The 90th percentile is the 10th ordered value, which is $x_{(10)} = 88$. The minimum is 45, and the maximum is 99. The sample range is $99 - 45 = 54$, and the interquartile range is $Q_3 - Q_1 = 78 - 53 = 25$. To find outliers, we calculate:

$$Q_1 - 1.5(\text{IQR}) = 53 - 1.5(25) = 15.5$$

$$Q_3 + 1.5(\text{IQR}) = 78 + 1.5(25) = 115.5$$

Because none of the scores in the data is less than 15.5 or greater than 115.5, there are no outliers.

Side-by-side boxplots are used when distributions of two or more data sets are compared. Figure 1.16 displays side-by-side boxplots comparing quality indices of products manufactured at four plants. We find that Plant 2 shows the largest variation among the four. It needs to reduce its variability. Overall, Plants 2 and 4 manufacture lower-quality products than Plants 1 and 3. Both Plant 2 and Plant 4 need to improve their quality levels. Products from Plants 1 and 3 show high overall quality with small variations.

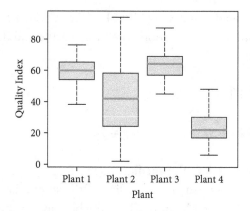

FIGURE 1.16   Side-by-side boxplots comparing qualities of products from four plants.

In R, the $100p$-th percentile of data "a" can be obtained as

```
>quantile(a, p)
```

More than one percentile, say n percentiles, can be obtained as

>quantile(a, c(p1, p2, …,pn))

For the data in Example 1.11, the quartiles can be obtained as

>a=c(81,57,85,84,99,90,69,76,76,83)

>quantile(a, c(0.25, 0.50, 0.75))

and the output is given below.

  25%   50%   75%

76.00 82.00 84.75

Note that the third quartile is slightly different from the 85 that we obtained in Example 1.11. This is because R uses a slightly different algorithm. The minimum and maximum can be obtained by

>min(a)

>max(a)

respectively. They can also be obtained by

>range(a)

The output of this is

57 99

which are the minimum and the maximum of the data, respectively. The interquartile range can be obtained by

>IQR(a)

Make sure that you use capital letters for "IQR."

Summary data can be obtained by

>summary(a)

The output is

Min.	1st Qu.	Median	Mean	3rd Qu.	Max.
57.00	76.00	82.00	80.00	84.75	99.00

A boxplot can be obtained by

>boxplot(a)

Side-by-side boxplots of a and b can be obtained by

>boxplot(a, b)

We can obtain the side-by-side boxplots of the data in Example 1.11 and Example 1.12 after entering the values of the data from Example 1.12 to b as below in addition to the existing data set a.

>b=c(78,99,47,53,71,69,60,57,45,88,59)

>boxplot(a, b)

The plots are shown in Figure 1.17. Here, the first boxplot denotes the outliers as points.

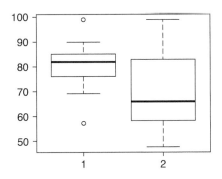

FIGURE 1.17   Side-by-side boxplots
generated by R for the data
in Examples 1.11 and 1.12.

The boxplot in Figure 1.16 is obtained using R as follows:

>boxplot(a1,a2,a3,a4,ylab="Quality index",xlab="Plant",names=c("Plant 1","Plant 2","Plant
3","Plant 4"))

Here a1 contains the data from Plant 1, a2 contains the data from Plant 2, a3 contains the data
from Plant 3, and a4 contains the data from Plant 4. Alternatively, you can combine the data
arrays in a matrix, say "a," and obtain the same plot by

>boxplot(a,ylab="Quality index",xlab="Plant",names=c("Plant 1","Plant 2","Plant 3","Plant 4"))

The y label is obtained by ylab="Quality index", and the x label is obtained by xlab="Plant".
The names of the plants are obtained by "names=c("Plant 1","Plant 2","Plant 3","Plant 4"). The
labels and variable names for other types of plots, such as histograms or time plots, can be
obtained similarly.

If the distribution is a bell-shaped curve, then roughly 68% of the data fall within one standard
deviation from the mean, 95% of the data fall within two standard deviations, and 99.7% of
the data fall within three standard deviations. This is called the *empirical rule*. This will be
further discussed in Chapter 4.

4. *Relationship Between Two Variables*

When we want to measure the relationship between two variables, *covariance* and *correlation coefficient* are used. Covariance measures how two variables are related. A positive covariance indicates that a variable tends to increase when the other variable increases, and a negative covariance indicates that a variable tends to decrease when the other variable increases. Zero covariance indicates that the change of a variable is random when the other variable increases or decreases. Figure 1.18 illustrates the relationship between two variables and the covariance.

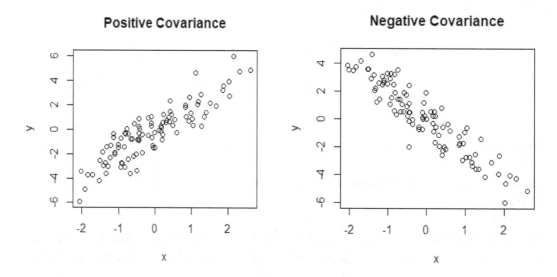

FIGURE 1.18   Covariance corresponding to each scatter plot.

A scatter plot (see Figure 1.18) of n pairs of observations $(x_1, y_1), (x_2, y_2), \cdots, (x_n, y_n)$ provides a rough idea about how the two variables are related. To measure the relationship of the variables, we use the following formula for sample covariance:

$$Cov(x, y) = \frac{\sum_{i=1}^{n} (x_i - \bar{x})(y_i - \bar{y})}{n - 1} \tag{1-3}$$

EXAMPLE 1.13 The following table shows the initial speed (miles per hour) and stopping distance (feet) of a car.

Initial speed	11	22	32	41	51
Stopping distance	8.2	32.8	82.0	144.4	236.2

Find the covariance between the initial speed and the stopping distance.

Here, $\bar{x} = 31.4$ and $\bar{y} = 100.72$. The covariance can be obtained as

i	x_i	y_i	$x_i - \bar{x}$	$y_i - \bar{y}$	$(x_i - \bar{x})(y_i - \bar{y})$
1	11	8.2	−20.4	−92.52	1887.41
2	22	32.8	−9.4	−67.92	638.45
3	32	82.0	0.6	−18.72	−11.23
4	41	144.4	9.6	43.68	419.33
5	51	236.2	19.6	135.48	2655.41
Total					5589.36

Thus,

$$Cov(x, y) = \frac{\sum_{i=1}^{n}(x_i - \bar{x})(y_i - \bar{y})}{n-1} = \frac{5589.36}{5-1} = 1397.34$$

THEOREM 1.2 The sample covariance in (1-3) can also be written as the following:

$$Cov(x, y) = \frac{\sum_{i=1}^{n} x_i y_i - n\bar{x}\,\bar{y}}{n-1} \tag{1-4}$$

PROOF

$$Cov(x, y) = \frac{\sum_{i=1}^{n}(x_i - \bar{x})(y_i - \bar{y})}{n-1} = \frac{\sum_{i=1}^{n}(x_i y_i - x_i \bar{y} - y_i \bar{x} + \bar{x}\bar{y})}{n-1}$$

$$= \frac{\sum_{i=1}^{n}x_i y_i - \bar{y}\sum_{i=1}^{n}x_i - \bar{x}\sum_{i=1}^{n}y_i + n\bar{x}\bar{y}}{n-1} = \frac{\sum_{i=1}^{n}x_i y_i - n\bar{x}\bar{y}}{n-1}$$

Calculation of the sample covariance is usually simpler with the above formula than the formula in (1-3).

Let's calculate the sample variance for the data given in Example 1.13 using the formula given in (1-4). Since $\sum_{i=1}^{n}x_i y_i = 21,402.4$, we get

$$Cov(x, y) = \frac{\sum_{i=1}^{n}x_i y_i - n\bar{x}\bar{y}}{n-1} = \frac{21402.4 - 5(31.4)(100.72)}{5-1} = 1397.34$$

which is the same as the value we obtained in Example 1.13.

Covariance describes how two variables are related. It indicates whether the variables are positively or negatively related. However, the covariance cannot be an efficient measure of the relationship between two variables, because it is also affected by the magnitude of the variables. If the magnitudes of the two variables are large, then the covariance may be large even if they are not highly correlated. If the magnitudes of the two variables are very small, then the covariance cannot be large even if they are perfectly correlated. To obtain a measure of relationship with a standard unit of measurement, we use the *correlation coefficient*. The correlation coefficient is a scaled version of covariance. The correlation coefficient ranges from −1 to 1. The correlation coefficient is obtained by dividing covariance by standard deviations of the two variables. If the correlation coefficient is one, the two variables have a perfect positive correlation. If the correlation coefficient is −1, the two variables have a perfect negative correlation. If the correlation coefficient is zero, then there is no relationship between the two variables.

Figure 1.19 displays scatter plots of 100 points each. Plot (a) shows a strong positive correlation indicating that y tends to increase as x increases, and plot (b) shows a strong negative correlation indicating that y tends to decrease as x increases. Plots (c) and (d) show correlation near zero. In plot (c), y appears to move randomly as x changes. Plot (d) shows a pattern that y has a quadratic relationship with x. However, it does not show a linear relationship. The actual correlation coefficients are (a) 0.87, (b) −0.90, (c) 0.012, and (d) is 0.023.

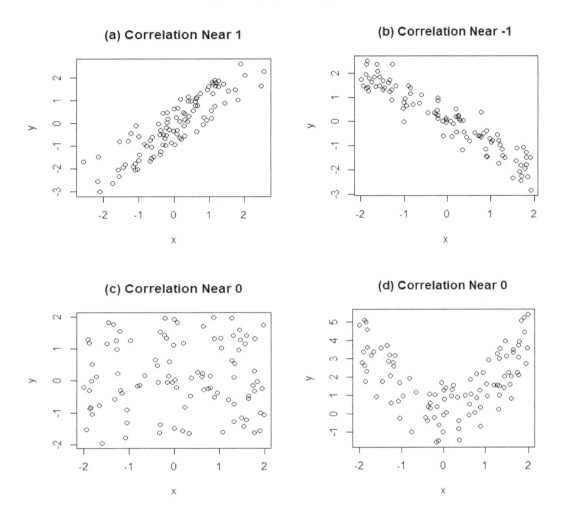

FIGURE 1.19   Correlation corresponding to each scatter plot.

The sample correlation coefficient is defined as follows:

$$r = \frac{Cov(x, y)}{s_x s_y}$$

where s_x and s_y are the sample standard deviations of x and y, respectively.

EXAMPLE 1.14 Let's calculate the sample correlation coefficient of the data given in Example 1.13. The sample standard deviations are:

$$s_x = \sqrt{\frac{\sum_{i=1}^{n} x_i^2 - n(\bar{x})^2}{n-1}} = \sqrt{\frac{5,911 - 5(31.4^2)}{5-1}} = 15.66$$

$$s_y = \sqrt{\frac{\sum_{i=1}^{n} y_i^2 - n(y)(\bar{y})^2}{n-1}} = \sqrt{\frac{84,508.88 - 5(100.72^2)}{5-1}} = 91.91$$

Thus, the sample correlation coefficient is

$$r = \frac{Cov(x, y)}{s_x s_y} = \frac{1,397.34}{(15.66)(91.91)} = 0.97$$

This implies that the initial speed and stopping distance of a car are highly correlated.

EXAMPLE 1.15 The following table shows the midterm and final exam scores of a statistics class of 20 students.

Student	1	2	3	4	5	6	7	8	9	10	11	12	13	14	15	16	17	18	19	20
Midterm	82	80	88	64	69	79	97	75	98	88	93	87	86	83	85	84	98	100	77	77
Final	81	55	22	57	89	87	82	85	78	89	91	76	81	80	85	80	83	99	39	52

A scatter plot of the exam scores is given in Figure 1.20.

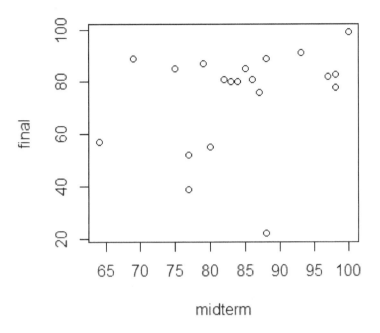

FIGURE 1.20   Scatter plot of the midterm and the final exam
scores given in Example 1.15.

Scores of the two exams do not appear to be highly correlated according to the above plot.
Some students who didn't perform well on the midterm got high scores on the final, while
some students who got high scores on the midterm did poorly on the final. Let's find the
correlation coefficient of the midterm and final exam scores.

$$Cov(x, y) = \frac{\sum_{i=1}^{n} x_i y_i - n\bar{x}\,\bar{y}}{n-1} = \frac{127{,}112 - 20(84.50)(74.55)}{20-1} = 59.08$$

$$s_x = \sqrt{\frac{\sum_{i=1}^{n} x_i^2 - n(\bar{x})^2}{n-1}} = \sqrt{\frac{144{,}594 - 20(84.5^2)}{20-1}} = 9.70$$

$$s_y = \sqrt{\frac{\sum_{i=1}^{n} y_i^2 - n(\bar{y})^2}{n-1}} = \sqrt{\frac{118{,}321 - 20(74.55^2)}{20-1}} = 19.42$$

Hence, the correlation coefficient is

$$r = \frac{Cov(x, y)}{s_x s_y} = \frac{59.08}{(9.70)(19.42)} = 0.31$$

In R, the scatter plot is obtained by

>plot(x, y, main="(title of the plot)")

The covariance is obtained by

>cov(x, y)

and the correlation is obtained by

>cor(x, y)

if the variable names are "x" and "y".

For Example 1.15, the scatter plot is obtained by

>plot(midterm, final, main="Exam Scores")

SUMMARY OF CHAPTER 1

1. Frequency Distribution, Basic steps:
   a. Find the minimum and the maximum values in the data set.
   b. Determine class intervals.
   c. Find the frequency in each interval.
   d. Relative frequency: (class frequency) / (total number of observations)
2. Histogram: A pictorial representation of a frequency distribution.
3. Stem-and-Leaf Plot: List one or more leading digits for the stem values. The trailing digit(s) becomes the leaves.
4. Sample Mean: $\bar{x} = \sum_{i=1}^{n} x_i / n$

5. Sample Median: Middle value.

   Calculating the median:

   a.   Order the data.

   b.   If n is odd, take the middle value.

   If n is even, take the average of the two middle values.

6. Mode: The number that occurs most often.

7. Possible Shapes of a Distribution

   a.   Symmetric: mean = median

   b.   Skewed to the left: mean < median

   c.   Skewed to the right: mean > median

8. Sample Variance:

$$s^2 = \frac{\sum_{i=1}^{n}(x_i - \bar{x})^2}{n-1} = \frac{\sum_{i=1}^{n}x_i^2 - \left(\sum_{i=1}^{n}x_i\right)^2 / n}{n-1}$$

9. Sample Standard Deviation: $s = \sqrt{s^2}$

10. Percentile: The points for division into hundreds.

    Calculating the sample $100p$-th percentile:

    a.   Order the n observations from smallest to largest.

    b.   Find np.

    c.   If np is not an integer, round it up to the next integer and find the corresponding ordered value.

    If np is an integer, say k, calculate $(x_{(k)} + x_{(k+1)})/2$.

11. Quartiles: The points for division into quarters.

    a.   First quartile (Q_1): 25th percentile

    b.   Second quartile (Q_2): 50th percentile = median

    c.   Third quartile (Q_3): 75th percentile

12. Sample Range: $x_{(n)} - x_{(1)} = \text{max} - \text{min}$.

    Interquartile Range (IQR) $= Q_3 - Q_1$

    Outlier: An observation that is less than $Q_1 - 1.5(\text{IQR})$ or greater than $Q_3 + 1.5(\text{IQR})$.

13. Boxplot: A display of data based on the five number summary: minimum, Q_1, median, Q_3, maximum.

14. Covariance:

$$Cov(x, y) = \frac{\sum_{i=1}^{n}(x_i - \bar{x})(y_i - \bar{y})}{n-1} = \frac{\sum_{i=1}^{n}x_i y_i - n\bar{x}\bar{y}}{n-1}$$

15. Correlation Coefficient:

$$r = \frac{Cov(x, y)}{S_x S_y}$$

EXERCISES

1.1 The following are ages of 62 people who live in a certain neighborhood:

2, 5, 6, 12, 14, 15, 15, 16, 18, 19, 20, 22, 23, 25, 27, 28, 30, 32, 33, 35, 36, 36, 37, 38, 39, 40, 40, 41, 42, 43, 43, 44, 44, 45, 45, 46, 47, 47, 48, 49, 50, 51, 56, 57, 58, 59, 59, 60, 62, 63, 65, 65, 67, 69, 71, 75, 78, 80, 82, 84, 90, 96

a. Display the data in a frequency table.
b. Display the data in a histogram.
c. Describe the shape of the distribution.
d. Display the data using a stem-and-leaf plot.

1.2 Illustrate the following heights (in inches) of 10 people:

64 66 67 68 69 70 70 71 71 74

with
a. a histogram
b. a stem-and-leaf plot

1.3   The following figure shows a histogram of 56 observations generated by R.

Each class interval includes the left endpoint but not the right.

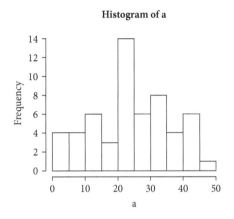

Histogram of a

a.   Find the relative frequency of each interval.
b.   Find the class interval containing the median.
c.   Describe the shape of the distribution.
d.   Which of the following can be the sample mean?
     (i) 16.50   (ii) 24.64   (iii) 33.50   (iv) 46.00

1.4   Use the histogram of 90 observations generated by R given below to answer the following questions.

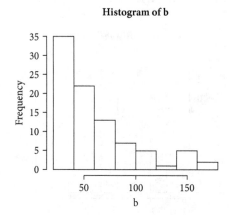

a.   Match the name in the left column with the correct number in the right column.

sample median	57.5
sample mean	37.7
sample standard deviation	67.3

b.   Describe the shape of the distribution.

1.5   Below are weights (in pounds) of male athletes in a high school.

Weight	Frequency
130–150	13
150–170	18
170–190	23
190–210	32
210–230	11

a.   Draw a histogram.
b.   Draw a pie chart.
c.   Find the class interval containing the median.

1.6   Following are the weights of 12 young goats, in pounds:

56, 32, 60, 59, 74, 65, 44, 51, 58, 51, 66, 49

    a.   Illustrate the observations with a frequency table.
    b.   Illustrate the observations with a histogram.
    c.   Compute the sample mean and the sample variance.
    d.   What proportion of the measurements lie in the interval $\bar{x} \pm 2s$?

1.7   The following are the ages of a sample of students: {9, 8, 12, 5, 3, 6, 13}.
    a.   Find the sample mean.
    b.   Find the sample standard deviation.

1.8   Cathy has obtained a sample mean of 0.8 and $\sum_{i=1}^{n} x_i^2 = 25$, but she cannot remember if the sample size was 30 or 40. She accidentally erased the data file. How can she figure out the correct sample size?

1.9   Compute the mean and standard deviation of the following grouped data.

x_i	8	12	16	20	24	28	32
Frequency	2	9	13	24	16	8	2

1.10   The monthly average New York City temperatures in Fahrenheit are given below.

Month	Jan	Feb	Mar	Apr	May	Jun	Jul	Aug	Sep	Oct	Nov	Dec
High	38	42	50	61	71	79	84	83	75	64	54	43
Low	27	29	35	45	54	64	69	68	61	50	42	28

    a.   Compute the sample variance of the high temperatures using the formula (1-1).
    b.   Compute the sample variance of the high temperatures using the formula (1-2).
    c.   Draw a time plot of the average high temperatures.
    d.   Compute the sample variance of the low temperatures using the formula (1-1).
    e.   Compute the sample variance of the low temperatures using the formula (1-2).
    f.   Draw a time plot of the average low temperatures.

1.11   Midterm exam scores in a math class are given below. Find the following percentiles:

81 89 69 72 91 58 69 66 60 67 95 83 84 68 53 76 63 74 72 68
79 81 81 86 72 79 83 73 58 73 81 77 92 87 48 49 89 88 97 80

a. 25th percentile
b. 50th percentile
c. 64th percentile
d. 75th percentile
e. 82nd percentile
f. 90th percentile

1.12   The weights (in pounds) of six people are: 113, 127, 131, 149, 174, 248. Compute the following sample statistics for the given weights:
a. median
b. 70th percentile
c. mean
d. variance and standard deviation
e. range

1.13   All fifth-grade children on Long Island are given an examination on mathematical achievement. A random sample of 50 fifth graders are selected. Their examination results are summarized in the following stem-and-leaf plot.

6	5
6	
6	9
7	011
7	
7	4
7	66677777
7	888889
8	0001111111
8	333
8	44445
8	667
8	89
9	01
9	33
9	455

a.   Compute the sample median.
b.   Compute the sample lower quartile and upper quartile.
c.   Compute the sample mean.
d.   Compute the sample standard deviation.
e.   What proportion of the measurements lie in the interval $\bar{x} \pm 2s$?
f.   Find the sample range and the interquartile range.
g.   Compute the 90th percentile.
h.   Construct a boxplot.
i.   Are there outliers?

1.14   The following stem-and-leaf plot shows scores on a statistics final exam.

2	9
3	38
4	3688
5	116
6	14678
7	7889
8	234556699
9	001347
10	0

a. Compute the sample median and quartiles.
b. Compute the sample mean and sample variance.
c. Compute the 66th percentile.
d. If a grade of B was given to the students with scores between the 36th percentile and the 72nd percentile (inclusive), find the number of students who received B's.
e. How many students obtained scores above the 80th percentile?
f. Compute the sample range and the interquartile range.
g. Construct a boxplot.
h. Find outliers, if there are any.

1.15   The final exam scores given in Example 1.2 are displayed using the stem-and-leaf plot.

```
 3 | 04
 3 | 8
 4 | 4
 4 | 5678
 5 | 00123334
 5 | 55566778899
 6 | 0000112222333333344
 6 | 555556677788888999999999
 7 | 0000001112233333333333444
 7 | 5556666667777788888999
 8 | 000011112222233333444444444
 8 | 55667777888888899999
 9 | 00012334444
 9 | 555566677889
10 | 0
```

a.   Describe the shape of the distribution.

b.   Which one is greater? The mean or the median?

c.   The instructor decided to give an A to the 80th percentile or above. What is the lowest score with an A grade?

d.   Construct a boxplot.

1.16   The average working hours of full-time office workers in a week in 97 countries are given below.

Number of working hours per week	Number of countries
33	1
34	2
35	4
36	5
37	7
38	12
39	15
40	18
41	13
42	9
43	6
44	3
45	2

a.   Find the sample median and quartiles.
b.   Compute the sample mean and standard deviation.
c.   Find the 90th percentile.
d.   Find outliers, if there are any.

1.17   Octane levels for various gasoline blends are given below:

87.9 84.2 86.9 87.7 91.7 88.8 95.3 93.5 94.3 88.1 90.2 91.4 91.3 93.9

a.   Construct a stem-and-leaf plot by using the tens digit as the stem.
b.   Construct a stem-and-leaf plot by using the first two digits (tens and ones digits) as the stem.
c.   Which of the above stem-and-leaf plots describes the distribution more efficiently?
d.   Draw a dot diagram.

1.18   The following are heights of female students in inches from a college in 2018.

67 67 67 60 68 64 69 71 67 67 66 63 67 62 66 70 67 61 68 67 68 64 69 67 70
72 61 67 69 68 69 72 66 67 66 67 69 64 64 63 68 66 65 60 70 65 68 66 61 65

    a.   Compute the median and the quartiles.
    b.   Find the range and interquartile range.
    c.   Construct a boxplot.
    d.   Draw a dot diagram.
    e.   Is there an outlier?

1.19   The following are amounts of total snowfall (in inches) in different northeastern cities in the United States in a certain year.

24 39 7 48 16 29 34 20 43 18 12 19 22 27 29 10 37 16 23 32

    a.   Construct a stem-and-leaf plot.
    b.   Compute the sample mean and standard deviation.
    c.   Compute the sample median and quartiles.
    d.   Construct a boxplot.
    e.   Are there outliers?

1.20   The following are the amounts of radiation received at a greenhouse.

6.4 7.2 8.5 8.9 9.1 10.0 10.1 10.2 10.6 10.8 11.0 11.2 11.3 11.4 12.0 12.3 13.2

    a.   Construct a stem-and-leaf plot.
    b.   Compute the sample mean and sample standard deviation.
    c.   Compute the sample median and quartiles.
    d.   Compute the 70th percentile.
    e.   Find the sample range and interquartile range.
    f.   Construct a boxplot.
    g.   Draw a dot diagram.

1.21 In a study of a parasite in humans and animals, researchers measured the lengths (in μm) of 90 individual parasites of certain species from the blood of a mouse. The measures are shown in the following table.

Length	19	20	21	22	23	24	25	26	27	28	29
Frequency	1	2	11	9	13	15	13	12	10	2	2

    a. Find the sample median and quartiles.
    b. Compute the sample mean and sample standard deviation.
    c. What percentage of the observations fall within one standard deviation of the sample mean?
    d. Compute the sample range and interquartile range.
    e. Find the 85th percentile.

1.22 The following side-by-side boxplots display the first and second midterm scores of an introductory statistics course.

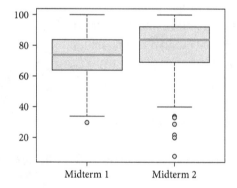

    a. Compare the two distributions by describing the shapes.
    b. Overall, did the students perform better on the second midterm?
    c. Which exam has a larger difference between the mean and the median? For this exam, is the mean or median larger? Why?
    d. Which exam has a larger standard deviation of the scores?

1.23   With the data given in Example 1.1, answer the following questions using R.
   a.   Draw a histogram.
   b.   Construct a stem-and-leaf plot.
   c.   Find the sample mean and sample standard deviation.
   d.   Find the median, Q_1, and Q_3.
   e.   Find the 80th percentile.
   f.   Find the interquartile range.
   g.   Draw a boxplot.

1.24   With the data given in Example 1.3, answer the following questions using R.
   a.   Draw a histogram.
   b.   Find the sample mean and sample variance.
   c.   Find the median, Q_1, and Q_3.
   d.   Find the 65th percentile.
   e.   Find the interquartile range.
   f.   Draw a boxplot.
   g.   Find summary statistics of the data.

1.25   The following table shows the results of the 100 m dash final in the 2016 Olympics and the heights of the runners.

Rank	Athlete	Result (sec)	Height (cm)
1	Usain Bolt	9.81	195
2	Justin Gatlin	9.89	185
3	Andre De Grasse	9.91	176
4	Yohan Blake	9.93	180
5	Akani Simbine	9.94	176
6	Ben Youssef Meite	9.96	180
7	Jimmy Vicaut	10.04	184
8	Trayvon Bromell	10.06	173

   a.   Draw boxplots of time and height separately and comment on them.
   b.   Draw a scatter plot of time and height.
   c.   Compute the covariance of time and height.
   d.   Compute the correlation coefficient of time and height.
   e.   Exclude Usain Bolt and then compute the correlation coefficient.
   f.   Comment on the relationship between the speed and height based on the above data.

1.26 The following data show the median annual income of Americans at every age bracket by gender according to the Bureau of Labor Statistics for the second quarter of 2017.

Age Group	Annual Salary	
	Men	Women
16 to 19	$22,880	$21,008
20 to 24	$28,548	$26,416
25 to 34	$43,056	$37,804
35 to 44	$55,380	$45,604
45 to 54	$56,888	$44,252
55 to 64	$55,016	$45,188
65 years and older	$52,260	$41,600

a. Draw side-by-side boxplots of annual salaries for the two gender groups, and comment on it.
b. Draw a scatter plot of women's versus men's salaries.
c. Find the covariance of the men's and women's salaries.
d. Find the correlation coefficient of the men's and women's salaries and comment on it.

1.27 Answer the questions in Exercise 1.26 using R and compare the answers obtained in 1.26.

1.28 The following data given in an article (G.A. Smith, W.C. Lenahan, and D.S. MacLeod, "Hydrogen, Oxygen, and Nitrogen in Cobalt Metal", Metallurgia, 1969, 121-127) contain pressure of extracted gas (microns) and extraction time (minutes).

Pressure	40	130	155	160	260	275	325	370	420	480
Time	2.5	3.0	3.1	3.3	3.7	4.1	4.3	4.8	5.0	5.4

a. Find the sample covariance of pressure and time.
b. Find the sample standard deviations of pressure and time.
c. Find the correlation coefficient of the two variables.

1.29   A study has been conducted to find the relationship between the shelf spaces (in square feet) to predict the monthly sales (in thousand dollars) of milk.
A random sample of 11 grocery stores is selected and the results are given below.

Shelf space (x)	5	7	8	9	10	12	13	15	16	18	20
Monthly sales (y)	3.2	4.4	2.8	3.8	4.7	5.2	4.6	5.4	5.6	5.4	6.1

a.   Draw a scatter plot of the shelf space and monthly sales.
b.   Find the correlation coefficient of the two variables.

Probability

2

1. Sample Spaces and Events

Define an *experiment* to be any process that generates observations. The set of all possible observations, or *outcomes*, of the experiment is called the *sample space*, usually denoted S. An *event* is a set of outcomes contained in the sample space S.

EXAMPLE 2.1

a. Toss a coin twice. The sample space S for this experiment can be written as follows:

$$S = \{HH, HT, TH, TT\}$$

The notation HT means the outcome of the first flip was a head (H) and the second flip a tail (T). HH means that both flips result in heads, and so on. Let A be the event that exactly one of the flips results in a head, and let B be the event that at least one of the flips results in a head. Then:

$$A = \{HT, TH\}, \quad B = \{HH, HT, TH\}$$

b. Toss a six-sided die. The sample space S consists of the numerical values of the faces of the die:

$$S = \{1, 2, 3, 4, 5, 6\}$$

Some events in this sample space are:
    a. the outcome is an odd number: {1, 3, 5}
    b. the outcome is less than 4: {1, 2, 3}
    c. the outcome is a 3: {3}

    c. Roll two dice, and record the sum. Then the sample space is:

$$S = \{2, 3, 4, 5, 6, 7, 8, 9, 10, 11, 12\}$$

Note the distinction between an outcome and an event. No two outcomes may occur simultaneously, whereas two events can occur at the same time. In the middle example above, the outcome of rolling a die can be both an odd number and less than 4, but it cannot be both a 3 and a 5.

The notation we are introducing here is that of sets, with a couple of new words for familiar concepts. The individual outcomes are otherwise known as *elements*, and events are otherwise known as *subsets* of the sample space. A set can, of course, consist of only one element. The sample space, which is the set of all possible outcomes, is an event. An event consisting of no outcomes is called the *empty set*, denoted $\varnothing$.

Sample space can be discrete or continuous, finite or infinite. A continuous sample space has an uncountable number of outcomes and is thus always infinite. The sample spaces given above are all discrete and finite. The number of leaves on a randomly selected tree is an outcome in a sample space that is discrete but infinite (disregarding the physical limitations on the number of leaves a tree can possess). Suppose that we pick two cities at random and record the air distance between them. This outcome would fall in a continuous interval bounded by zero and half the circumference of the earth; the sample space is therefore a finite interval, but it is continuous and therefore contains uncountably many outcomes.

A *Venn diagram* is a useful pictorial method for analyzing sample spaces and events. Denote the sample space S as the region inside a rectangle, and any event as the region inside a circle or ovum (see Figures 2.1–2.3). Since A is a subset of S, written $A \subset S$, its picture is drawn within the rectangle representing S. The idea here is that the relationship between sets is expressed by the geometry of the shapes in the Venn diagram.

The event consisting of all outcomes that are contained in A or B (or both) is called the *union* of A and B. This set is represented in a Venn diagram as the combined region occupied by A and B, i.e., the shaded region in Figure 2.1. In the notation of sets, it is denoted $A \cup B$, or equivalently, A or B.

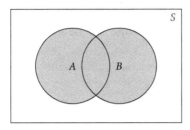

FIGURE 2.1   Union: $A \cup B$ (A or B).

The event consisting of all outcomes that are in both A and B is called the intersection of A and B. This set is represented in a Venn diagram as the region of overlap between A and B, as shown in Figure 2.2. In the notation of sets, it is denoted $A \cap B$, or equivalently, A and B.

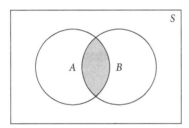

FIGURE 2.2   Intersection: $A \cap B$ (A and B).

The event consisting of all outcomes that are not in A is called the complement of A (or not A). This set is represented in a Venn diagram as the region of S outside of A (Figure 2.3). In the notation of sets, it is denoted A^C.

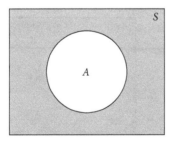

FIGURE 2.3   Complement of A: A^C (not A).

EXAMPLE 2.2   In tossing a coin twice, define A as the event that a tail occurs at the second toss and B as the event that at least one head occurs. Then:

$$A = \{HT, TT\}, \; B = \{HH, HT, TH\}, \; \text{and}$$

$$A \cup B = \{HH, HT, TH, TT\} = S, \quad A \cap B = \{HT\}, \quad A^C = \{HH, TH\}, \quad B^C = \{TT\}$$

THEOREM 2.1   De Morgan's law: Let A and B be any two events in a sample space. Then:
     a. $(A \cup B)^C = A^C \cap B^C$
     b. $(A \cap B)^C = A^C \cup B^C$

The proof of each statement is a straightforward exercise in the use of a Venn diagram. The shaded region of Figure 2.4 is $(A \cup B)^C$. The shaded region of Figure 2.5 (a) is A^C, and the shaded region of Figure 2.5 (b) is B^C. We can see that the intersection of A^C and B^C in Figure 2.5 is the same as $(A \cup B)^C$ in Figure 2.4. From Example 2.2, $(A \cup B)^C = \varnothing = A^C \cap B^C$ and $(A \cap B)^C = \{HH, TH, TT\} = A^C \cup B^C$. This is not a formal proof, but we can see that De Morgan's law works here.

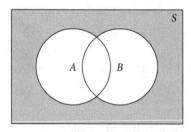

FIGURE 2.4 $(A \cup B)^C$.

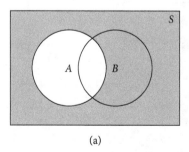

(a)

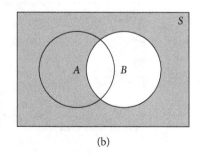

(b)

FIGURE 2.5   (a) A^C, (b) B^C.

Events A and B are *mutually exclusive* (or *disjoint*) if the two events have no outcomes in common. In set notation, $A \cap B = \varnothing$. Figure 2.6 shows two mutually exclusive events.

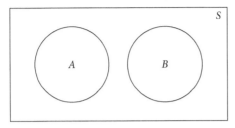

FIGURE 2.6   Mutually exclusive (disjoint) events A and B.

EXAMPLE 2.3   In flipping a coin twice, define A as the event that the second flip results in a tail and D as the event that both flips result in heads. Then $A = \{HT, TT\}$ and $D = \{HH\}$. Here, A and D are disjoint.

2. *Counting*

The number of elements in a sample space or one of its subsets is often so large that it is impractical to simply list them. Furthermore, it is often not important what the outcomes in a set are; only the number of outcomes in the set matters. It is therefore appropriate to develop enumerating techniques.

A. TREE DIAGRAMS

EXAMPLE 2.4   A three-course dinner at a restaurant consists of a soup, a side dish, and the main dish. The choices of soup are clam chowder and cream of broccoli; the choices of side dish are french fries and salad; and the choices of the main dish are chicken, beef, and pork. Enumerate, using a tree diagram, the number of possible three-course dinners at this restaurant. Let's use the following abbreviations for the food choices:

Soup: clam chowder (CC), broccoli (BR)

Side dish: french fries (F), salad (S)

Main dish: chicken (C), beef (B), pork (P)

In selecting a three-course meal, the order in which the selections are made is not important, so let's assume that a customer first orders a soup, then a side dish, and then main dish. A meal selection is then given as one of the paths through the tree diagram shown in Figure 2.7. The number of possible meals is the number of branches on the right hand side of the tree diagram, which in this example is 12.

This result could also be obtained from straight multiplications of the number of possibilities within each course. Since the selection of, for example, a soup does not change the possible choices of side dish, we have:

$$2 \text{ soups} \times 2 \text{ side dishes} \times 3 \text{ main dishes} = 12 \text{ dinner combinations}$$

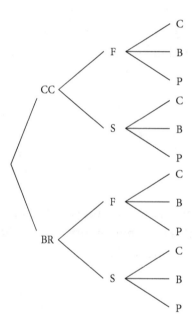

FIGURE 2.7   Counting: $2 \times 2 \times 3 = 12$ possible choices.

B. PRODUCT RULE

Example 2.4 is a realization of the following result.

THEOREM 2.2   Product rule: Let the sets $A_1, \cdots, A_k$ contain, respectively, $n_1, \cdots, n_k$ elements. The number of ways of choosing one element from A_1, one element from A_2, $\cdots$, one element from A_k is $n_1 n_2 \cdots n_k$.

EXAMPLE 2.5   A test consists of 12 true-false questions. In how many different ways can the questions be answered?

Let $A_j = \{T, F\}$, $j = 1, \cdots, 12$. Then the answer follows the product rule and number of possible responses $= n_1 \cdots n_{12} = 2 \cdots 2 = 2^{12} = 4{,}096$.

EXAMPLE 2.6   A license plate has exactly six characters, each of which can be a digit or a letter excluding letters O and I. How many possible license plates are possible?

Let the six characters occupy slots 1 through 6, and let A_i be the set of characters that may occupy slot i. Then $A_i = \{0, 1, 2, \cdots 9, A, B, \cdots H, J, \cdots, N, P, \cdots, Z\}, i = 1, \cdots, 6$. There are

$$n_1 \cdots n_6 = 34 \times 34 \times 34 \times 34 \times 34 \times 34 = 34^6 = 1{,}544{,}804{,}416$$

possible license plates.

What if the first three characters must be digits, and the last three must be letters excluding O and I?

$$\text{number of possible license plates} = 10 \times 10 \times 10 \times 24 \times 24 \times 24 = 13{,}824{,}000.$$

Factorial: To further discuss counting, we need to know the factorial. For a positive integer n, the factorial of n is denoted as $n!$ and it is defined as follows:

$$n! = n(n-1)(n-2)\cdots 2 \cdot 1$$

One factorial ($1!$) is 1 according to the definition, and zero factorial ($0!$) is defined to be 1.

C. PERMUTATIONS: ORDER IS IMPORTANT

THEOREM 2.3   Permutation: An ordered sequence of k objects from a set of n distinct objects.

$$_nP_k = n(n-1)(n-2)\cdots(n-k+1) = \frac{n!}{(n-k)!}$$

EXAMPLE 2.7   A committee consists of 10 members. How many ways are there of selecting, from this group, a chair, vice chair, and secretary? Assume that no person can fill more than one position.

The order in which these positions are filled is not important. However, the choice of one position affects the choice of subsequent positions. Suppose we choose, in order, the secretary, chair, and vice chair. There are 10 possible candidates for the secretary position, but once we make this selection, then there are only 9 possible candidates for the chair position, and once the chair is selected, then there are only 8 possible candidates for the vice chair position. Using the multiplication rule, we get:

$$_{10}P_3 = \frac{10!}{(10-3)!} = \frac{10!}{7!} = 10 \cdot 9 \cdot 8 = 720$$

In permutation, the order in which the k selections are made is not important, but the order of *arrangement* of the k objects is important. In Example 2.7, the event that the secretary is Andrew, the vice chair is Bob, and the chair is Cathy is distinct from the event that the secretary is Andrew, the vice chair is Cathy, and the chair is Bob. If all the positions were labeled "secretary," the answer would be different, as we will see below.

D. COMBINATIONS: IGNORE THE ORDER

THEOREM 2.4   Combination: Any unordered subset of k objects from a set of n distinct objects.

$$_nC_k = \binom{n}{k} = \frac{n(n-1)(n-2)\cdots(n-k+1)}{k!} = \frac{n!}{k!(n-k)!}$$

EXAMPLE 2.8 A committee consists of 10 members. How many ways are there of selecting 3 representatives from this group?

$$\binom{10}{3} = \frac{10!}{3!(10-3)!} = \frac{10 \cdot 9 \cdot 8}{3!} = \frac{720}{6} = 120$$

The order of arrangement of the selections is not important in the above example.

EXAMPLE 2.9 Pick 3 cards in succession from a full deck of 52. Let A_i be the face value of the ith card chosen ($i = 1, 2, 3$). Consider the following cases:

a. Suppose the order of arrangement of the selection is important. For example, the outcomes

$$A_1 = Q\spadesuit, \quad A_2 = 2\heartsuit, \quad A_3 = 5\diamondsuit$$

and

$$A_1 = 2\heartsuit, \quad A_2 = Q\spadesuit \quad A_3 = 5\diamondsuit$$

are distinct. In this case, there are

$$_{52}P_3 = \frac{52!}{(52-3)!} = \frac{52!}{49!} = 52 \cdot 51 \cdot 50 = 132{,}600$$

possible outcomes.

b. Suppose the order of arrangement of the selections is not important. For example, the two outcomes given above are identical. In this case, $_{52}P_3$ counts each outcome multiple times, so we have to apply a correction. Again, consider the outcome where the three cards chosen are the queen of spades ($Q\spadesuit$), two of hearts ($2\heartsuit$), and five of diamonds ($5\diamondsuit$). In how many ways could these three cards have been selected? There are $_3P_3 = 3! = 6$ different ways of selecting these three cards. Since the three cards we are using are not special, it is apparent that *every combination* of three cards has been overcounted by a factor of $3!$. Thus, to arrive at the correct answer, we need to divide $_{52}P_3$ by $3!$. Therefore, the number of possible selections is:

$$\binom{52}{3} = \frac{_{52}P_3}{3!} = \frac{52!}{(52-3)!3!} = \frac{52 \cdot 51 \cdot 50}{3 \cdot 2} = \frac{132{,}600}{6} = 22{,}100$$

3. Relative Frequency (Equally Likely Outcomes)

If an experiment has n equally likely outcomes, and s of these outcomes are labeled success, then the probability of a successful outcome is s / n.

EXAMPLE 2.10   In tossing a fair die, the sample space is $S = \{1, 2, 3, 4, 5, 6\}$. Let A be the event that the toss is an odd number. Then $A = \{1, 3, 5\}$. The probability of getting an odd number in this experiment is:

$$\text{Probability of } A = \frac{\#\text{outcomes in } A}{\#\text{outcomes in } S} = \frac{3}{6} = \frac{1}{2}$$

EXAMPLE 2.11   The probability of drawing a king from a well-shuffled deck of 52 playing cards is:

$$\frac{s}{n} = \frac{4}{52} = \frac{1}{13}$$

EXAMPLE 2.12   Toss a balanced die twice and record the outcome of each toss. The sample space S has $6 \times 6 = 36$ outcomes. Let A be the event that the sum of the numbers in two tosses is 6. Then:

$$A = \{\text{sum of the numbers is } 6\} = \{(1, 5), (2, 4), (3, 3), (4, 2), (5, 1)\}$$

The probability of A is:

$$\text{Probability of } A = \frac{\#\text{elelments in } A}{\#\text{elements in } S} = \frac{5}{36}$$

4. Probability

Let S be a sample space and A an arbitrary event in S. The probability of A is denoted $P(A)$ and satisfies the following properties.

Probability axioms:

(i) $0 \leq P(A) \leq 1$

(ii) $P(S) = 1$

(iii) If A and B are any mutually exclusive events in S, then
$P(A \cup B) = P(A) + P(B)$.

The first two axioms are consistent with our commonsense notion of the probability of an event being a fraction or percentage. If $P(A) = 0$, then it is absolutely certain that A will not occur. If $P(A) = 1$, then it is absolutely certain that A will occur. Any uncertainty in whether A will occur is reflected by a probability of A intermediate between 0 and 1. Since S contains all possible outcomes, and it is absolutely certain that one of these outcomes will result, we must have $P(S) = 1$.

EXAMPLE 2.13. The third axiom of probability is called the *additive property* of probability. It concerns the probability of union of two disjoint events. Does a similar formula hold for the probability of the union of three mutually exclusive events A, B, and C? Note that the events $A \cup B$ and C are disjoint. Therefore, we can apply the third axiom to these two events:

$$P(A \cup B \cup C) = P((A \cup B) \cup C) = P(A \cup B) + P(C)$$

Applying the third axiom a second time, this time to the disjoint events A and B, leads to the following result:

$$P(A \cup B \cup C) = P(A) + P(B) + P(C)$$

It can be generalized to the union of n disjoint events, as follows:

THEOREM 2.5 If $A_1, A_2, \cdots, A_n$ are mutually exclusive events in S, then:

$$P(A_1 \cup A_2 \cup \cdots \cup A_n) = P(A_1) + P(A_2) + \cdots + P(A_n) = \sum_{i=1}^{n} P(A_i)$$

EXAMPLE 2.14 In tossing a fair coin, the sample space is $S = \{H, T\}$. Since the outcomes are mutually exclusive, we have:

$$P(H) = P(T) = \frac{1}{2}$$

$$P(S) = P(H) + P(T)$$

EXAMPLE 2.15 In flipping two balanced coins, the sample space is $S = \{HH, HT, TH, TT\}$. The individual outcomes constituting S are themselves mutually exclusive events, so that

$$P(S) = P(HH) + P(HT) + P(TH) + P(TT) = \frac{1}{4} + \frac{1}{4} + \frac{1}{4} + \frac{1}{4} = 1$$

confirming the second axiom of probability. Let A be the event of obtaining at least one tail, i.e., $A = \{HT, TH, TT\}$. Then:

$$P(A) = P(HT) + P(TH) + P(TT) = \frac{1}{4} + \frac{1}{4} + \frac{1}{4} = \frac{3}{4}$$

EXAMPLE 2.16 In an experiment of tossing a fair coin until a tail is obtained, the sample space is $S = \{T, HT, HHT, \cdots\}$ and

$$P(S) = P(T) + P(HT) + P(HHT) + \cdots = \frac{1}{2} + \frac{1}{4} + \frac{1}{8} + \cdots + \frac{1}{2^n} + \cdots = \frac{\frac{1}{2}}{1 - \frac{1}{2}} = 1$$

THEOREM 2.6 Law of complementation: Let A be any event. Then $P(A^C) = 1 - P(A)$.

PROOF Since A and A^C are disjoint, $A \cup A^C = S$. Applying (ii) and (iii) of the probability axioms, we have:

$$1 = P(S) = P(A \cup A^C) = P(A) + P(A^C)$$

Thus, $P(A^C) = 1 - P(A)$.

EXAMPLE 2.17 In Example 2.15, the sample space is $S = \{HH, HT, TH, TT\}$. Let A be the event that both the flips result in heads. Then $A = \{HH\}$ and

$$A^C = \{\text{at least one is tail}\} = \{HT, TH, TT\}$$

The probabilities of A and A^C are

$$P(A) = \frac{1}{4}, \quad P(A^C) = \frac{3}{4}$$

which shows that $P(A^C) = 1 - P(A)$.

If $A \subset B$, i.e., A is a subset of B, then $P(A) \leq P(B)$.

THEOREM 2.7   If the events A and B are mutually exclusive, then $P(A \cap B) = 0$.

PROOF $A \cap B = \varnothing$ and $(A \cap B)^C = S$

$$1 = P(S) = P((A \cap B)^C) = 1 - P(A \cap B)$$

Hence, $P(A \cap B) = 0$.

THEOREM 2.8   Let A and B be any events. Then $P(A \cup B) = P(A) + P(B) - P(A \cap B)$.

PROOF   Since $A \cup B = (A \cap B) \cup (A \cap B^C) \cup (A^C \cap B)$,

$$P(A \cup B) = P(A \cap B) + P(A \cap B^C) + P(A^C \cap B)$$

$$= [P(A \cap B) + P(A \cap B^C)] + [P(A \cap B) + P(A^C \cap B)] - P(A \cap B)$$

$$= P(A) + P(B) - P(A \cap B)$$

If the events A and B are mutually exclusive, then $P(A \cup B) = P(A) + P(B)$.

EXAMPLE 2.18 Let 80% of freshmen at a college take statistics, 50% take physics, and 40% take both statistics and physics.

a. What is the probability that a freshman at this college takes at least one of these courses?

Let $A = \{\text{taking statistics}\}$ and $B = \{\text{taking physics}\}$. Then:

$$P(A) = 0.8, \ P(B) = 0.5, \ P(A \cap B) = 0.4$$

$$P(A \cup B) = P(A) + P(B) - P(A \cap B) = 0.8 + 0.5 - 0.4 = 0.9$$

b. What is the probability that a freshman at this college takes only one of these courses? Figure 2.8 illustrates this event. The probability of this event is:

$$P(A \cup B) - P(A \cap B) = 0.9 - 0.4 = 0.5$$

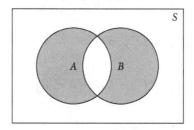

FIGURE 2.8   The event that a freshman takes only one of the two courses in Example 2.18.

THEOREM 2.9   The probability of the union of three events is given below.

$$P(A \cup B \cup C) = P(A) + P(B) + P(C) - P(A \cap B) - P(B \cap C) - P(C \cap A) + P(A \cap B \cap C)$$

PROOF   We can derive a rule for the probability of the union of three events by applying the two-set formula repeatedly, as follows:

$$P(A \cup B \cup C) = P[(A \cup B) \cup C] = P(A \cup B) + P(C) - P[(A \cup B) \cap C]$$

$$= [P(A) + P(B) - P(A \cap B)] + P(C) - P[(A \cap C) \cup (B \cap C)]$$

$$= P(A) + P(B) + P(C) - P(A \cap B) - [P(A \cap C) + P(B \cap C) - P\{(A \cap C) \cap (B \cap C)\}]$$

$$= P(A) + P(B) + P(C) - P(A \cap B) - P(B \cap C) - P(C \cap A) + P(A \cap B \cap C)$$

It is easy to see that the more events in the union, the more tedious the derivation and resulting formula. Even the rule for three events is somewhat tedious; in problems involving three events, it is usually easier to figure out the probabilities using a Venn diagram than by applying a long formula.

EXAMPLE 2.19 A committee has 8 male members and 12 female members. Choose 5 representatives in this committee at random and find:

a.   the probability that exactly 3 of the 5 representatives are females.

Let $D_i = P(\text{exactly } i \text{ of the 5 students are females})$, $i = 0, 1, 2, 3, 4, 5$. Then:

$$P(D_3) = \frac{\#\text{committees with 3 females and 2 males}}{\#\text{committees in total}} = \frac{\binom{12}{3}\binom{8}{2}}{\binom{20}{5}} = \frac{\frac{12!}{3!9!} \cdot \frac{8!}{2!6!}}{\frac{20!}{5!15!}} = 0.3973$$

b.   the probability that at least 3 of the 5 representatives are females.

$$P(D_3 \cup D_4 \cup D_5) = \frac{\binom{12}{3}\binom{8}{2}}{\binom{20}{5}} + \frac{\binom{12}{4}\binom{8}{1}}{\binom{20}{5}} + \frac{\binom{12}{5}\binom{8}{0}}{\binom{20}{5}}$$

$$= 0.3973 + 0.2554 + 0.0511 = 0.7038$$

EXAMPLE 2.20 Among a group of 200 students, 137 are enrolled in a math class, 50 in history, and 124 in art; 33 in both history and math, 29 in both history and art, and 92 in both math and art; 18 in all three classes. Pick a student at random from this group. What is the probability that he or she is enrolled in at least one of these classes?

Define the following events: $A = \{\text{enrolled in art}\}$, $H = \{\text{enrolled in history}\}$, $M = \{\text{enrolled in math}\}$. Then:

$$P(A) = \frac{124}{200} = 0.62,\ P(H) = \frac{50}{200} = 0.25,\ P(M) = \frac{137}{200} = 0.685,$$

$$P(A \cap H) = \frac{29}{200} = 0.145,\ P(H \cap M) = \frac{33}{200} = 0.165,\ P(M \cap A) = \frac{92}{200} = 0.46,$$

$$P(A \cap H \cap M) = \frac{18}{200} = 0.09$$

The question is asking for $P(A \cup H \cup M)$ and the answer is:

$$P(A \cup H \cup M) = P(A) + P(H) + P(M) - P(A \cap H) - P(H \cap M) - P(M \cap A) + P(A \cap H \cap M)$$

$$= 0.62 + 0.25 + 0.685 - 0.145 - 0.165 - 0.46 + 0.09 = 0.875$$

5. *Conditional Probability*

The conditional probability of an event A given event B is the probability that A will occur given the knowledge that B has already occurred. This probability is written as $P(A|B)$, which is defined as follows.

Conditional probability of A given B: $P(A|B) = \dfrac{P(A \cap B)}{P(B)}$ if $P(B) > 0$

Equivalently,

$$P(A \cap B) = P(A)P(B|A)\ \text{if}\ P(A) > 0$$

$$= P(B)P(A|B)\ \text{if}\ P(B) > 0$$

Figure 2.9 illustrates the conditional probability of A given B.

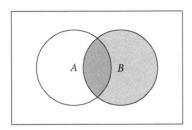

FIGURE 2.9   Conditional probability of A given B is the probability of the intersection of A and B given that B has already occurred.

EXAMPLE 2.21 The following table shows the distribution of body mass index (BMI) and age groups among male adults in a certain country.

	Normal or low BMI	Overweight	Obese	Total
Age <30	0.09	0.06	0.05	0.20
Age ≥ 30	0.20	0.32	0.28	0.80
Total	0.29	0.38	0.33	1.00

a.   What is the probability that a person selected at random from the group will be obese?

Let $A = \{\text{obese}\}$. Then $P(A) = 0.33$.

b.   A person, selected at random from this group, is found to be obese. What is the probability that this person is younger than age 30?
Let $B = (\text{younger than age 30})$. Then:

$$P(B|A) = \frac{P(A \cap B)}{P(A)} = \frac{0.05}{0.33} = 0.15$$

THEOREM 2.10 The conditional probability is a probability.

PROOF   We need to show that the conditional probability satisfies the three axioms of probability. Let A and D be events with $P(D) > 0$. Then:

(i) $P(A|D) = \dfrac{P(A \cap D)}{P(D)}$ and $0 \le \dfrac{P(A \cap D)}{P(D)} \le 1$ because $P(A \cap D) \le P(D)$

(ii) $P(A^C|D) = \dfrac{P(A^C \cap D)}{P(D)} = \dfrac{P(D) - P(A \cap D)}{P(D)} = 1 - \dfrac{P(A \cap D)}{P(D)} = 1 - P(A|D)$

(iii) Let B be an event such that A and B are mutually exclusive. Then

$$P(A \cup B|D) = \frac{P((A \cup B) \cap D)}{P(D)} = \frac{P((A \cap D) \cup (B \cap D))}{P(D)} = \frac{P(A \cap D) + P(B \cap D)}{P(D)}$$

$$= \frac{P(A \cap D)}{P(D)} + \frac{P(B \cap D)}{P(D)} = P(A|D) + P(B|D)$$

6. Independence

If two events are not related, then they are independent. Two events are independent if the probability of one event is not affected by the occurrence of the other event. In other words, two events A and B are independent if the conditional probability of A given B is the same as the probability of A.

> A and B are independent if $P(A|B) = P(A)$.

EXAMPLE 2.22. In rolling a fair die, let $A = \{2, 4, 6\}$, $B = \{1, 2, 3\}$ and $C = \{1, 2, 3, 4\}$. Then:

$$P(A) = \frac{1}{2}, \ P(A|B) = \frac{1}{3}, \ P(A|C) = \frac{1}{2}$$

Hence, A and B are dependent, and A and C are independent.

It follows that the probability of the intersection of two independent events is the product of the probabilities of the two events.

> A and B are independent if $P(A \cap B) = P(A)P(B)$.

It can be proved by using the definition of the conditional probability. If A and B are independent, then:

$$P(A \cap B) = P(A \mid B)P(B) \text{ and by independence } P(A \mid B) = P(A)$$

$$\text{Therefore, } P(A \cap B) = P(A)P(B)$$

THEOREM 2.11 If two events A and B are independent, then the following are true.

a. $P(A \cap B^C) = P(A)P(B^C)$
b. $P(A^C \cap B) = P(A^C)P(B)$
c. $P(A^C \cap B^C) = P(A^C)P(B^C)$

PROOF

(a) $P(A \cap B^C) = P(A) - P(A \cap B) = P(A) - P(A)P(B) = P(A)[1 - P(B)] = P(A)P(B^C)$

Parts (b) and (c) can be proved similarly.

EXAMPLE 2.22 (CONTINUED)   From Example 2.22, $A \cap B = \{2\}$. Therefore:

$$P(A \cap B) = \frac{1}{6} \neq P(A)P(B) = \frac{1}{2} \cdot \frac{1}{2} = \frac{1}{4}$$

Thus, A and B are dependent. For A and C, $A \cap C = \{2, 4\}$.

$$P(A \cap C) = \frac{1}{3} = P(A)P(C) = \frac{1}{2} \cdot \frac{2}{3}$$

Thus, A and C are independent.

The independence can be extended to more than two events, as follows:

$A_1, A_2, \cdots, A_n$ are independent if, for any k, and every subset of indices $i_1, i_2, \cdots, i_k$,
$$P(A_{i_1} \cap A_{i_2} \cap \cdots \cap A_{i_k}) = P(A_{i_1})P(A_{i_2}) \cdots P(A_{i_k}).$$

EXAMPLE 2.23 Suppose components A, B_1, and B_2 operate independently in an electronic system shown in Figure 2.10. Let the probability that each of the components will operate for 10 days without failure be $P(A) = 0.9$, $P(B_1) = 0.8$, and $P(B_2) = 0.7$. The system works if A works and either B_1 or B_2 works. Find the probability that the entire system will operate without failure for 10 days. Assume that all the components in the system start running at the same time and a component does not work again once it fails.

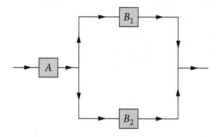

FIGURE 2.10   An electronic system described in Example 2.23.

$$P(\text{system working}) = P[A \cap (B_1 \cup B_2)] = P(A)P(B_1 \cup B_2) \quad \text{(by independence)}$$
$$= 0.9[P(B_1) + P(B_2) - P(B_1 \cap B_2)]$$
$$= 0.9[P(B_1) + P(B_2) - P(B_1)P(B_2)]$$
$$= 0.9[0.8 + 0.7 - (0.8)(0.7)]$$
$$= 0.9(0.94) = 0.846$$

7. *Bayes' Theorem*

A sequence of events A_1, A_2, $\cdots$, A_n are *mutually exclusive* if $A_i \cap A_j = \varnothing$ for all $i \neq j$. The sequence of events are called *exhaustive* if $A_1 \cup A_2 \cup \cdots \cup A_n = S$.

A_1, A_2, $\cdots$, A_n are mutually exclusive if $A_i \cap A_j = \varnothing$ for all $i \neq j$.

A_1, A_2, $\cdots$, A_n are exhaustive if $A_1 \cup A_2 \cup \cdots \cup A_n = S$.

THEOREM 2.12 If A_1, A_2, $\cdots$, A_n are mutually exclusive and exhaustive events and B is an event (see Figure 2.11), then:

$$
\begin{aligned}
P(B) &= P(B \cap A_1) + P(B \cap A_2) + \cdots + P(B \cap A_n) \\
&= P(A_1)P(B|A_1) + P(A_2)P(B|A_2) + \cdots + P(A_n)P(B|A_n) \\
&= \sum_{i=1}^{n} P(A_i)P(B|A_i)
\end{aligned}
$$

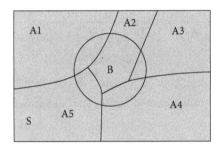

FIGURE 2.11   Mutually exclusive and exhaustive events A_1, A_2, $\cdots$, A_n and an event B.

THEOREM 2.13 (BAYES' THEOREM)   Let A_1, A_2, $\cdots$, A_n be mutually exclusive and exhaustive events and B an event. Then for some integer k such that $1 \le k \le n$:

$$
P(A_k|B) = \frac{P(A_k)P(B \mid A_k)}{\sum_{i=1}^{n} P(A_i)P(B|A_i)}
$$

PROOF

$$
P(A_k|B) = \frac{P(A_k \cap B)}{P(B)} = \frac{P(A_k)P(B \mid A_k)}{\sum_{i=1}^{n} P(A_i)P(B|A_i)}
$$

EXAMPLE 2.24 Approximately 0.4% of the people in the United States are living with an HIV infection. Modern HIV testing is highly accurate. The HIV screening detects 99.7% of the people who have HIV and shows a negative result for

98.5% of the people who do not have HIV. Suppose a person had a positive test result. What is the probability that the individual has an HIV infection?

Let $A_1 = \{$individual has HIV$\}$, $A_2 = \{$individual does not have HIV$\}$, and

$B = \{$positive HIV test result$\}$. Then $P(A_1) = 0.004$, $P(A_2) = 0.996$, $P(B \mid A_1) = 0.997$ and

$P(B \mid A_2) = 0.015$. Using Bayes' theorem:

$$P(A_1 \mid B) = \frac{P(A_1)P(B \mid A_1)}{P(A_1)P(B \mid A_1) + P(A_2)P(B \mid A_2)} = \frac{(0.004)(0.997)}{(0.004)(0.997) + (0.996)(0.015)} = 0.21$$

The false positive rate is very high because the HIV infection is rare. To reduce the false positive rate to a reasonable number for such a rare disease, the test should be very accurate.

SUMMARY OF CHAPTER 2

1. Experiment: Any action or process that generates observations.
   a. Sample Space: The set of all possible outcomes of an experiment.
   b. Event: The set of outcomes contained in the sample space.
2. Set Operations
   a. Union of two events $(A \cup B)$: The event consisting of all outcomes that are either in A or B or both.
   b. Intersection of two events $(A \cap B)$: The event consisting of all outcomes that are in both A and B.
   c. Complement (A^C): Not A.
   d. De Morgan's Law: $(A \cup B)^C = A^C \cap B^C$ and $(A \cap B)^C = A^C \cup B^C$.
   e. A and B are <u>mutually exclusive (disjoint)</u> if $A \cap B$ is empty.
3. Counting
   a. Product Rule: If $A_1, \cdots, A_k$ contain, respectively, $n_1, \cdots, n_k$ elements, the number of ways of choosing one element from A_1, one element from A_2, $\cdots$, one element from A_k is $n_1 n_2 \cdots n_k$.
   b. Permutation: An ordered sequence of k objects from a set of n distinct objects. $n! / (n-k)!$
   c. Combination: Any unordered subset of k objects from a set of n distinct objects. $n! / [k!(n-k)!]$

4. Probability
   a. Probability Axioms:
      i. $0 \leq P(A) \leq 1$ for any subset A of the sample space S
      ii. $P(S) = 1$
      iii. If $A_1, A_2, \cdots, A_n$ are mutually exclusive events in S, then
         $$P(A_1 \cup A_2 \cup \cdots \cup A_n) = P(A_1) + P(A_2) + \cdots + P(A_n).$$
   b. If A is a subset of B, then $P(A) \leq P(B)$.
   c. $P(A^C) = 1 - P(A)$ for any event A.
   d. For events A and B, $P(A \cup B) = P(A) + P(B) - P(A \cap B)$.
   e. If A and B are <u>mutually exclusive,</u> then $P(A \cup B) = P(A) + P(B)$.
   f. For events A, B, and C, $P(A \cup B \cup C) = P(A) + P(B) + P(C) - P(A \cap B) - P(B \cap C) - P(C \cap A) + P(A \cap B \cap C)$.

5. Conditional Probability: For events A and B with $P(B) > 0$, the conditional probability of A given B is:

$$P(A|B) = \frac{P(A \cap B)}{P(B)}$$

or equivalently, $P(A \cap B) = P(A|B)P(B)$.

6. Two events A and B are <u>independent</u> if $P(A|B) = P(A)$, or equivalently, $P(A \cap B) = P(A)P(B)$.
   a. If A and B are independent, then $P(A \cap B^C) = P(A)P(B^C)$, $P(A^C \cap B) = P(A^C)P(B)$ and $P(A^C \cap B^C) = P(A^C)P(B^C)$.
   b. $A_1, A_2, \cdots, A_n$ are independent if, for any k, and every subset of indices, $i_1, i_2, \cdots, i_k$,
   $$P(A_{i_1} \cap A_{i_2} \cap \cdots \cap A_{i_k}) = P(A_{i_1})P(A_{i_2}) \cdots P(A_{i_k}).$$

7. If events $A_1, A_2, \cdots, A_n$ are mutually exclusive and exhaustive with $P(A_i) > 0$ for $i = 1, 2, \cdots, n$, then for any event B in S, $P(B) = \sum_{i=1}^{n} P(A_i)P(B|A_i)$.

8. Bayes' Theorem: If $A_1, A_2, \cdots, A_n$ and B are the same as in 7., then:

$$P(A_r|B) = \frac{P(A_r)P(B \mid A_r)}{\sum_{i=1}^{n} P(A_i)P(B|A_i)} \quad \text{for} \quad r = 1, 2, \cdots, n$$

EXERCISES

2.1   Pick a card from a full deck of 52. Let E be the event that a card drawn at random from the deck is a spade, and let F be the event that the card is a face card (J, Q, or K). Find the following events:
   a. $E \cup F$
   b. $E \cap F$
   c. F^C

Let G be the event that the card is a 3. Are the following events disjoint?
   d. E and F
   e. F and G

2.2   If A is the solution set of the equation $x^2 - 9 = 0$ and B is the solution set of the equation $x^2 - 4x + 3 = 0$, find $A \cup B$.

2.3   Let A be the multiples of 3 that are less than 10, and B the set of odd positive integers less than 10. Find $A \cup B$ and $A \cap B$.

2.4   A child is putting white clothes on two dolls. There are three shirts with different shapes, and three pants with different shapes. The shirts and pants that one doll wore should not be put on to the other doll. After putting clothes to two dolls, she colors the shirts and pants with either gray or blue. The colors of the shirts and pants on each doll should be different. How many different possible outcomes are possible?

2.5   Compute the following:

   a. $\binom{8}{3}$

   b. $\binom{10}{4}$

   c. $_9P_3$

   d. $_7P_4$

2.6   Minnie has homework from 6 subjects including calculus and statistics this week. Among them, she wants to finish homework for 4 subjects including calculus and statistics today. She wants to finish statistics before calculus. How many ways are there for her to select the 4 subjects, and then decide the order of doing homework?

2.7   After a fair die is tossed 4 times, the numbers appeared on top arc arranged to form
a four digit integer. How many ways do the numbers become a multiple of 5?

2.8   Answer the following questions.
   a.   In the general education course requirement at a college, a student needs to choose
        one each from social sciences, humanities, natural sciences, and foreign languages.
        There are 5 social science courses, 4 humanity courses, 4 natural science courses,
        and 3 foreign language courses available for general education. How many different
        ways can a student choose general education courses from these 4 areas?
   b.   Four people are chosen from a 25-member club for president, vice president,
        secretary, and treasurer. In how many different ways can this be done?
   c.   In how many different ways can 5 tosses of a coin yield 2 heads and 3 tails?

2.9   In a simulation study, a statistical model has 3 components: mean, variance, and
number of variables. Four different means, three different variances, and five
different variables are considered. For each model, a statistician chooses one value
from each component. How many simulation models are needed if the two lowest
means, the lowest variance, and three out of the five variables are selected?

2.10   A student is randomly choosing the answer to each of five true-false questions in a
test. How many possible ways can the student answer the five questions?

2.11   In a poker game, how many possible ways can a hand of five cards be dealt?

2.12   The Department of Applied Mathematics and Statistics (AMS) of a state university
has 24 full-time faculty members, of whom 4 are women and 20 are men. A
committee of 3 faculty members is to be selected from this group in order to study
the issue of purchasing a new copier.
   a.   How many ways are there to select a committee?
   b.   How many ways are there to select a committee that has more women than men?
   c.   If the positions in the committee are titled chair, vice chair, and secretary, in how
        many ways can this committee be formed among the 24 AMS faculty members?

2.13   A real estate agent is showing homes to a prospective buyer. There are 10 homes in
the desired price range listed in the area. The buyer wants to visit only 3 of them.
   a.   In how many ways could the 3 homes be chosen if the order of visiting is
        considered?
   b.   In how many ways could the 3 homes be chosen if the order of visiting is not
        important?

2.14 How many batting orders are possible for a baseball team consisting of 9 players?

2.15 In a box containing 36 strawberries, 2 of them are rotten. Kyle randomly picked 5 of these strawberries.
  a. What is the probability of having at least 1 rotten strawberry among the 5?
  b. How many strawberries should be picked so that the probability of having exactly 2 rotten strawberries among them equals 2/35?

2.16 The same kind of 5 cookies are given to 3 kids. At least one cookie is given to each child. Also, the same kind of 5 candies are given to the kids who received only one cookie. How many ways are there to give cookies and candies to these children?

2.17 Among a group of 100 people, 68 can speak English, 45 can speak French, 42 can speak German, 27 can speak both English and French, 25 can speak both English and German, 16 can speak both French and German, and 9 can speak all 3 languages. Pick a person at random from this group. What is the probability that this person can speak at least 1 of these languages?

2.18 A ball numbered 1, two balls numbered 2, and three balls numbered 3 are in a jar. A ball is randomly chosen from the jar twice and the numbers written on the balls are recorded. Find the probability that the total of the two numbers is 4 if
  a. the ball from the first pick is returned to the jar before the second pick.
  b. the ball from the first pick is not returned to the jar before the second pick.

2.19 If an integer is randomly chosen from the first 50 positive integers, what is the probability that the chosen integer will be a two-digit number?

2.20 A fair coin is tossed four times.
  a. Find the probability that the outcome is HTTH in that order.
  b. Find the probability that exactly two heads are obtained.

2.21 Suppose stocks of companies A, B, and C are popular among investors. Suppose 18% of the investors own A stocks, 49% own B stocks, 32% own C stocks, 5% own all three stocks, 8% own A and B stocks, 10% own B and C stocks, and 12% own C and A stocks.
  a. What proportion of the investors own stocks of only one of these companies?
  b. What proportion of the investors do not invest in any of these three companies?

2.22   In a group of k people, where $2 \leq k \leq 365$, what is the probability that at least 2 people in this group have the same birthday if
   a. $k = 23$?
   b. $k = 30$?
   c. $k = 50$?

   Ignore February 29.

2.23   A company in New Jersey is hiring 5 people for a position. Of the applicants, 7 are from New Jersey, 8 are from New York, and 5 are from Pennsylvania. What is the probability that of the 5 people hired, exactly 2 are from New Jersey?

2.24   A salesperson at a car dealership is showing cars to a prospective buyer. There are 9 models in the dealership. The customer wants to test-drive only 3 of them.
   a.   In how many ways could the 3 models be chosen if the order of test-driving is considered?
   b.   In how many ways could the 3 models be chosen if the order of test-driving is not important?
   c.   Suppose 6 of the models are new and the other 3 models are used. If the 3 cars to test-drive are randomly chosen, what is the probability that all 3 are new?
   d.   Is the answer to part (c) different depending on whether or not the order is considered?

2.25   A health club has 300 members and operates a gym that includes a swimming pool and 10 exercise machines. According to a survey, 60% of the members regularly use the exercise machines, 50% regularly use the swimming pool, and 25% use both of these facilities regularly.
   a.   What is the probability that a randomly chosen member regularly uses exercise machines or the swimming pool or both?
   b.   What is the probability that a randomly chosen member does not use any of these facilities regularly?
   c.   A randomly chosen member is known to use the swimming pool regularly. What is the probability that this member uses the exercise machines regularly?

2.26   For two events A and B, the probability that A occurs is 0.6, the probability that B occurs is 0.5, and the probability that both occur is 0.3. Given that B occurred, what is the probability that A also occurred?

2.27 When two events A and B are mutually exclusive, $P(A) = P(B)$, and $P(A)P(B) = 1/16$, find $P(A \cup B)$.

2.28 The following table shows the number of bedrooms and bathrooms 90 students in a class have in their house.

	A_2	A_3	A_4	Total
B_1	3	7	7	17
B_2	8	16	9	33
B_3	14	11	15	40
Total	25	34	31	90

Here, $A_2 = 2$ bedrooms, $A_3 = 3$ bedrooms, $A_4 = 4$ bedrooms, $B_1 = 1$ bathroom, $B_2 = 2$ bathrooms, and $B_3 = 3$ bathrooms. Find the following probabilities.

a. $P(A_3)$
b. Probability that a randomly selected student has a total of 6 bedrooms and bathrooms combined
c. $P(B_2 | A_3)$
d. $P(A_3 \cup B_2)$
e. $P(A_4^C)$
f. $P(A_2 \cap B_2)$

2.29 Given that $P(A) = \frac{1}{2}$ and $P(A \cap B) = \frac{1}{6}$, find $P(B^C | A)$.

2.30 If event A is a subset of B and $3P(A) = P(B) = 3/4$, find $P(A^C \cap B)$.

2.31 Given that $P(A) = 0.4$, $P(B) = 0.7$, and $P(A \cap B) = 0.3$, find
a. $P(B^C)$
b. $P(A \cup B)$
c. $P(A | B)$

2.32 Given that $P(A) = 0.6$, $P(A \cap B) = 0.2$, and $P(B) = 0.3$, find
a. $P(A | B)$
b. $P(A^C \cap B^C)$
c. $P(A \cap B^C)$
d. $P(A | B^C)$

2.33   Given that $P(A) = 0.5$, $P(B) = 0.3$, and $P(A \cup B) = 0.7$, find
   a. $P(A \mid B)$
   b. $P(A^C \cap B^C)$
   c. $P(A^C \mid B)$
   d. $P(A^C \cap B)$
   e. $P(A^C \cup B)$

2.34   Among the 640 employees in a company, 60% of males and 50% of females are married. If a randomly selected employee is married, the probability that this person is a male is double the probability that this person is a female. What is the number of female employees in this company?

2.35   In a mall, three neighboring stores offer discount coupons. Store A has 300 envelopes, Store B has 600 envelopes, and Store C has 900 envelopes. Each envelope contains one, two, or three coupons. The following table shows the number of envelopes according to the number of coupons in them in each store.

Store\#Coupons	1	2	3	Total
A	200	100	0	300
B	300	200	100	600
C	400	300	200	900

Each customer selects an envelope from each store and gets coupons in it. Rebecca went to the mall when it opened, and became the first customer for all three stores. She randomly selected an envelope from each store. After visiting the three stores, Rebecca found that she received 4 coupons. What is the probability that she received two coupons from Store A?

2.36   Brian is interested in the result of a baseball game whether the ace starting pitcher of his team pitches or not. He is given the following probabilities. The probability that the pitcher pitches and his team wins is 0.18; the probability that his team wins is 0.55. The probability that the pitcher pitches in a game is 0.25.
   a.   Given that Brian's team wins, what is the probability that the pitcher pitched?
   b.   Given that the pitcher pitched, what is the probability that Brian's team wins?
   c.   What is that probability that Brian's team loses or the pitcher did not pitch?
   d.   What is the probability that the pitcher did not pitch and Brian's team loses?

2.37 A boy has color blindness and has trouble distinguishing blue and green. There are 60 blue pens and 40 green pens mixed together in a box. Given that he picks up a blue pen, there is a 60% chance that he thinks it is a blue pen and a 40% chance that he thinks it is a green pen. Given that he picks up a green pen, there is an 80% chance that he thinks it is a green pen and a 20% chance that he thinks it is a blue pen. Assume that the boy randomly selects one of the pens from the box.
   a. What is the probability that he picks up a blue pen and recognizes it as a blue pen?
   b. What is the probability that he chooses a pen and thinks it is blue?
   c. Given that he thinks he chose a blue pen, what is the probability that he actually chose a blue pen?

2.38 The sides of a cube show numbers 2, 3, 3, 4, 4, 4. Alice is rolling this cube three times. Find the probability that the first roll results in an even number, and the sum of the numbers obtained from the second and third rolls is six.

2.39 In selecting a card from a deck of 52 cards, answer the following questions:
   a. What is the probability of selecting a diamond or a queen?
   b. Are the events selecting a diamond and selecting a queen independent?

2.40 Tags are attached to the left and right hind legs of a cow in a pasture. Let A_1 be the event that the left leg tag is lost and A_2 the event that the right leg tag is lost. Suppose these two events are independent and $P(A_1) = P(A_2) = 0.3$.
   a. Find the probability that at least one leg tag is lost.
   b. Find the probability that exactly one tag is lost, given that at least one tag is lost.
   c. Find the probability that exactly one tag is lost, given that at most one tag is lost.

2.41 In a high school, 60% of the students live east of the school and 40% live west of the school. Among the students who live east of the school, 30% are in the math club, and among the students who live west of the school, 20% are in the math club.
   a. Find the probability that a randomly selected student is from east of the school and in the math club.
   b. Given that a student is from west of the school, what is the probability that he or she is in the math club?
   c. Is participation in the math club independent from whether a student lives east or west of the school? Justify your answer.

2.42 Find $P(A \cup B)$, given that $P(A) = 0.4$, $P(B) = 0.3$, and
   a. A and B are independent
   b. A and B are mutually exclusive

2.43   Let A and B be events such that $P(A) = 0.2$ and $P(B) = 0.6$.
   a.   What is the largest possible value of $P(A \cap B)$?
   b.   What is the largest possible value of $P(A \cup B)$?
   c.   What is the smallest possible value of $P(A \cap B)$?
   d.   What is the smallest possible value of $P(A \cup B)$?

2.44   Let A and B be events such that $P(A) = 0.7$ and $P(A \cup B) = 0.6$.
   a.   What is the largest possible value of $P(A \cap B)$?
   b.   What is the largest possible value of $P(A \cup B)$?
   c.   What is the smallest possible value of $P(A \cap B)$?
   d.   What is the smallest possible value of $P(A \cup B)$?

2.45   Let $P(A) = 0.7$, $P(B^C) = 0.4$, and $P(B \cap C) = 0.48$.
   a.   Find $P(A \cup B)$ when A and B are independent.
   b.   Is it possible that A and C are mutually exclusive if they are independent?

2.46 A and B are events such that $P(A) = 0.4$ and $P(A \cup B) = 0.6$. Find $P(B)$ in each of the following cases.
   a. A and B are mutually exclusive
   b. A and B are independent
   c. $P(A|B) = 0.2$
   d. $P(A \cap B) = 0.3$

2.47   Answer the following questions.
   a.   Suppose $P(B|A) = 0.2$, $P(B \mid A^C) = 0.4$, and $P(A) = 0.7$. What is $P(B)$?
   b.   If $P(A) = 0.4$, $P(B) = 0.3$, and A and B are independent, are they mutually exclusive?

2.48   Answer the following questions.
   a.   Suppose $P(A|B) = 0.6$, $P(A|B^C) = 0.4$, and $P(A) = 0.5$. What is $P(B)$?
   b.   If $P(A) = 0.4$, $P(B) = 0.3$, and A and B are mutually exclusive, are they independent?

2.49   In a certain college class, 55% of the admitted students were in the top 10% of their high school class, 30% were in the next 10% , and the remaining 15% were below the top 20%. Of these students, 95%, 80%, and 20% were passing this course, respectively. If a randomly selected student is failing, then what is the probability that this student was below 20% of his or her high school class?

2.50   Box 1 contains 2 yellow and 4 green balls, whereas Box 2 contains 1 yellow and 1 green ball. A ball is randomly chosen from Box 1 and then transferred to Box 2, and a ball is then randomly selected from Box 2.
   a.   What is the probability that the ball selected from Box 2 is yellow?
   b.   What is the conditional probability that the transferred ball was yellow, given that a yellow ball is selected from Box 2?

2.51   Box 1 contains 2 yellow and 5 green balls, whereas Box 2 contains 1 yellow and 2 green balls. A ball is drawn from a randomly selected box.
   a.   What is the probability that this ball is yellow?
   b.   Given that the ball is yellow, what is the probability that it came from Box 2?
   c.   Let A be the event that the ball is yellow, and B the event that the ball came from Box 2. Are they independent?

2.52   Suppose components A, B_1, and B_2 operate independently in the system shown in Figure 2.10 (see Example 2.23), and the probabilities that any one of the components will operate for one week without failure are $P(A) = P(B_1) = 0.9$ and $P(B_2) = 0.8$. The system works (i.e., operates without failure) if A works and either B_1 or B_2 works. Assume that all the components in the system start operating at the same time and a component does not work again once it fails.
   a.   Find the probability that the entire system will operate without failure for one week.
   b.   Suppose B_1 failed within a week. Find the probability that the entire system will operate without failure for one week.

2.53   Suppose components A_1, A_2, B_1, and B_2 operate independently in the system shown in Figure 2.12. Assume that all the components in the system start operating at the same time and a component does not work again once it fails. The probability of functioning for each of the components is 0.8. The entire system works (i.e., operates without failure) if A_1 or A_2 works and B_1 or B_2 works. Find the probability that the entire system works.

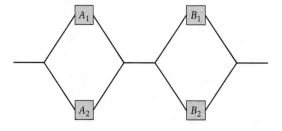

FIGURE 2.12.   System described in Exercise 2.36.

2.54   Julie is taking statistics and physics courses on Mondays. She is late for the physics class with probability 0.4 and late for the statistics class with probability 0.3. Suppose the two events are independent.
   a.   What is the probability that Julie is late for at least one of the classes?
   b.   What is the probability that Julie is on time to both the classes?
   c.   What is the probability that Julie is on time to exactly one of the classes?

2.55   In each problem below, compute the probability if the information given is sufficient to answer the question. Otherwise, answer "not enough information."
   a.   Find $P(A^C)$ when $P(A) = 0.4$.
   b.   Find $P(A \cap B)$ when $P(A) = 0.4$ and $P(B) = 0.2$.
   c.   Find $P(A|B)$ when $P(A) = 0.3$ and A and B are independent.
   d.   Find $P(A \cap B^C)$ when $P(A) = 1/4$ and A and B are mutually exclusive.
   e.   Find $P(A \cup B)$ when $P(A) = 0.23$, $P(B|A) = 0.11$, and $P(A|B) = 0.253$.
   f.   Find $P(B)$ when $P(A) = 0.62$, $P(A \cup B) = 0.67$, and $P(A|B) = 0.4$.
   g.   Find $P(A|B)$ when $P(A \cap B) = 1/8$ and A and B are independent.
   h.   Find $P(A \cup B)$ when $P(A^C \cap B^C) = 0.2$.

2.56   Bowl 1 contains 3 red chips and 7 blue chips. Bowl 2 contains 6 red chips and 4 blue chips. A single chip is drawn from a randomly selected bowl.
   a.   What is the probability that this chip is red?
   b.   Given that the chip is red, what is the probability that it came from bowl 2?
   c.   Let $A = \{$the chip is red$\}$ and $B = \{$the chip came from bowl 2$\}$.   Are   they independent?

2.57   My next-door neighbor George's car didn't start this morning. He was not sure if it was because of the battery or a damaged starter. If he replaces the battery, the engine will run with probability 0.8. If he replaces the starter, it will run with probability 0.15. He can try one of these in an hour. I was 90% certain that he would replace the battery.
   a.   What is the probability that the engine runs in an hour?
   b.   If the engine doesn't run in an hour, what is the probability that George replaced the starter?

2.58   An automobile company has three different production sites. Four percent of the cars from Site 1, 6% from Site 2, and 8% from Site 3 have been recalled due to a faulty brake system. Suppose that 50% of the cars are produced at Site 1, 30% at Site 2, and 20% at Site 3. If a randomly selected car has been recalled, what is the probability that it came from

    a.   Site 1?

    b.   Site 2?

    c.   Site 3?

2.59   Suppose two different methods are available for shoulder surgery. The probability that the shoulder has not recovered in a month is 0.001 if method A is used. When method B is used, the probability that the shoulder has not recovered in a month is 0.01. Assume that 30% of shoulder surgeries are done with method A and 70% are done with Method B in a certain hospital.

    a.   What is the probability that the shoulder has not recovered within a month after surgery?

    b.   If a shoulder is recovered within a month after surgery is done in the hospital, what is the probability that method B was performed?

2.60   A department store sells refrigerators from three manufacturers: 50% from Company A, 30% from Company B, and 20% from Company C. Suppose 10% of the refrigerators from Company A, 20% from Company B, and 5% from Company C last more than 15 years.

    a.   What is the probability that a refrigerator purchased from this department store lasts more than 15 years?

    b.   If a refrigerator purchased from this department store lasts more than 15 years, what is the probability that it was made by Company B?

2.61   Seventy percent of clay pots are produced by Machine 1 and 30% by Machine 2. Among all the pots produced by Machine 1, 4% are defective, and of those produced by Machine 2, 8% are defective.

    a.   What percentage of the total production of pots is defective?

    b.   If a pot is found to be defective, what is the probability that it was produced by Machine 2?

2.62   In a certain company, 40% of the employees are females. Suppose 60% of the male workers are married and 40% of the female workers are married. What is the probability that a married worker is male?

2.63   Sean went on a vacation for a week and asked his brother Mike to feed his 13-year-old dog Huxley. But Mike is forgetful, and Sean is 75% sure Mike will forget to feed his dog. Without food, Huxley will die with probability 0.45. With food, he will die with probability 0.01.

   a.   Find the probability that Huxley will die while Sean is on vacation.

   b.   Sean came back from vacation and found Huxley dead. What is the probability that Mike forgot to feed Huxley?

2.64   Ten percent of emails Sonya receives include the word "present." Fifty percent of emails containing the word "present" are advertisements, and twenty percent of emails not containing the word "present" are advertisements. If Sonya just received an advertising email, what is the probability that this email contains the word "present?"

Discrete Distributions

1. Random Variables

We assign a unique number to every outcome in a sample space. The outcome of an experiment is then described as a single numerical value X, which is called a *random variable*.

EXAMPLE 3.1 Let X be the number of heads obtained in 3 tosses of a fair coin. The following tables show how a number is assigned to X from each outcome of an experiment.

Outcome	X
HHH	3
HHT	2
HTH	2
HTT	1
THH	2
THT	1
TTH	1
TTT	0

Value of X	Event
0	{TTT}
1	{HTT, THT, TTH}
2	{HHT, HTH, THH}
3	{HHH}

The probabilities of X are given below.

$$P(X = 0) = \frac{1}{8}, \ P(X = 1) = \frac{3}{8}, \ P(X = 2) = \frac{3}{8}, \ P(X = 3) = \frac{1}{8}$$

A random variable is *discrete* if its set of possible values is a discrete set. In the above example, X is a discrete random variable. A random variable is *continuous* if it represents some measurement on a continuous scale. For example, the amount of precipitation produced by a storm is a continuous random variable. Continuous random variables are discussed in Chapter 4.

2. Probability Distribution

If a random variable X has a discrete set of possible values (as in Example 3.1), then its probability distribution, denoted $f(x)$, is defined as follows:

$$f(x) = P(X = x)$$

Based on this definition, we note the following necessary and sufficient conditions for $f(x)$ to be a probability distribution:

i. $0 \le f(x) \le 1$ for all x
ii. $\sum_{all\ x} f(x) = 1$

EXAMPLE 3.2

a.   Let $f(x) = \frac{x-1}{3}$ for $x = 0, 1, 2, 3$.
     Since $f(0) = -\frac{1}{3} < 0$, this function is not a valid probability distribution.
b.   Let $f(x) = \frac{x^2}{12}$ for $x = 0, 1, 2, 3$.
     Then $0 \le f(x) \le 1$ for all x. However,

$$\sum_{x=0}^{3} f(x) = 0 + \frac{1}{12} + \frac{4}{12} + \frac{9}{12} = \frac{14}{12} = \frac{7}{6} > 1$$

Thus, f is not a valid probability distribution.

EXAMPLE 3.3   Thirty percent of the automobiles in a certain city are foreign made. Four cars are selected at random. Let X be the number of cars sampled that are foreign made. Let F: foreign made, and D: domestic. The following table displays all possible outcomes for each value of X.

X – 0	X = 1	X = 2	X = 3	X = 4
DDDD	DDDF	DDFF	DFFF	FFFF
	DDFD	DFDF	FDFF	
	DFDD	DFFD	FFDF	
	FDDD	FDDF	FFFD	
		FDFD		
		FFDD		

The probability of each value of X is given below (see Figure 3.1).

$$P(X = 0) = P(DDDD) = 0.7^4 = 0.2401$$

$$P(X = 1) = 4 \cdot 0.7^3 \cdot 0.3 = 0.4116$$

$$P(X = 2) = 6 \cdot 0.7^2 \cdot 0.3^2 = 0.2646$$

$$P(X = 3) = 4 \cdot 0.7 \cdot 0.3^3 = 0.0756$$

$$P(X = 4) = 0.3^4 = 0.0081$$

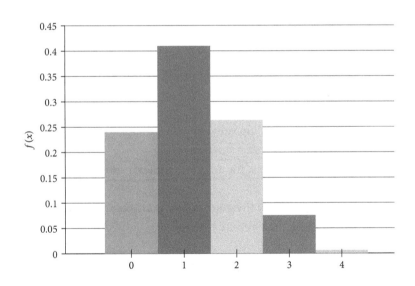

FIGURE 3.1   Probability distribution of X.

An assumed form of the probability distribution that describes the chance behavior for a random variable is called a *probability model*. Probabilities are expressed in terms of relevant population quantities, called *parameters*.

EXAMPLE 3.4 Bernoulli trials (p: parameter) have the following properties:

  a. Each trial yields one of two outcomes: success (S) or failure (F).
  b. $P(S) = P(X = 1) = p$, $P(F) = P(X = 0) = 1 - p$

$$\text{or} \ \ f(x) = \begin{cases} p, & \text{if } x = 1 \\ 1 - p, & \text{if } x = 0 \\ 0, & \text{otherwise} \end{cases}$$

where p is the probability of success in a single trial.
  c. Each trial is independent.

A Bernoulli random variable X is denoted as $X \sim \text{Bernoulli}(p)$. For a discrete random variable X, the *cumulative distribution function* (cdf) for a probability distribution $f(x)$ is denoted $F(x)$ and is defined as follows:

$$F(x) = P(X \le x) = \sum_{y \le x} f(y)$$

If the range of a random variable X consists of the values $x_1 < x_2 < \cdots < x_n$, then $f(x_1) = F(x_1)$ and $f(x_i) = F(x_i) - F(x_{i-1})$ for $i = 2, 3, \cdots, n$. The cdf is a nondecreasing function.

For a discrete random variable:
  i. A cdf has a jump at each possible value equal to the probability of that value.
  ii. The graph of the cdf will be a step function.
  iii. The graph increases from a minimum of 0 to a maximum of 1.

EXAMPLE 3.3 (CONTINUED)   From the automobile example in Example 3.3, the distribution of X was obtained as follows:

$$f(0) = 0.2401, \ f(1) = 0.4116, \ f(2) = 0.2646, \ f(3) = 0.0756, \ f(4) = 0.0081$$

and thus the cdf is given as follows (see Figure 3.2):

$$F(0) = P(X \leq 0) = \text{probability of no foreign made car} = f(0) = 0.2401$$
$$F(1) = P(X \leq 1) = \text{probability that at most 1 car is foreign made} = f(0) + f(1)$$
$$= 0.2401 + 0.4116 = 0.6517$$

$$F(2) = P(X \leq 2) = \text{probability that at most 2 cars are foreign made}$$
$$= f(0) + f(1) + f(2) = 0.9163$$

$$F(3) = P(X \leq 3) = f(0) + f(1) + f(2) + f(3) = 0.9919$$
$$F(4) = P(X \leq 4) = 1$$

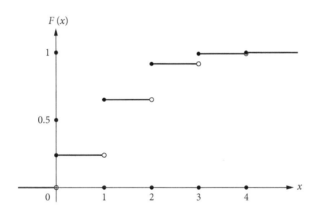

FIGURE 3.2   The cdf of X in Example 3.3.

The probability of a discrete distribution varies depending on the inclusion and exclusion of the boundary values. Figure 3.3 shows the probability expressed in terms of cdf's for $a < b$ in each case. In this figure, a^- denotes a number that is less than a by an infinitesimally small value.

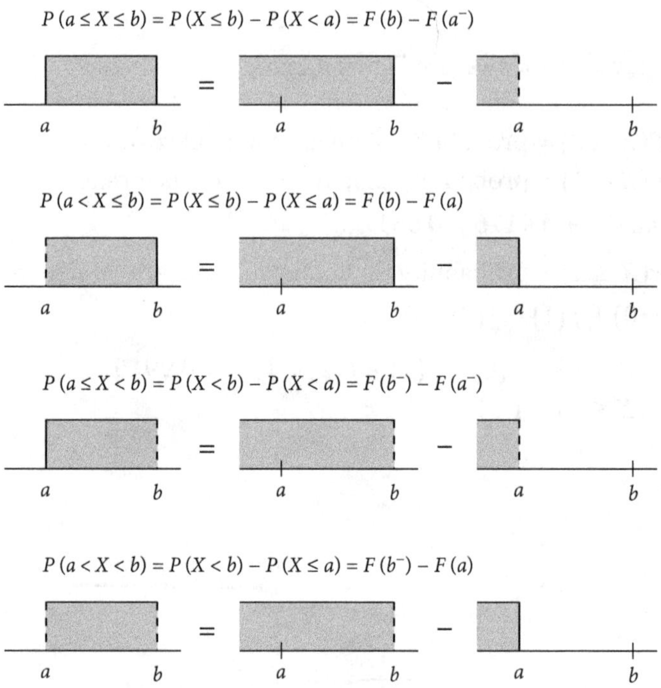

$$P(a \leq X \leq b) = P(X \leq b) - P(X < a) = F(b) - F(a^-)$$

$$P(a < X \leq b) = P(X \leq b) - P(X \leq a) = F(b) - F(a)$$

$$P(a \leq X < b) = P(X < b) - P(X < a) = F(b^-) - F(a^-)$$

$$P(a < X < b) = P(X < b) - P(X \leq a) = F(b^-) - F(a)$$

FIGURE 3.3   Probability of X in terms of the cdf.

EXAMPLE 3.5   Let X have the following distribution.

x	f(x)	F(x)
0	0.1	0.1
1	0.2	0.3
2	0.3	0.6
3	0.2	0.8
4	0.2	1

$$P(1 \leq X \leq 3) = F(3) - F(0) = 0.8 - 0.1 = 0.7$$

$$P(1 < X \leq 3) = F(3) - F(1) = 0.8 - 0.3 = 0.5$$

$$P(1 \leq X < 3) = F(2) - F(0) = 0.6 - 0.1 = 0.5$$

$$P(1 < X < 3) = F(2) - F(1) = 0.6 - 0.3 = 0.3$$

3. *The Mean and Variance of Discrete Random Variables*

The *mean* (*expected value*) of a discrete random variable X is defined as follows:

$$E(X) = \mu = \sum_{\text{all } x} xf(x)$$

(3-1)

EXAMPLE 3.6   In flipping a balanced coin 3 times, let X be the number of heads.

x	f(x)	xf(x)
0	1/8	0
1	3/8	3/8
2	3/8	3/4
3	1/8	3/8
Total	1	$\mu = 1.5$

In this example, $E(X) = 1.5$. This means that if this experiment were repeated an extremely large number of times, the average number of heads obtained per experiment would be very close to 1.5.

EXAMPLE 3.7   Let X be a Bernoulli random variable with success probability p. Then the probability distribution is:

$$f(x) = \begin{cases} p, & \text{if } x = 1 \\ 1 - p, & \text{if } x = 0 \\ 0, & \text{otherwise} \end{cases}$$

and the expectation is:

$$\mu = E(X) = \sum_x xf(x) = 0 \cdot f(0) + 1 \cdot f(1) = p$$

EXAMPLE 3.8   In a state lottery, a player picks 5 different integers between 1 and 50. If all 5 of these numbers are drawn, the prize is $1,000,000. If 4 of the 5 match, the prize is $1,000. If 3 of the 5 match, the prize is a free lottery ticket (worth $1). Matching 2 or fewer of the numbers earns no prize. For an individual player, what is the expected prize?

Assume that there is no sharing of the top prize. Let the random variable X be the monetary value of the prize in dollars. We want to compute $E(X)$.

$$P(X = 1,000,000) = \frac{1}{\binom{50}{5}} = \frac{1}{2,118,760}$$

$$P(X = 1,000) = \frac{\binom{5}{4}\binom{45}{1}}{\binom{50}{5}} = \frac{225}{2,118,760}$$

$$P(X = 1) = \frac{\binom{5}{3}\binom{45}{2}}{\binom{50}{5}} = \frac{9,900}{2,118,760}$$

$$P(X = 0) = 1 - P(X = 1) - P(X = 1,000) - P(X = 1,000,000)$$

$$E(X) = 0 \cdot P(X = 0) + 1 \cdot P(X = 1) + 1,000 P(X = 1,000) + 1,000,000 P(X = 1,000,000)$$

$$= 0 + \frac{9,900}{2,118,760} + \frac{1,000 \cdot 225}{2,118,760} + \frac{1,000,000}{2,118,760} \approx 0.583$$

The expected prize, which is also the value of a ticket, is 58.3 cents. Since a ticket costs $1, playing the lottery is a losing proposition.

Note that the expected prize must be less than the price of a ticket for the state to make money on the lottery. Because there are typically millions of players, the state can predict its profit very accurately based on the number of tickets it sells. If, for example, 10,000,000 tickets are sold, then the expected profit of the state is

10,000,000 tickets $\times$ $1/ticket $-$ 10,000,000 tickets $\times$ $0.583/ticket = $4,171,591.

Let $h(x)$ be any function of a real number x. Let X be a random variable with a probability distribution $f(x)$. Then $h(X)$ is also a random variable (on some other sample space), and we can compute its expectation in the obvious way:

$$E(h(X)) = \sum_{\text{all } x} h(x) f(x)$$

EXAMPLE 3.9   In flipping 3 balanced coins, find $E(X^3 - X)$.

x	$x^3 - x$	$f(x)$	$(x^3 - x)\,f(x)$
0	0	1/8	0
1	0	3/8	0
2	6	3/8	9/4
3	24	1/8	3
Total	30	1	21/4

$$E(X^3 - X) = 0 + 0 + \frac{9}{4} + 3 = \frac{21}{4} = 5.25$$

Now we can define the variance of a probability distribution as follows.

$$Var(X) = \sigma^2 = E((X - \mu)^2) = \sum_{all\ x}(x - \mu)^2 f(x) \qquad \textbf{(3-2)}$$

The standard deviation σ is the square root of the variance:

$$sd(X) = \sigma = \sqrt{Var(X)}$$

EXAMPLE 3.10   For the following distribution, the mean and variance of X can be calculated as follows.

x	$f(x)$	$xf(x)$	$(x - \mu)^2$	$(x - \mu)^2\,f(x)$
1	0.3	0.3	4	1.2
2	0.4	0.8	1	0.4
5	0.2	1.0	4	0.8
9	0.1	0.9	36	3.6
Total	1	$\mu = 3$		$\sigma^2 = 6$

$$\sigma^2 = Var(X) = \sum_{x}(x - \mu)^2 f(x) = 6$$

THEOREM 3.1   Let X be a random variable, and let Y be a random variable related to X as follows: $Y = aX + b$, where a and b are constants. Then $E(Y) = aE(X) + b$.

PROOF

$$E(Y) = E(aX + b)$$
$$= \sum_x (ax + b)f(x) = \sum_x axf(x) + \sum_x bf(x) = a\sum_x xf(x) + b\sum_x f(x) = aE(X) + b$$

THEOREM 3.2   The variance formula in (3-2) has the following alternative form.

$$Var(X) = E(X^2) - \mu^2 = \sum_{all\ x} x^2 f(x) - \left[\sum_{all\ x} xf(x)\right]^2$$

PROOF

$$Var(X) = E[(X - \mu)^2] = \sum_{all\ x} (x - \mu)^2 f(x) = \sum_{all\ x} (x^2 - 2\mu x + \mu^2)f(x)$$
$$= \sum_{all\ x} x^2 f(x) - 2\mu \sum_{all\ x} xf(x) + \mu^2 \sum_{all\ x} f(x)$$
$$= \sum_{all\ x} x^2 f(x) - 2\mu^2 + \mu^2 = \sum_{all\ x} x^2 f(x) - \mu^2$$
$$= E(X^2) - \mu^2$$

Compare the formula of Theorem 3.2 to the alternative formula for the sample variance given in Theorem 1.1 of Chapter 1.

EXAMPLE 3.11   The variance of X in Example 3.10 can be calculated using the alternative formula in Theorem 3.2.

x	x^2	$f(x)$	$xf(x)$	$x^2f(x)$
1	1	0.3	0.3	0.3
2	4	0.4	0.8	1.6
5	25	0.2	1.0	5.0
9	81	0.1	0.9	8.1
Total		1	$\mu = 3$	$E(X^2) = 15$

$$Var(X) = E(X^2) - \mu^2 = 15 - 3^2 = 6$$

EXAMPLE 3.12 The evaluation of the variance of the Bernoulli random variable X with the success probability p can be done as follows.

$$f(x) = \begin{cases} p, & \text{if } x = 1 \\ 1-p, & \text{if } x = 0 \\ 0, & \text{otherwise} \end{cases}$$

$\mu = E(X) = p$ as obtained in Example 3.11.

$$E(X^2) = \sum_x x^2 f(x) = 0 \cdot f(0) + 1 \cdot f(1) = p$$

$$Var(X) = E(X^2) - \mu^2 = p - p^2 = p(1-p)$$

4. The Binomial Distribution

In n independent Bernoulli trials, we define a random variable X as the number of successes. Here, X has the binomial distribution with success probability p with the following properties.

$X \sim \text{Bin}(n, p)$, where
n: a fixed number of Bernoulli trials
p: the probability of success in each trial
X: number of successes in n trials

Let's revisit the automobile example in Example 3.3. Thirty percent of the automobiles in a certain city are foreign made. Four cars are selected at random. Let X be the number of cars sampled that are foreign made. Let F: foreign made, and D: domestic. The following table displays all possible outcomes for each value of X.

	$X = 0$	$X = 1$	$X = 2$	$X = 3$	$X = 4$
	DDDD	DDDF	DDFF	DFFF	FFFF
		DDFD	DFDF	FDFF	
		DFDD	DFFD	FFDF	
		FDDD	FDDF	FFFD	
			FDFD		
			FFDD		
Probability of each outcome	$(1-p)^4$	$p(1-p)^3$	$p^2(1-p)^2$	$p^3(1-p)$	p^4
Number of outcomes	$\binom{4}{0} = 1$	$\binom{4}{1} = 4$	$\binom{4}{2} = 6$	$\binom{4}{3} = 4$	$\binom{4}{4} = 1$

Here, X has a binomial distribution with $n = 4$ and $p = 0.3$. The probability of $X = x$, where $x = 0, 1, \cdots, 4$ is given as:

$$f(x) = P(X = x) = \binom{4}{x} p^x (1-p)^{4-x}$$

Using the above formula, we can find the following probabilities.

$P(X = 3) = f(3) = 4(0.3)^3(0.7) = 0.0756$

$P(X \geq 3) = \sum_{x=3}^{4} f(x) = \binom{4}{3}(0.3)^3(0.7) + \binom{4}{4}(0.3)^4 = 0.0756 + 0.0081 = 0.0837$

$P(X \leq 1) = \sum_{x=0}^{1} f(x) = \binom{4}{0}(0.7)^4 + \binom{4}{1}(0.3)(0.7)^3 = 0.2401 + 0.4116 = 0.6517$

$P(X < 2) = P(X \leq 1) = 0.6517$

Let's generalize this case, so that in an experiment of n independent Bernoulli trials the probability of success in any individual trial is p. Then we obtain the binomial distribution given as follows:

$$P(X = x) = \binom{n}{x} p^x (1-p)^{n-x}, \qquad x = 0, 1, \cdots, n$$

Bin(n, p) is a probability distribution because $f(x) \geq 0$ and:

$$\sum_{x=0}^{n} f(x) = \sum_{x=0}^{n} \binom{n}{x} p^x (1-p)^{n-x} = [p + (1-p)]^n = 1$$

Binomial tables are given in Table A.2 in the Appendix. The tables display the following cdf of $X \sim$ Bin(n, p) for various values of n and p.

$$P(X \leq x) = \sum_{k=0}^{x} \binom{n}{k} p^k (1-p)^{n-k}, \qquad x = 0, 1, \cdots, n$$

The binomial distribution and cdf can also be obtained using the statistical software R. For Bin(n, p), the probability distribution can be obtained as:

>dbinom(x, n, p)

and the cdf can be obtained as:

>pbinom(x, n, p)

EXAMPLE 3.13 For a binomial distribution, if $n = 12$ and $p = 0.3$, then $X \sim$ Bin$(12, 0.3)$.

1. $P(X \leq 7) = \sum_{x=0}^{7} \binom{12}{x} p^x (1-p)^{12-x} = 0.9905$
    Using R, it can be obtained as:
    >pbinom(7, 12, 0.3)
2. $P(X = 7) = P(X \leq 7) - P(X \leq 6) = 0.9905 - 0.9614 = 0.0291$
    Using R, it can be obtained as:
    >dbinom(7, 12, 0.3)
3. $P(X \geq 7) = 1 - P(X < 7) = 1 - P(X \leq 6) = 1 - 0.9614 = 0.0386$
    Using R, it can be obtained as:
    >1-pbinom(6, 12, 0.3)
4. $P(4 \leq X \leq 7) = P(X \leq 7) - P(X \leq 3) = 0.9905 - 0.4925 = 0.4980$
    Using R, it can be obtained as:
    >pbinom(7, 12, 0.3) – pbinom(3, 12, 0.3)

EXAMPLE 3.14 If the probability is 0.1 that a certain device fails a comprehensive safety test, what are the probabilities among 15 of such devices that

a. at most 2 will fail?

$$P(X \le 2) = 0.8159$$

b. at least 3 will fail?

$$P(X \ge 3) = 1 - P(X \le 2) = 1 - 0.8159 = 0.1841$$

The binomial distribution has special formulas for the mean and variance. For $X \sim \text{Bin}(n, p)$,

$$E(X) = np \text{ and } Var(X) = np(1-p)$$

$$\text{Hence, } \sigma = \sqrt{np(1-p)}$$

Because the binomial random variable is the sum of n independent Bernoulli random variables each having the mean of p, the mean of a binomial random variable is np. Since the variance of a Bernoulli random variable is $p(1-p)$, the variance of a binomial random variable is $np(1-p)$. We can also derive the mean and variance of a binomial random variable by using the equations (3-1) and (3-2) given in Section 3.

EXAMPLE 3.15 Find the mean and variance of the probability distribution of the number of heads obtained in 3 flips of a balanced coin.

$X \sim \text{Bin}(3, 0.5)$

$E(X) = \mu = np = 3(0.5) = 1.5$

$Var(X) = \sigma^2 = np(1-p) = 3(0.5)(0.5) = 0.75$

The probability distribution is skewed to the right when $p < 0.5$, skewed to the left when $p > 0.5$, and symmetric when $p = 0.5$. The variance $np(1-p)$ is the largest when $p = 0.5$. Figure 3.4 illustrates the shape of binomial distributions depending on the value of p.

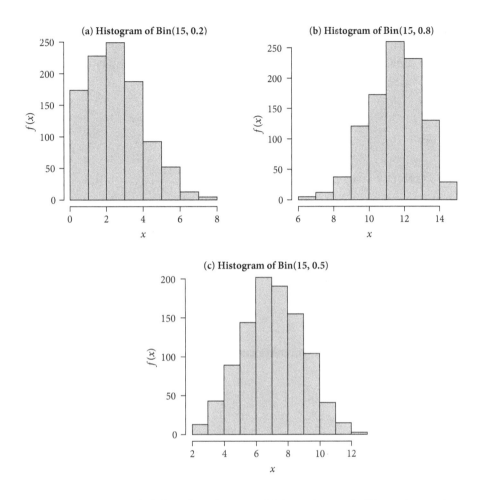

FIGURE 3.4   Probability distribution of (a) Bin(15, 0.2), (b) Bin(15, 0.8), and (c) Bin(15, 0.5).

5. *The Hypergeometric Distribution*

The binomial distribution is applied to an infinite population or describes success/failure in repeated trials of *sampling with replacement* in a finite population. What about sampling *without* replacement in a finite population?

EXAMPLE 3.16 Pick a card at random from a full deck, record its face value, and then put it back in the deck and reshuffle. What is the probability distribution for selecting jacks if we perform this process 3 times?

$$P(\text{select a jack in 1 trial}) = \frac{4 \text{ jacks}}{52 \text{ cards}} = \frac{1}{13}$$

$$P(\text{select } x \text{ jacks in 3 trials}) = \binom{3}{x}\left(\frac{1}{13}\right)^x \left(\frac{12}{13}\right)^{3-x}, x = 0, 1, 2, 3$$

Now consider the following alternative: Select 1 card at random 3 times, but do not put each card back in the deck prior to making the next selection. This is sampling without replacement, as in dealing a hand in a card game. Note that the individual trials are *not* independent.

$$P(\text{select no jack}) = \frac{\binom{4}{0}\binom{48}{3}}{\binom{52}{3}} \qquad\qquad P(\text{select 1 jack}) = \frac{\binom{4}{1}\binom{48}{2}}{\binom{52}{3}}$$

$$P(\text{select 2 jacks}) = \frac{\binom{4}{2}\binom{48}{1}}{\binom{52}{3}} \qquad\qquad P(\text{select 3 jacks}) = \frac{\binom{4}{3}\binom{48}{0}}{\binom{52}{3}}$$

The distribution can be written as:

$$f(x) = P(\text{select } x \text{ jacks}) = \frac{\binom{4}{x}\binom{48}{3-x}}{\binom{52}{3}}, \qquad x = 0, 1, 2, 3$$

Now let's generalize this example. Suppose that we have a set of N objects, of which a are considered "desirable." We select n of these objects at random without replacement. What is the probability of obtaining exactly x successes? In the example above, we have $N = 52$ cards, $a = 4$ jacks, $n = 3$ cards drawn, and the random variable X is the number of jacks selected.

The probability distribution for this random variable depends on the parameters n, a, and N, and is given as, using the example above as a guide, as follows:

$$f(x) = \frac{\binom{a}{x}\binom{N-a}{n-x}}{\binom{N}{n}}, \quad \max\{0, n - N + a\} \le x \le \min\{n, a\}.$$

The constraint of x comes from $x \leq a$ and $n - x \leq N - a$.

The hypergeometric distribution is defined for the following setup:

    i.   Finite population with N individuals (binomial: infinite population)
    ii. Success or failure. There are a successes in the population.
    iii. Sample size: n. Each subset of size n is equally likely to be chosen.

The hypergeometric distribution and cdf can also be obtained using R. For the parameters shown above, the probability distribution can be obtained as:

>dhyper(x, a, N-a, n)

and the cdf can be obtained as:

>phyper(x, a, N-a, n)

EXAMPLE 3.17 A shipment of 25 compact discs contains 5 defective ones. Ten are selected at random. What is the probability that two of them will be defective?

$$x = 2, n = 10, a = 5, N = 25$$

$$f(2) = P(X = 2) = \frac{\binom{5}{2}\binom{20}{8}}{\binom{25}{10}} = \frac{10 \cdot 1{,}511{,}640}{49{,}031{,}400} = 0.385$$

Using R, it can be obtained as:

>dhyper(2, 5, 20, 10)

The mean and variance of the hypergeometric distribution can be obtained using the equations (3-1) and (3-2) as:

$$E(X) = \frac{na}{N}, \qquad Var(X) = \frac{na(N-a)(N-n)}{N^2(N-1)}$$

The mean can be derived rigorously from the expression for the distribution function and the equations, but it should make perfect sense anyway: a/N is the proportion of the objects that are desirable. If we select n objects at random, we should expect, on average, to obtain na/N "hits."

The hypergeometric distribution accounts for the finite size of the set of objects that is being sampled. Thus, we should expect the hypergeometric distribution to approach the binomial distribution in the limit as the number of objects becomes extremely large. If we let $a/N = p$, then:

$$E(X) = np \text{ and } Var(X) = \frac{na}{N}\left(1 - \frac{a}{N}\right)\left(\frac{N-n}{N-1}\right) = np(1-p)\left(\frac{N-n}{N-1}\right)$$

$$\text{If } N \rangle\rangle n, \text{ then } Var(X) \approx np(1-p)$$

Here, $(N-n)/(N-1)$ is called the finite population correction factor. In the limit $N \to \infty$ and $a \to \infty$ (such that the proportion of desirable objects remains intact as we take this limit), the hypergeometric distribution function converges to the binomial distribution function. The binomial approximation to the hypergeometric distribution is adequate for an experiment without replacement with $n/N \leq 0.1$ and p not too close to either 0 or 1.

EXAMPLE 3.18 For Example 3.17, suppose a shipment of 100 compact discs contains 20 defective ones. Ten are selected at random. Find the probability that 2 of them will be defective using:

a. The formula for the hypergeometric distribution:

$$x = 2, n = 10, a = 20, N = 100$$

$$P(X = 2) = \frac{\binom{20}{2}\binom{80}{8}}{\binom{100}{10}} = 0.318$$

Using R, it can be obtained as:
>dhyper(2, 20, 80, 10)

b. The binomial approximation to the hypergeometric distribution:

$$x = 2, n = 10, p = \frac{a}{N} = \frac{20}{100} = 0.2, \frac{n}{N} = 0.1$$

$$P(X = 2) = \binom{10}{2}(0.2)^2(0.8)^8 = 0.302$$

Using R, it can be obtained as:

>dbinom(2, 10, 0.2)

Note that the answers to part (a) and part (b) are pretty close.

6. *The Poisson Distribution*

Let us first state the Poisson distribution, and then explore its meaning and applications. If the random variable X is distributed as Poisson with mean λ, then we denote it as $X \sim \text{Poisson}(\lambda)$. The Poisson distribution is given as follows:

$$f(x) = \frac{\lambda^x e^{-\lambda}}{x!}, \qquad x = 0, 1, 2, \cdots; \lambda > 0$$

Is the above function a valid probability distribution? It is clearly positive for all x. The factorial in the denominator guarantees that f decreases rapidly as x increases, for any given value of λ. Thus, a quick calculator test, using small values of x, will show that f is between 0 and 1 for all x. Is $\sum_x f(x) = 1$? Recall that

$$e^\lambda = 1 + \lambda + \frac{\lambda^2}{2!} + \frac{\lambda^3}{3!} + \cdots = \sum_{x=0}^{\infty} \frac{\lambda^x}{x!}$$

is the Maclaurin series of e^λ. Thus, we can conclude that

$$\sum f(x) = \sum \frac{\lambda^x e^{-\lambda}}{x!} = e^{-\lambda} e^{\lambda} = 1$$

and therefore, $f(x)$ is a probability distribution.

What is the use of the Poisson distribution? Consider an experiment in which a measuring device (possibly human) is set up to count the number of occurrences of some repeating event—i.e., the number of "hits"—over a given time interval. Common examples are:

- The number of calls received at a telephone switchboard.
- The number of atomic particles emitted by a radioactive substance. The emission of each particle is recorded as a click on a Geiger counter.
- The number of cars passing by a mileage marker on an uncrowded highway, as recorded by a person or camera.

In all of these examples, the events recorded are infrequent, in the sense that it is possible to divide the total time of the experiment T into n *small* intervals of length Δt ($\Delta t = T/n$) so that:

1. The probability of a hit during *any* of these intervals is the same and proportional to the length of the interval, i.e., Δt.
2. The probability of more than one hit during this interval is negligible.
3. The hits are independent events.

Poisson tables are given in Table A.3. The tables display the following cdf of $X \sim$ Poisson(λ) for various values of λ.

$$P(X \le x) = \sum_{k=0}^{x} \frac{\lambda^k e^{-\lambda}}{k!}, \qquad x = 0, 1, \cdots$$

The Poisson distribution and cdf can also be obtained using R. For Poisson(λ), the probability distribution can be obtained as:

>dpois(x, λ)

and the cdf can be obtained as:

>ppois(x, λ)

The Poisson distribution is highly skewed if the value of λ is small. As λ increases, the distribution becomes more symmetric. Figure 3.5 illustrates the shape change of Poisson distributions as λ varies.

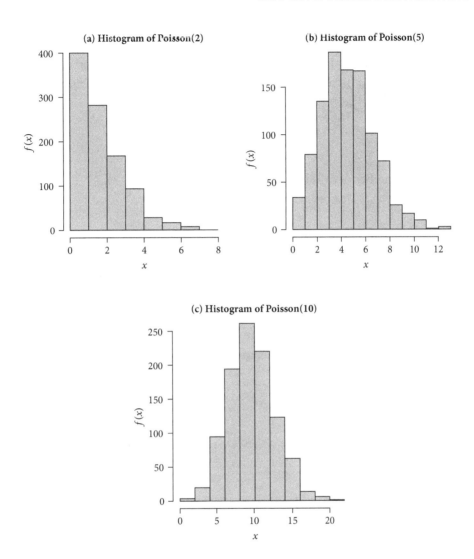

FIGURE 3.5   Probability distribution of (a) Poisson(2), (b) Poisson(5), and (c) Poisson(10).

EXAMPLE 3.19 A 911 operator receives, on average, 4 calls every 3 hours.

a.   Find the probability of receiving no calls in the next hour.

Let X be the number of calls received in the next hour. Then X is distributed as Poisson, and we need to know the value of the parameter λ. Recall that λ is the mean of the Poisson distribution, which in the example is simply the number of calls expected over the next hour. We know that 4 calls are received every 3 hours on average, so we should therefore expect 4/3 calls over the next hour. This can be obtained as below.

$$E(X) = \lambda = 4/3$$

If X is the number of calls in an hour, then $X \sim$ Poisson($4/3$).

$$P(X = 0) = \frac{(4/3)^0 e^{-4/3}}{0!} = e^{-4/3} = 0.264$$

The same answer can be obtained using R as:
>dpois(0, 4/3)

b. Find the probability of receiving at most 2 calls in the next hour.

$$P(X \le 2) = \frac{(4/3)^0 e^{-4/3}}{0!} + \frac{(4/3)e^{-4/3}}{1!} + \frac{(4/3)^2 e^{-4/3}}{2!}$$

$$= e^{-4/3}\left(1 + \frac{4}{3} + \frac{16}{18}\right) = \frac{29}{9}e^{-4/3} = 0.849$$

The same answer can be obtained using R as:
>ppois(2, 4/3)

EXAMPLE 3.20 Suppose, on average, 12 cars pass a tollbooth per minute during rush hours. Find the probability that:

a. One car passes the booth in 3 seconds.

$$\lambda = (0.2 \text{ car/sec})(3 \text{ seconds}) = 0.6 \text{ car}$$

The probability can be found using the Poisson table as:

$$P(X = 1) = P(X \le 1) - P(X \le 0) = 0.878 - 0.549 = 0.329$$

The same answer can be obtained using R as:
>dpois(1, 0.6)

b.   At least 2 cars pass in 5 seconds.

$$\lambda = (0.2)5 = 1$$

$$P(X \geq 2) = 1 - P(X < 2) = 1 - P(X \leq 1) = 1 - 0.736 = 0.264$$

The same answer can be obtained using R as:
>1-ppois(1, 1)

c.   At most 1 car passes in 10 seconds.

$$\lambda = (0.2)10 = 2$$

$$P(X \leq 1) = 0.406$$

The same answer can be obtained using R as:
>ppois(1, 2)

Poisson approximation to the binomial

Let X be a random variable from a binomial distribution. If we are taking the limits $n \to \infty$ and $p \to 0$ such that np remains constant (np is a moderate number), then the Poisson distribution is obtained. In other words, when n is very large and p is very small, then the binomial distribution is approximated by the Poisson distribution with $\lambda = np$. A rule of thumb is that the approximation is acceptable when $n \geq 20$ with $p \leq 0.05$. The approximation is excellent when $n \geq 100$ and $p \leq 0.01$ with $np \leq 20$.

n is large, p is small $\Rightarrow$ Bin(n, p) $\approx$ Poisson(λ), where $\lambda = np$

If we apply the same limit used above to the mean and variance of the binomial distribution, then we can derive the mean and variance of the Poisson distribution. The mean of the binomial distribution is np, which is a constant, equal to λ, in this limit. Thus, λ must be the mean of the Poisson distribution. The variance is almost as easy:

$$\sigma^2 = np(1 - p) = \lambda(1 - p) \to \lambda$$

as $p \to 0$. Thus, the variance of the Poisson distribution is also equal to λ.

$X \sim$ Poisson(λ) $\Rightarrow$ $E(X) = \lambda$ and $Var(X) = \lambda$

EXAMPLE 3.21 A publisher of mystery novels tries hard to keep its books free of typographical errors (typos), to the extent that if we select a page at random from any book published by this firm, the probability that we will find at least 1 typo is 0.003. Assume that the occurrence of typos on different pages are independent events.

a. What is the approximate probability that a 500-page novel has exactly 1 page with typos?

Let X be the number of pages in the 500 pages that contain typos. Then $X \sim \text{Bin}(n, p)$, where $n = 500$, $p = 0.003$. Therefore, the exact answer is:

$$P(X = 1) = \binom{500}{1}(0.003)^1(0.997)^{499} = 0.3349$$

The same answer can be obtained using R as:
```
>dbinom(1, 500, 0.003)
```

Let's use the Poisson distribution to approximate this result. We have $\lambda = np = (500)(0.003) = 1.5$ pages with typos expected. Thus:

$$P(X = 1) \approx \frac{1.5e^{-1.5}}{1!} = 0.3347$$

The same answer can be obtained using R as:
```
>dpois(1, 1.5)
```

We see that the Poisson distribution gives a highly accurate approximation. In fact, for all practical purposes, one can consider this result to be exact, since the error in the approximation is far less than the error in the problem data. In other words, the value of p, 0.003, has only 1 significant digit, whereas we need several digits to distinguish the approximation from the exact answer!

b. What is the probability that the novel has at most 2 pages with typos?

$$P(X \leq 2) \approx \sum_{x=0}^{2} \frac{1.5^x e^{-1.5}}{x!} = 0.223 + 0.335 + 0.251 = 0.809$$

The same answer can be obtained using R as:
```
>ppois(2, 1.5)
```

7. *The Geometric Distribution*

Again, we consider a problem consisting of a number of Bernoulli trials, where the probability of success in any single trial is p, but this time we do not fix the number of trials. Instead, we perform one trial after another, stopping when a success is obtained. If we define the random variable X to be the number of trials performed, then X has a geometric distribution ($X \sim \text{Geometric}(p)$) of the form

$$f(x) = p(1-p)^{x-1}, \qquad x = 1, 2, \cdots$$

This distribution can be understood by using the example of coin flips. Here, the experiment is to flip a balanced coin repeatedly until the first tail is obtained. Half of the time, we expect to get tails on the first flip, in which case $X = 1$. Thus, $P(X = 1) = 1/2$. In the other half of the cases, we proceed to a second flip, and half of these trials will result in tails, in which case we stop at $X = 2$. Thus, $P(X = 2) = (1/2)(1/2) = 1/4$. Continuing in this fashion, we obtain $P(X = x) = (1/2)^x$, $x = 1, 2, \cdots$. If we now generalize this example to the case where there is no equal probability of success or failure (i.e., $p \neq 1/2$), then we require, for an experiment with $X = x$, a sequence of $x - 1$ consecutive failures followed by a success. The probability of this event is therefore $p(1-p)^{x-1}$, $x = 1, 2, \cdots$.

In summary, the geometric distribution has the following properties:
  i.   Each trial yields either success (S) or failure (F).
  ii.   The trials are independent.
  iii. $P(S) = p$
  iv.   The process is continued until the first success is observed.

The mean and variance of the geometric distribution are:

$$E(X) = \frac{1}{p}, \quad Var(X) = \frac{1-p}{p^2}$$

EXAMPLE 3.22   Consider observing the sex of each newborn child at a certain hospital until a girl is born. Let G be the event that a girl is born and B the event that a boy is born. Let X be the number of births observed and $p = P(G)$. Then:

$$f(1) = P(X = 1) = P(G) = p$$

$$f(2) = P(X = 2) = P(BG) = p(1-p)$$

$$f(3) = P(X = 3) = P(BBG) = p(1-p)^2$$
$$\vdots$$
$$f(x) = P(X = x) = P(\underbrace{BB\cdots B}G) = p(1-p)^{x-1}$$

$x-1$ times

Probability of getting the first success on the xth trial is of the form

$$f(x) = p(1-p)^{x-1}, \qquad x = 1, 2, \cdots$$

EXAMPLE 3.23 A fair die is tossed until a certain number appears. What is the probability that the first 6 appears at the fifth toss?

Let X be the number of tosses. Then X has a geometric distribution with $p = 1/6$. The answer is:

$$P(X = 5) = \frac{1}{6}\left(\frac{5}{6}\right)^{5-1} = 0.080$$

The geometric distribution and cdf can also be obtained using R. For Geometric(p), the probability distribution can be obtained as:

```
>dgeom(x-1, p)
```

and the cdf can be obtained as:

```
>pgeom(x-1, p)
```

Note that the number of failures $(x-1)$ should be entered in R. For Example 3.23, the same answer is obtained using R by:

```
>dgeom (4,1/6)
```

8. *Chebyshev's Inequality*

If a probability distribution has mean μ and standard deviation σ, the probability of getting a value which deviates from μ by at least $k\sigma$ is at most $1/k^2$.

$$P(|X - \mu| \geq k\sigma) \leq \frac{1}{k^2}$$

An equivalent form of the Chebyshev's inequality can be obtained as:

$$P(|X - \mu| < k\sigma) \geq 1 - \frac{1}{k^2}$$

using the law of complementation.

EXAMPLE 3.24 The bearings made by a certain process have a mean radius of 18 mm with a standard deviation of 0.025 mm. With what minimum probability can we assert that the radius of a bearing will be between 17.9 mm and 18.1 mm?

$$17.9 = \mu - 0.1, 18.1 = \mu + 0.1 \Rightarrow |X - \mu| \leq 0.1$$

$$0.1 = k\sigma = 0.025k \Rightarrow k = \frac{0.1}{0.025} = 4$$

$$P(|X - \mu| \geq 0.1) = P(|X - \mu| \geq 4\sigma) \leq \frac{1}{4^2} = \frac{1}{16}$$

$$P(|X - \mu| < 0.1) = 1 - P(|X - \mu| \geq 0.1) \geq 1 - \frac{1}{16} = \frac{15}{16}$$

See Figure 3.6 for an illustration.

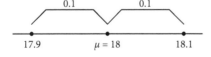

0.1   0.1

17.9 $\mu = 18$ 18.1

FIGURE 3.6   Illustration of Chebyshev's inequality in Example 3.24.

Therefore, the probability is at least 15/16 that the radius of a bearing will be between 17.9 mm and 18.1 mm.

EXAMPLE 3.25 What can you say about the probability that a random variable falls within two standard deviations of its mean?

$$P(-2\sigma < X - \mu < 2\sigma) = P(|X - \mu| < 2\sigma) \geq 1 - \frac{1}{2^2} = 1 - \frac{1}{4} = \frac{3}{4}$$

Therefore, the probability is at least 75% that a random variable falls within two standard deviations of its mean. This is valid for any distribution. Recall that about 95% of the observations fall within two standard deviations from the mean for a bell-shaped distribution.

9. The Multinomial Distribution

Let's consider a generalization of the binomial problem, whereby each trial can have more than 2 possible outcomes. Specifically, let each trial have k possible outcomes, where k is any positive integer. Label the possible outcomes with an index ($i = 1, \cdots, k$), and let p_i be the probability that a single trial results in outcome i, so that $\sum_{i=1}^{k} p_i = 1$. We now want to consider a set of k random variables $X_1, \cdots, X_k$, where X_i is the number of times outcome i occurs in an experiment of n trials. The *joint* probability distribution for this problem is:

$$f(x_1, \cdots, x_k) = P(X_1 = x_1, \cdots, X_k = x_k) = \frac{n!}{x_1! \cdots x_k!} p_1^{x_1} \cdots p_k^{x_k}$$
where

$$p_i = P(\text{outcome } i \text{ on any trial})$$
$$X_i: \text{number of trials resulting in outcome } i$$
$$x_i = 0, 1, \cdots, n, \quad x_1 + x_2 + \cdots + x_k = n$$

This function is known as the *multinomial* distribution. Here, x_k can be expressed in terms of $x_1, \cdots, x_{k-1}$ as $x_k = n - (x_1 + \cdots + x_{k-1})$ and p_k can be expressed in terms of $p_1, \cdots, p_{k-1}$ as $p_k = 1 - (p_1 + \cdots + p_{k-1})$. When $k = 2$, the distribution reduces to the binomial distribution. Note that the form of the multinomial distribution function gives the individual terms in the multinomial expansion of $(p_1 + \cdots + p_k)^n$. The factor multiplying $p_1^{x_1} \cdots p_k^{x_k}$ is a multinomial coefficient.

EXAMPLE 3.26 Consider an experiment of throwing a drum 10 times. Let X_1 be the number of heads, X_2 the number of sides, and X_3 the number of tails. Let $p_1 = 1/4$, $p_2 = 1/2$, $p_3 = 1/4$. What is the probability of having 2 heads, 5 sides, and 3 tails?

Here we have $k = 3$ outcomes for each of $n = 10$ trials. The answer is:

$$P(X_1 = 2, X_2 = 5, X_3 = 3) = \frac{10!}{2!5!3!}(1/4)^2(1/2)^5(1/4)^3 = \frac{10!}{2!5!3!}(1/4)^5(1/2)^5 = 0.0769$$

The multinomial distribution with $k = 3$ is called a *trinomial* distribution.

SUMMARY OF CHAPTER 3

1. Random Variable: A real-valued function defined over the elements of the sample space.
2. A random variable is <u>discrete</u> if its set of possible values is a discrete set.
3. Probability Distribution Function: $f(x) = P(X = x)$ for each x within the range of X.
    a. $f(x) \geq 0$ for each value within its domain
    b. $\sum_{\text{all } x} f(x) = 1$
4. Cumulative Distribution Function (cdf) of a discrete random variable X:

$$F(x) = P(X \leq x) = \sum_{y \leq x} f(y)$$

5. If the range of a random variable X consists of the values $x_1 < x_2 < \cdots < x_n$, then $f(x_1) = F(x_1)$ and $f(x_i) = F(x_i) - F(x_{i-1})$ for $i = 2, 3, \cdots, n$.
6. For any numbers a and b with $a < b$
    a. $P(a \leq X \leq b) = F(b) - F(a^-)$
    b. $P(a < X \leq b) = F(b) - F(a)$
    c. $P(a \leq X < b) = F(b^-) - F(a^-)$
    d. $P(a < X < b) = F(b^-) - F(a)$
7. Expected Value (Mean): $E(X) = \mu = \sum_{\text{all } x} xf(x)$

8. Variance
    a. $Var(X) = \sigma^2 = E((X - \mu)^2) = \sum_{\text{all } x}(x - \mu)^2 f(x)$
    b. Alternative formula: $Var(X) = E(X^2) - \mu^2 = \sum_x x^2 f(x) - \left[\sum_x xf(x)\right]^2$
    c. Standard deviation: $\sigma = \sqrt{\sigma^2}$
9. The Bernoulli Trial (success and failure; p: probability of success): $X \sim \text{Bernoulli}(p)$
    a. $f(x) = p^x(1 - p)^{1-x}$, $x = 0, 1$
    b. $E(X) = p$, $Var(X) = p(1 - p)$
10. The Binomial Distribution (n independent Bernoulli trials, with replacement): $X \sim \text{Bin}(n, p)$
    a. $f(x) = \binom{n}{x} p^x(1 - p)^{n-x}$, $x = 0, 1, \cdots, n$
    b. $E(X) = np$, $Var(X) = np(1 - p)$
11. The Hypergeometric Distribution (without replacement, trials are not independent)

    a. $f(x) = \binom{a}{x}\binom{N-a}{n-x} \Big/ \binom{N}{n}$, $\max\{0, n - N + a\} \le x \le \min\{n, a\}$

    b. $E(X) = \frac{na}{N}$, $Var(X) = \frac{na(N-a)(N-n)}{N^2(N-1)}$
12. The Geometric Distribution: $X \sim \text{Geometric}(p)$
    a. Each trial yields success (S) or failure (F)
    b. Independent trials until the first success is observed
    c. $P(S) = p$
    d. $f(x) = p(1 - p)^{x-1}$, $x = 1, 2, \cdots$
    e. $E(X) = 1/p$
    f. $Var(X) = (1 - p)/p^2$
13. The Poisson Distribution with mean λ (e.g., number of occurrences in a given duration): $X \sim \text{Poisson}(\lambda)$
    a. $f(x) = \frac{\lambda^x e^{-\lambda}}{x!}$, $x = 0, 1, \cdots$
    b. $E(X) = Var(X) = \lambda$
14. Poisson Approximation to the Binomial: Let X follow the binomial distribution. If n is large, p is small, and the value of np is moderate, then X is approximately Poisson with $\lambda = np$.
15. Chebyshev's Inequality: $P(|X - \mu| < k\sigma) \ge 1 - \frac{1}{k^2}$
16. The Multinomial Distribution
    a. Each trial results in any one of k possible outcomes.
    b. $p_i = P(\text{outcome } i \text{ on any trial})$
    c. X_i: number of trials resulting in outcome i
    d. $f(x_1, \cdots, x_k) = \frac{n!}{x_1! \cdots x_k!} p_1^{x_1} \cdots p_k^{x_k}$, $x_i = 0, 1, \cdots, n$, $x_1 + x_2 + \cdots + x_k = n$

EXERCISES

3.1   A fair coin is flipped 3 times. Consider a random variable X, which is the number of runs. The number of runs is number of changes of letter H and T. For example, HHH has one run, TTH has two runs, and THT has three runs. Find the probability distribution of the random variable X.

3.2   Determine if the following are legitimate probability distributions.
a. $f(x) = \frac{x-1}{5}$, $x = 0, 1, 2, 3, 4$
b. $f(x) = \frac{x+1}{15}$, $x = 0, 1, 2, 3, 4$
c.

x	−2	0	1	3	4
$f(x)$	0.2	0.5	0.1	0.05	0.15

d.

x	1	2	3	4	5
$f(x)$	0.3	0.2	−0.1	0.4	0.2

e.

x	0	1	2	3	4
$f(x)$	0.15	0.45	0.3	0.05	0.1

3.3   There are four balls of the same size in a box. The balls are numbered 1, 2, 3, and 4. Connie randomly picks up one ball, puts it back, and then randomly picks up one ball from the box. Let X denote the sum of the numbers on the two balls she picked up.
a.   Find the probability distribution and events corresponding to the values of X.
b.   Obtain the cumulative distribution function of X.
c.   Find $P(3 \leq X < 7)$.

3.4   Consider the following probability distribution:

x	1	2	3	5	6
$f(x)$	0.05	0.20	0.45	0.20	0.10

a.   Find $P(X \geq 3)$.
b.   Find the mean.

3.5   Consider the following probability distribution:

x	0	1	2
$f(x)$	2/7	3/7	2/7

a. Find the mean of $7X$.
b. Find the variance of $7X$.

3.6   Consider the following probability distribution:

x	0	1	2	3
$f(x)$	0.05	0.30	0.20	0.45

a. Find the probability that X is greater than 2.
b. Find the mean.
c. Find the standard deviation.

3.7   Consider the following probability distribution:

x	1	2	4
$f(x)$	0.5	0.3	0.2

a. Find $P(X = 4)$.
b. Find $P(X < 2)$.
c. Find the mean.
d. Find the variance.

3.8   Consider randomly selecting a student who is among the 12,000 registered for the current semester in a college. Let X be the number of courses the selected student is taking, and suppose that X has the following probability distribution:

x	1	2	3	4	5	6	7
$f(x)$	0.01	0.02	0.12	0.25	0.42	0.16	0.02

a. Find the cdf of X.
b. Find the expected number of courses a student is taking this semester.
c. Find the variance of X.
d. Find the third quartile of this distribution.

3.9   Let X be a random variable with cdf

$$F(x) = \begin{cases} 0, & x < -2 \\ 1/8, & -2 \le x < -1 \\ 3/8, & -1 \le x < 0 \\ 5/8, & 0 \le x < 1 \\ 7/8, & 1 \le x < 2 \\ 1 & x \ge 2 \end{cases}$$

a. Find the probability distribution of X.
b. Find the expected value and standard deviation of X.
c. Find the 70th percentile.
d. Find $P(-1 \le X \le 1)$.

3.10   Let X be a random variable with cdf

$$F(x) = \begin{cases} 0, & x < 0 \\ 0.05, & 0 \le x < 1 \\ 0.3, & 1 \le x < 2 \\ 0.6, & 2 \le x < 3 \\ 0.9, & 3 \le x < 4 \\ 1 & x \ge 4 \end{cases}$$

a. Find the probability distribution of X.
b. Find the expected value and standard deviation of X.
c. Find $P(1.5 < X \le 3)$.
d. Find $P(X < 2)$.
e. Find $P(X = 2)$.
f. Find $P(X > 2)$.
g. Find $F(2.5)$.
h. Find $P(X \ge 3)$.
i. Find the median.

3.11   Let X denote the sum of the numbers in two tosses of a fair four-sided die with four equilateral triangle-shaped faces. Each of the dice's faces shows a different number from 1 to 4.
   a.   Find $P(X \leq 3 \text{ or } X \geq 6)$.
   b.   Find $P(3 < X \leq 6)$.
   c.   Find $E(X)$.
   d.   Find $Var(X)$.

3.12   You are told that $E(X) = 1$, $Var(X^2) = 4$, and $E(X^4) = 20$. What is $Var(X)$?

3.13   On a standardized test there are 20 multiple-choice questions. On each question there are five answer choices, but only one is correct. Steve guesses on each question. Find the probability that he answers between 4 and 8 (inclusive) questions correctly.

3.14   Which of the following experiments are Bernoulli trials?
   a.   A die is tossed 4 times and we record the number obtained for each trial.
   b.   A die is tossed 4 times and we record whether or not the number obtained is odd for each trial.
   c.   A box contains 4 red and 6 blue balls. A ball is picked up from the box 3 times with replacement.
   d.   A box contains 4 red and 6 blue balls. A ball is picked up from the box 3 times without replacement.

3.15   Let $X \sim Bin(20, 0.6)$. Find the following probabilities.
   a. $P(X < 7)$
   b. $P(3 < X < 11)$
   c. $P(X > 8)$
   d. $P(X \leq 7)$
   e. $P(3 < X \leq 11)$
   f. $P(X \geq 9)$

3.16   A fair coin is flipped four times. Let H denote a head is obtained and T denote a tail is obtained in a flip.
   a.   Find the probability that the outcome is $HHTH$ in that order.
   b.   Find the probability that exactly 3 heads are obtained in 4 flips.

3.17 When a fair coin is tossed 6 times, what is the probability that the product of the number of heads and the number of tails is 8?

3.18 David didn't study for his introduction to logic exam consisting of 15 true-false questions. He did blind guessing on each question.
a. If he needs to score 10 or more correct to pass, what is the probability that he will fail the exam?
b. Find the probability that he answers 6 to 11 (inclusive) questions correctly.

3.19 Let the random variable X represent the number of children in a randomly selected family in a town. The following is the probability distribution of X:

x	0	1	2	3
$f(x)$	0.05	0.25	0.40	0.30

a. Find the cdf of X.
b. Find the expected number of children in a family.
c. Find the standard deviation of X.
d. Find the probability that a family has at most one child.
e. If four families are randomly chosen, find the probability that two families have at most one child.

3.20 Let X have a binomial distribution with $n = 5$ and $p = 0.2$.
a. Find $P(X = 1)$.
b. Find $P(X < 3)$.
c. Find the mean of X.
d. Find the standard deviation of X.

3.21 Let X have probability distribution

$$f(x) = \frac{2!}{x!(2-x)!}\left(\frac{1}{5}\right)^x \left(\frac{4}{5}\right)^{2-x}, \ x = 0, 1, 2.$$

a. Find the cdf $F(x)$ of X.
b. Find the probability that X is odd.
c. Find $E(X)$.
d. Find Var(X).

3.22   Let X have the following probability distribution:

$$f(x) = \frac{3!}{x!(3-x)!}\left(\frac{1}{3}\right)^x\left(\frac{2}{3}\right)^{3-x}, x = 0, 1, 2, 3$$

    a.   Find the cdf of X.
    b.   Find the probability that X is odd.
    c.   Find the mean of X.
    d.   Find the variance of X.

3.23   Find the probability that number 5 appears only once when a fair die is tossed 4 times.

3.24   Ten percent of an airline's current customers qualify for an executive traveler's club membership.
    a.   Find the probability that 2 to 5 (inclusive) out of 20 randomly selected customers qualify for the membership.
    b.   Find the expected value and the standard deviation of the number who qualify in a randomly selected sample of 50 customers.

3.25   A company is interested in evaluating its current inspection procedure on large shipments of identical items. The procedure is to take a sample of 5 items and pass the shipment if no more than 1 item is found to be defective. It is known that items are defective at a 10% rate overall.
    a.   What is the probability that the inspection procedure will pass the shipment?
    b.   What is the expected number of defectives in this process of inspecting 5 items?
    c.   If items are defective at a 20% rate overall, what is the probability that you will find 4 defectives in a sample of 5?

3.26   It is known that 30% of the 10-year-old children in a city have exactly two siblings. Twenty 10-year-old children are selected at random. Find the probability that more than 12 of these selected children have exactly two siblings.

3.27   In a high school statistics class, the teacher formed 10 groups of project teams, each consisting of 3 girls and 2 boys. The teacher is randomly selecting two presenters of the results from each group. If X is the random variable of the number of teams having only girl presenters, what is $E(X)$? Assume that no student is in more than one team.

3.28   In textbooks published by a certain publisher, the number of typographical errors (typos) and the corresponding percent frequency are given in the table below.

Number of typos	Percent frequency
0	20
1	32
2	25
3	16
4	4
5	3

a. Find the mean and variance of the number of typos in a textbook published by this publisher.
b. Suppose that 10 textbooks from this publisher are chosen at random. Find the probability that 6 of these texts contain at most one typo and the remaining 4 contain more than one typo.
c. Suppose that 10 textbook titles from this publisher are chosen at random. Find the mean and variance of the number of texts that contain at most one typo.

3.29   The human sex ratio at birth is commonly thought to be 107 boys to 100 girls. Suppose 5 infants are chosen at random.
a. What is the probability that 3 are males and 2 are females?
b. What is the probability that at least 1 is a male?

3.30   Two dice are tossed. Let X be the random variable that shows the maximum of the two tosses.
a. Find the distribution of X.
b. Find $P(X \leq 3)$.
c. Find $E(X)$.

3.31   In a classroom of 30 students, 3 of the students wear wrist watches.
a. If 14 students are selected *with replacement*, what is the probability that exactly 2 of them wear wrist watches?
b. If 14 students are selected *without replacement*, what is the probability that exactly 2 of them wear wrist watches?

3.32   In a college classroom some chairs are designed for left-handers. Suppose this classroom has 20 chairs, and 8 of them are made for lefties. If 5 students randomly select chairs and sit on them in this classroom, what is the probability that at least 4 of those selected will be seats for lefties?

3.33   In a box of 25 external hard disks, there are 2 defectives. An inspector examines 5 of these hard disks. Find the probability that there is at least 1 defective hard disk among the 5.

3.34   The school newspaper of a college has ten reporters. Three of the reporters are male students and the other seven are female students.
   a.   If you randomly select four reporters to work on an article about the 50-year anniversary of the school, in how many ways two of the chosen reporters are males?
   b.   What is the probability that at least two male students are selected among the 4 reporters chosen in part (a)?

3.35   In a class of 100 students, 25 students have hardcover and 75 students have paperback textbooks for the course. If you randomly choose 10 students in this class, find the probability that 2 of them have hardcover texts in the following ways:
   a.   the exact probability
   b.   approximate probability using a binomial distribution

3.36   Answer the questions in Exercise 3.35 using R.

3.37   Suppose a representative at a credit card customer service center receives a phone call every 5 minutes on average.
   a.   Find the probability that she receives 3 phone calls in 20 minutes.
   b.   Find the probability that she receives fewer than 3 phone calls in 20 minutes.
   c.   Find the expected number of phone calls she receives in 20 minutes.

3.38   In a certain apartment complex, the number of people who catch a cold in October is a random variable having a Poisson distribution with mean 2.
   a.   Find the probability that 2 people catch a cold in this apartment complex in October.
   b.   Find the probability that at least 1 person catches a cold here in October.

3.39 A certain online site has four visitors in three minutes on average. Let a Poisson random variable X denote the number of visitors per minute to this site.
  a. Find the probability that this site gets two visitors in a minute.
  b. Find the probability that this site gets at least two visitors in a minute.

3.40 Let X denote the number of bombs that hit per minute in an area of 1 square mile on a certain day during a war. Suppose X has a Poisson distribution with $\lambda = 5$.
  a. Find the probability that 2 bombs hit that area in a minute.
  b. Find the probability that at most 2 bombs hit that area in a minute.
  c. Find the expected number of bomb hits per minute in that area.
  d. Find the variance of the number of bomb hits per minute in that area.

3.41 In a certain community the number of persons who catch the flu each year is a random variable having a Poisson distribution with $\lambda = 3$.
  a. Find the probability that 3 people catch the flu in a given year.
  b. Find the probability that at least 1 person catches the flu in a given year.

3.42 Suppose the number of customer visits per minute in a convenient store follows a Poisson distribution with average 1 per two minutes in early afternoon.
  a. Find the probability that 2 customers visit the store in a given 5-minute time interval.
  b. Find the probability that at most one customer visits the store in a given 5-minute time interval.
  c. Find the expected number of customers visiting the store in a 10-minute interval.

3.43 One percent of a certain model of cars have defective mufflers. Suppose 400 cars in this model are ready to ship.
  a. Find an approximate probability that more than 5 cars in the shipment have defective mufflers.
  b. Find an approximate probability that 3 to 6 cars (inclusive) in the shipment have defective mufflers.

3.44 A chicken sexer makes 1 error in every 500 examinations when he detects the sexes of chicks.
  a. Find the exact probability that he makes no more than 2 errors in 1,000 examinations.
  b. Using the Poisson distribution, find an approximate probability that he makes no more than 2 errors in 1,000 examinations.

3.45 A certain drug causes liver damage in 0.5% of patients. Suppose the drug is to be tested on 100 patients. Answer the following questions.
a. Find an approximate probability that none of these patients will experience liver damage.
b. Find an approximate probability that 1 of these patients will experience liver damage.
c. Find an approximate probability that 2 or more of these patients will experience liver damage.
d. Find the exact probabilities of the above three questions and compare them with the approximate probabilities.

3.46 Answer the following questions from Example 3.21 using R.
a. Find $P(X = 2)$ using the exact distribution and using a Poisson approximation.
b. Find $P(X \leq 2)$ using the exact distribution and using a Poisson approximation.

3.47 The probability is 0.5 that an artist makes a craft item with satisfactory quality. Assume the production of each craft item by this artist is independent. What is the probability that at most 3 attempts are required to produce a craft item with satisfactory quality?

3.48 An apartment seeker in Manhattan estimates that 10% of the available apartments in her price range are in acceptable condition. Furthermore, she has time to look at only one apartment every weekend.
a. How many weeks should this person expect to spend for apartment hunting?
b. What is the probability that she will find an acceptable apartment in the first two weekends that she looks?
c. What is the probability that it will take more than 4 months (17 weeks) to find an acceptable apartment?

3.49 A certain city has an inept law enforcement agency, to the extent that only 10% of all thefts are solved (i.e., the culprit is arrested and convicted). A thief who knows some probability theory decides to permit no more than a 20% chance of being caught. Based on this criterion and the data given above, how many thefts may this criminal commit? (Presumably, he will then move to a different city.) What assumptions need to be made here in order to arrive at this estimate?

3.50   A certain model of a passenger car has two types: four-door and two-door cars. Suppose 60% of the cars in this model are four-door cars. The two types of cars in this model will be randomly delivered to a certain dealer one by one in a particular week. Zachary wants to buy a two-door car in the model from this dealer, so he asked the dealer to notify him when the first two-door car is delivered.
  a.   Find the probability that the first two-door car is the third one in this model delivered to this dealer this week.
  b.   Find the expected number of cars in this model until the first two-door car is delivered in this week.

3.51   Andrew rolled a fair die 24 times.
  a.   What is the probability that a 4 occurs no more than once?
  b.   Find the mean and variance of the probability distribution of the number of 4's occurred in this experiment.
  c.   What is the probability that the first 4 occurs at the 5th roll?
  d.   Find the expected number of rolls until the first 4 appears. Assume that Andrew continues rolling the die if the first 4 doesn't appear in first 24 trials.

3.52   Suppose the probability is 0.4 that a certain horse wins in each race.
  a.   Find the probability that this horse needs to run for 3 races until it wins a race.
  b.   Find the probability that this horse wins in at most 3 races.

3.53   Answer the questions in Exercise 3.52 using R.

3.54   Suppose the probability of an unsuccessful rocket launch is 0.2.
  a.   If 10 independent test launches are performed, find the probability of exactly 1 unsuccessful launch.
  b.   Find the probability of more than 2 unsuccessful launches in 10 independent launches.
  c.   If no more launches are performed after an unsuccessful launch occurs, what is the probability that exactly 5 total launches will be performed?

3.55   In a box there are r red pens and b blue pens. Pens are randomly selected, one at a time, until a red one is obtained. Assume that each selected pen is replaced before the next one is drawn.
  a.   What is the probability that you need to pick up a pen 3 times?
  b.   What is the probability that you need to pick up a pen at least 4 times?

3.56   When a balanced coin is flipped 10,000 times, find the lower bound of the probability that the proportion of heads obtained will fall between 0.45 and 0.55.

3.57   When a fair die is flipped 72,000 times, find the lower bound of the probability that the number 5 occurs between 11,500 and 12,500 times.

3.58   The number of passengers who arrive at the platform in an Amtrak train station for the 2:00 p.m. Saturday train is a random variable with a mean of 180 and a standard deviation of 25. Find the lower bound of the probability that there will be 80 to 280 passengers.

3.59   A certain manufacturing procedure produces items that have a mean weight of 55 pounds and a standard deviation of 3.2 pounds. With what minimum probability can we assert that the weight of a randomly selected item produced by this procedure is between 45.4 pounds and 64.6 pounds?

3.60   A die is rolled 10 times. What is the probability that 1, 2 and 3 occur once each, 4 and 5 occur twice each, and 6 occurs 3 times?

3.61   In 2000 the population of California consisted of approximately 80% white, 8% black, and 12% Asian. Perform the following calculations.
   a.   Find the probability that you met 3 white individuals, 1 black individual, and 1 Asian individual if you randomly met 5 Californians in 2000.
   b.   Find the probability that you met more white people than black and Asian people combined if you randomly met 5 Californians in 2000.

3.62   In a certain town, 30% of the eligible voters prefer the candidate from a conservative party, 20% prefer the candidate from a liberal party, and the remaining 50% have no preference in a presidential election. If you randomly sample 10 eligible voters, what is the probability that 3 will prefer the candidate from the conservative party, 2 will prefer the candidate from the liberal party, and the remaining 5 will have no preference?

Continuous Distributions

1. Probability Density

A random variable X is *continuous* if it represents some measurement on a continuous scale; for example, the weight of a newborn baby or the waiting time at a highway tollbooth. We saw in Chapter 3 that a discrete random variable has an associated probability distribution, a function that returns the probability of a given outcome. A continuous random variable X has an associated *probability density function* (pdf) $f(x)$, which is defined by the following properties:

a. $P(a \le X \le b) = \int_a^b f(x)\,dx$ for $a \le b$

b. $f(x) \ge 0$ for all x

c. $\int_{-\infty}^{\infty} f(x)\,dx = 1$

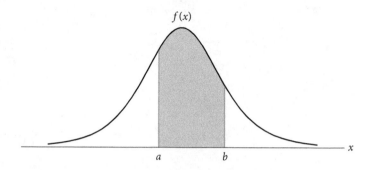

FIGURE 4.1   The area under the graph of $f(x)$ representing $P(a \leq X \leq b)$.

Figure 4.1 illustrates that the probability $P(a \leq X \leq b)$ of a continuous distribution is the area under the graph of the pdf. The probability that X takes on *some* value must be 1, as shown in Figure 4.2. Thus, condition (a) implies condition (c).

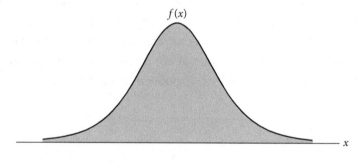

FIGURE 4.2   The area under the graph of $f(x)$ representing $\int_{-\infty}^{\infty} f(x)\,dx = 1$.

Condition (a) implies that for any number c,

$$P(X = c) = \int_{c}^{c} f(x)\,dx = 0$$

This is because the area of a line is zero.

If X is a continuous random variable, then for any two numbers a and b with $a < b$,

$$P(a \leq X \leq b) = P(a \leq X < b) = P(a < X \leq b) = P(a < X < b)$$

since the probability that a continuous random variable takes on a particular value is zero.

2. The Uniform Distribution

EXAMPLE 4.1   Suppose that a subway train arrives at a certain stop at 10-minute intervals starting at 6:00 a.m. Every time Rachel takes this train, she arrives at the station at a different time, so her waiting time is random within the 10-minute interval. Her waiting time X is a continuous random variable. What is the associated probability distribution?

A modeling assumption here is that the subway trains are reliable; i.e., it is absolutely certain that a train leaves every 10 minutes. Therefore, the probability that she will have to wait more than 10 minutes for a train is zero. A negative waiting time is certainly an impossible event, while waiting times between 0 and 10 minutes are equally likely. Thus, the pdf $f(x)$ must have the following form:

$$f(x) = \begin{cases} k, & 0 < x < 10 \\ 0, & \text{otherwise} \end{cases}$$

where k is a constant to be determined. How do we determine k? We know that it is absolutely certain that she will wait between 0 and 10 minutes for the train. Therefore,

$$1 = \int_{-\infty}^{\infty} f(x)\,dx = \int_{0}^{10} k\,dx = kx\Big|_{0}^{10} = k(10-0) = 10k$$

Thus, it follows that $k = 1/10$ and the pdf is

$$f(x) = \begin{cases} \dfrac{1}{10}, & 0 < x < 10 \\ 0, & \text{otherwise} \end{cases}$$

Having determined $f(x)$, we can compute any probability associated with waiting at the train stop; for example:

a.   The probability that she waits from 1 to 4 minutes is

$$P(1 < X < 4) = \int_{1}^{4} f(x)dx = \int_{1}^{4} \frac{1}{10} dx = \frac{x}{10}\Big|_{1}^{4} = \frac{1}{10}(4-1) = \frac{3}{10}$$

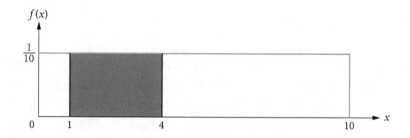

FIGURE 4.3   Probability for part (a) of Example 4.1.

We can find this probability in an alternative way. Because the probability is the area of the rectangle shown in Figure 4.3, it can be obtained by multiplying the base and the height as follows:

$$P(1 < X < 4) = \frac{1}{10}(4 - 1) = \frac{3}{10}$$

which is the last part of the previous calculation.

b.   The probability that she waits at least 6 minutes is

$$P(6 < X < 10) = \int_{6}^{10} f(x)dx = \int_{6}^{10} \frac{1}{10}dx = \frac{x}{10}\Big|_{6}^{10} = \frac{1}{10}(10 - 6) = \frac{2}{5}$$

or

$$P(6 < X < 10) = \frac{1}{10}(10 - 6) = \frac{2}{5}$$

Figure 4.4 illustrates this probability.

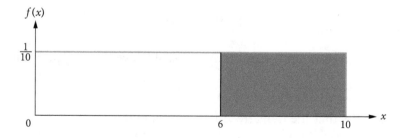

FIGURE 4.4   Probability for part (b) of Example 4.1.

The random variable in this example has a *uniform distribution*. To be general, consider the interval (a,b), and suppose that X can take on values only in this interval and with equal probability. The pdf then has the form

$$f(x) = \begin{cases} \dfrac{1}{b-a}, & a < x < b \\[2mm] 0, & \text{otherwise} \end{cases}$$

In Example 4.1, $a = 0$ and $b = 10$.

$X \sim \text{Uniform}(a, b)$ if

$$f(x) = \begin{cases} \dfrac{1}{b-a}, & a < x < b \\[2mm] 0, & \text{otherwise} \end{cases}$$

Probabilities of a uniform distribution can be obtained using the statistical software R. For $X \sim \text{Uniform}(a, b)$, $P(X < c)$, where $a < c < b$ can be obtained as follows:

>punif(c, a, b)

For Example 4.1, the answer to part (a) can be obtained by

>punif(4, 0, 10) – punif(1, 0, 10)

and the answer to part (b) can be obtained by

>1-punif(6, 0, 10)

When $a = 0$ and $b = 1$, i.e., for $X \sim \text{Uniform}(0, 1)$, $P(X < c)$, where $0 < c < 1$ can be obtained as follows:

>punif(c)

because the default is $a = 0$ and $b = 1$.

3. *The Exponential Distribution*

A large class of random variables is described with the exponential distribution with mean λ, which is defined as follows:

$$f(x) = \begin{cases} \dfrac{1}{\lambda} e^{-x/\lambda}, & x > 0 \\ 0, & \text{otherwise} \end{cases} \qquad \text{where } \lambda > 0$$

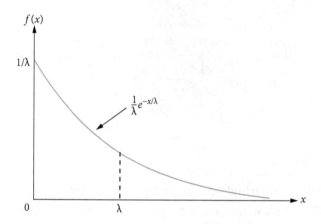

FIGURE 4.5 Exponential pdf with mean λ.

Figure 4.5 shows the pdf of an exponential distribution. We will show that $f(x)$ is a pdf. First, it is obvious that $f(x) \geq 0$ for all x. Next,

$$\int_{-\infty}^{\infty} f(x)dx = \int_{0}^{\infty} \frac{1}{\lambda} e^{-x/\lambda} dx = -e^{-x/\lambda} \Big|_{0}^{\infty} = -(0-1) = 1$$

so $f(x)$ is a pdf.

Probabilities of an exponential distribution can be obtained using R. For $X \sim \text{Exponential}(\lambda)$, $P(X < c)$ can be obtained as follows:

```
>pexp(c, 1/ λ)
```

EXAMPLE 4.2   The lifetime of a certain rodent species has the following distribution (in years).

$$f(x) = \begin{cases} \dfrac{1}{3}e^{-x/3}, & x \geq 0 \\ 0, & \text{otherwise} \end{cases}$$

a.   Find the probability that a rodent in this species lives longer than 2 years.

$$P(X > 2) = \int_2^\infty f(x)dx = \int_2^\infty \frac{1}{3}e^{-x/3}dx = -e^{-x/3}\Big|_2^\infty = e^{-2/3} = 0.5134$$

Using R, it can be obtained as follows:

>1-pexp(2, 1/3)

b.   Find the probability that a rodent in this species lives from 1 to 2 years.

$$P(1 < X < 2) = \int_1^2 \frac{1}{3}e^{-x/3}dx = -e^{-x/3}\Big|_1^2 = e^{-1/3} - e^{-2/3} = 0.2031$$

Using R, it can be obtained as follows:

>pexp(2, 1/3)-pexp(1, 1/3)

4. *The Cumulative Distribution Function*

The cumulative distribution function (cdf) of a continuous random variable is defined identically to the cdf of a discrete random variable, but of course, it is evaluated differently, as follows:

$$F(x) = P(X \leq x) = \int_{-\infty}^{x} f(y)\,dy$$

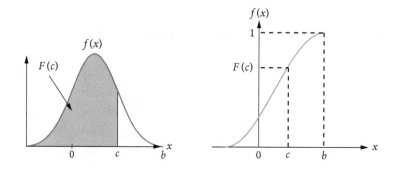

FIGURE 4.6   Cumulative distribution function of a continuous distribution.

Figure 4.6 illustrates the cdf. From the definition of $F(x)$, the following identity follows:

$$P(a < X < b) = F(b) - F(a)$$

This is illustrated in Figure 4.7.

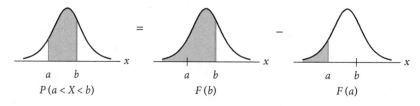

FIGURE 4.7 $P(a < X < b)$ in terms of two cumulative distribution functions.

The following probability can be expressed in terms of cdf using the law of complementation.

$$P(X \geq a) = 1 - P(X \leq a) = 1 - F(a)$$

Again, it does not matter whether the inequalities are strict or not, since the probability that X takes on any particular value is absolutely negligible. It follows from the fundamental theorem of calculus that the cdf is the antiderivative of the pdf.

If X is a continuous random variable, then $F'(x) = f(x)$ at every x at which $F'(x)$ exists.

EXAMPLE 4.3   The cdf of a uniform distribution: Let X have a uniform distribution on the interval $[a, b]$, where $a \le x \le b$. Then from Section 4.2,

$$f(x) = \begin{cases} \dfrac{1}{b-a}, & a \le x \le b \\ 0, & \text{otherwise} \end{cases}$$

Figure 4.8 shows the pdf of a uniform distribution.

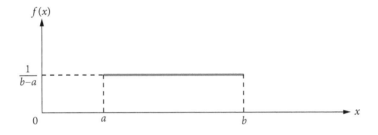

FIGURE 4.8   A uniform pdf.

For $a \le x \le b$,

$$F(x) = \int_{-\infty}^{x} f(y)dy = \int_{a}^{x} \frac{1}{b-a}dy = \frac{y}{b-a}\Big|_{a}^{x} = \frac{x-a}{b-a}$$

Thus,

$$F(x) = \begin{cases} 0, & x < a \\ \dfrac{x-a}{b-a}, & a \le x \le b \\ 1, & x > b \end{cases}$$

Figure 4.9 shows the cdf of a uniform distribution.

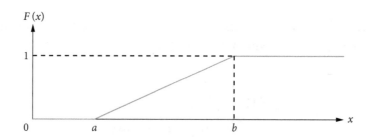

FIGURE 4.9   A uniform cdf.

EXAMPLE 4.4   The cdf of an exponential distribution: Let X have an exponential distribution with mean λ. Then the pdf is

$$f(x) = \begin{cases} \dfrac{1}{\lambda}e^{-x/\lambda}, & x > 0 \\ \\ 0, & \text{otherwise} \end{cases}$$

The cdf is

$$F(x) = \begin{cases} \displaystyle\int_{-\infty}^{x} f(y)dy = \int_{0}^{x} \frac{1}{\lambda}e^{-y/\lambda}dy = -e^{-\frac{y}{\lambda}}\Big|_{0}^{x} = 1 - e^{-\frac{x}{\lambda}}, & x > 0 \\ \\ 0, & x \leq 0 \end{cases}$$

The cdf is used for computing percentiles. The $100p$-th percentile for a distribution of X is the value of x such that, if we were to make a large number of measurements based on this distribution, we would expect a fraction p of the measurements to be no greater than x. Mathematically,

The $100p$-th percentile: x such that $F(x) = p$

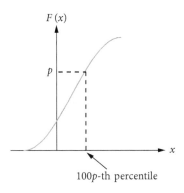

FIGURE 4.10   Finding the $100p$-th percentile.

Figure 4.10 illustrates how to find the $100p$-th percentile. The median is the 50th percentile; thus, it is x such that $F(x) = 1/2$.

EXAMPLE 4.5   Given the following pdf, find the 75th percentile and the median.

$$f(x) = \begin{cases} 3x^2, & 0 \le x \le 1 \\ 0, & \text{otherwise} \end{cases}$$

For $0 \le x \le 1$,

$$F(x) = \int_{-\infty}^{x} f(y)dy = \int_{0}^{x} 3y^2 dy = y^3 \Big|_{0}^{x} = x^3$$

The 75th percentile is obtained as follows:

$$F(x) = \frac{3}{4} \rightarrow x^3 = \frac{3}{4} \rightarrow x = \sqrt[3]{3/4}$$

The median is obtained as follows:

$$F(x) = \frac{1}{2} \rightarrow x^3 = \frac{1}{2} \rightarrow x = \frac{1}{\sqrt[3]{2}}$$

Percentiles for specific distributions can be obtained using R. For a Uniform(a,b) distribution, the $100p$-th percentile can be obtained by

```
>qunif(p, a, b)
```

From Example 4.1, the cdf of X is

$$F(x) = \begin{cases} 0, & x < 0 \\ \dfrac{x}{10}, & 0 \le x \le 10 \\ 1, & x > 10 \end{cases}$$

as obtained in Example 4.3. The 70th percentile of the waiting time can be obtained by finding x satisfying $F(x) = 0.7$, which is $\frac{x}{10} = 0.7$, and thus $x = 7$ minutes. Using R, it can be obtained by

>qunif(0.7, 0, 10)

For the exponential distribution with mean λ, the $100p$-th percentile can be obtained by

>qexp(p, 1/λ)

From Example 4.2, the cdf of X is

$$F(x) = \begin{cases} 1 - e^{-\frac{x}{3}}, & x > 0 \\ 0, & x \le 0 \end{cases}$$

as obtained in Example 4.4. The median lifetime can be obtained by finding x satisfying

$$\frac{1}{2} = F(x) = 1 - e^{-\frac{x}{3}}, \quad e^{-\frac{x}{3}} = \frac{1}{2}, \quad -\frac{x}{3} = \ln\left(\frac{1}{2}\right) = -\ln 2, \quad x = 3\ln 2 = 2.08$$

Using R, the same answer can be obtained by

>qexp(0.5, 1/3)

5. *Expectations*

The expected value (mean) of a continuous random variable is defined in the obvious way:

$$\mu = E(X) = \int_{-\infty}^{\infty} x f(x)\, dx$$

The variance of a continuous random variable is a special form of the mean defined as:

$$\sigma^2 = Var(X) = E[(X - \mu)^2] = \int_{-\infty}^{\infty} (x - \mu)^2 f(x)\, dx$$

The standard deviation of X is $\sigma = \sqrt{Var(X)}$.

THEOREM 4.1   Let X be any continuous random variable. Then Theorem 3.2 can be extended to continuous random variables as follows.

$$Var(X) = E(X^2) - \mu^2$$

The proof is similar to that of Theorem 3.2.

EXAMPLE 4.6   Let the pdf of a random variable X be given by

$$f(x) = \begin{cases} 6x(1-x), & 0 \le x \le 1 \\ 0, & \text{otherwise} \end{cases}$$

Compute the mean and standard deviation of this distribution.

$$\mu = \int_{-\infty}^{\infty} x f(x)\,dx = \int_0^1 6x^2(1-x)\,dx = 6\int_0^1 (x^2 - x^3)\,dx = 6\left(\frac{x^3}{3} - \frac{x^4}{4}\right)\Bigg|_0^1$$

$$= 6\left(\frac{1}{3} - \frac{1}{4}\right) = 6 \cdot \frac{1}{12} = \frac{1}{2}$$

$$E(X^2) = \int_{-\infty}^{\infty} x^2 f(x)dx = \int_0^1 6x^3(1-x)dx$$

$$= 6\int_0^1 (x^3 - x^4)dx = 6\left(\frac{x^4}{4} - \frac{x^5}{5}\right)\Bigg|_0^1 = 6\left(\frac{1}{4} - \frac{1}{5}\right) = 6 \cdot \frac{1}{20} = \frac{3}{10}$$

$$\sigma^2 = E(X^2) - \mu^2 = \frac{3}{10} - \left(\frac{1}{2}\right)^2 = \frac{1}{20}$$

EXAMPLE 4.7   The mean and variance of the uniform distribution on $[a, b]$ follow from straightforward integrations.

$$\mu = \int_{-\infty}^{\infty} xf(x)dx = \int_a^b \frac{x}{b-a}dx = \frac{x^2}{2(b-a)}\Bigg|_a^b = \frac{b^2 - a^2}{2(b-a)} = \frac{(b-a)(b+a)}{2(b-a)} = \frac{a+b}{2}$$

This result makes sense. If any measurement in the interval $[a, b]$ is equally likely, then the average measurement should be the midpoint $(a + b)/2$.

$$E(X^2) = \int_{-\infty}^{\infty} x^2 f(x)dx = \int_a^b \frac{x^2}{b-a}dx = \frac{x^3}{3(b-a)}\Bigg|_a^b = \frac{b^3 - a^3}{3(b-a)} = \frac{(b-a)(b^2 + ab + a^2)}{3(b-a)}$$

$$= \frac{a^2 + ab + b^2}{3}$$

$$\sigma^2 = E(X^2) - \mu^2 = \frac{a^2 + ab + b^2}{3} - \left(\frac{a+b}{2}\right)^2 = \frac{4(a^2 + ab + b^2) - 3(a+b)^2}{12}$$

$$= \frac{a^2 - 2ab + b^2}{12} = \frac{(a-b)^2}{12}$$

6. *The Normal Distribution*

By far the most important probability distribution is the *normal distribution*. When X has a normal distribution with mean μ and variance σ^2, we denote it as $X \sim N(\mu, \sigma^2)$, and the pdf of X is of the form

$$f(x) = \frac{1}{\sqrt{2\pi}\sigma}e^{-\frac{(x-\mu)^2}{2\sigma^2}}, \quad -\infty < x < \infty, \quad -\infty < \mu < \infty, \quad \sigma > 0$$

The normal distribution has been written in such a way that its two parameters μ and σ^2 are equal to its mean and variance, respectively. When one speaks of a bell-shaped curve, one is typically referring to the normal distribution, also called the *Gaussian* distribution. Note that the graph of the normal distribution is symmetric about its mean μ (i.e., for any $c > 0$, $P(X > \mu + c) = P(X < \mu - c)$, and that σ^2 is a measure of the width of the bell shape—a larger value of σ^2 implies a wider bell. See Figure 4.11.

$$X \sim N(\mu, \sigma^2) \Rightarrow E(X) = \mu, \quad Var(X) = \sigma^2$$

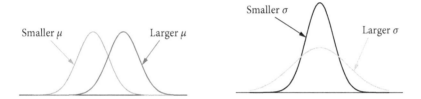

FIGURE 4.11   Normal pdf's with different means and variances.

The *standard normal* distribution is a normal distribution centered at the origin (i.e., $\mu = 0$) and with a standard deviation of 1 unit. The standard normal distribution is of the following form:

$$Z \sim N(0,1) \quad \leftarrow \mu = 0, \sigma^2 = 1$$

$$f(z) = \frac{1}{\sqrt{2\pi}}e^{-z^2/2}, \quad -\infty < z < \infty$$

The cdf of the standard normal distribution,

$$P(Z \le z) = \int_{-\infty}^{z} f(x)dx = \int_{-\infty}^{z} \frac{1}{\sqrt{2\pi}\sigma}e^{-x^2/2}dx = \Phi(z)$$

cannot be evaluated in a closed form, and so it is commonly evaluated using a table or computer program. A standard normal distribution table is given in Table A.1. Note that the symmetry of the standard normal distribution about $z = 0$ (see Figure 4.12) can be expressed as:

$$P(Z \geq z) = P(Z \leq -z)$$

for any z. Since $P(Z \geq z) = 1 - P(Z < z)$, it follows that

$$\Phi(z) = 1 - \Phi(-z)$$

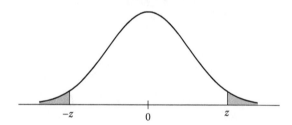

FIGURE 4.12   Standard normal probabilities
$P(Z \geq z) = P(Z \leq -z)$.

Normal probabilities can also be obtained using R. For $X \sim N(\mu, \sigma^2)$, $P(X \leq x)$ can be obtained as

>pnorm(x, μ, σ)

and for $Z \sim N(0,1)$, $P(Z \leq z)$ can be obtained as

>pnorm(z)

i.e., the default of the normal distribution in R is $\mu = 0$ and $\sigma = 1$.

EXAMPLE 4.8 Let the random variable Z have a standard normal distribution. Using Table A.1, find the following probabilities.

a. $P(Z < 1.96) = \Phi(1.96) = 0.9750$
    Using R, it can be obtained as
    >pnorm(1.96)

b. $P(Z > 1.96) = 1 - P(Z \leq 1.96) = 1 - \Phi(1.96) = 1 - 0.9750 = 0.025$
    Using R, it can be obtained as
    >1-pnorm(1.96)

c. $P(Z \leq -1.96) = \Phi(-1.96) = 0.025$
    Using R, it can be obtained as
    >pnorm(-1.96)

d. $P(0 \leq Z \leq 1.96) = \Phi(1.96) - \Phi(0) = 0.9750 - 0.5 = 0.4750$
    Using R, it can be obtained as
    >pnorm(1.96)-pnorm(0)

e. $P(Z < -1.96 \text{ or } Z > 2.0) = P(Z < -1.96) + P(Z > 2.0) = \Phi(-1.96) + 1 - \Phi(2.0)$
    $= 0.025 + 1 - 0.9772 = 0.0478$

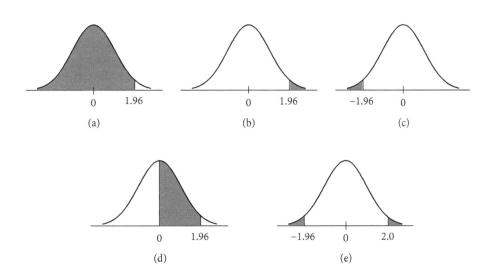

FIGURE 4.13.   Probabilities for Example 4.8.

What is the median of the standard normal distribution? It is symmetric about 0, so the median is equal to the mean $\mu = 0$. What about other percentiles? Recall that the $100p$-th percentile is found by solving $F(x) = p$ for x. Accordingly, we find the $100p$-th percentile of the standard normal distribution by solving $\Phi(z) = P(Z \leq z) = p$ for z using the standard normal distribution table. More generally, you may interpolate between neighboring values in a table or you can take the closest value in the table.

Normal percentiles can also be obtained by using R. The $100p$-th percentile of $X \sim N(\mu, \sigma^2)$ can be obtained as

>qnorm(p, μ, σ)

and the $100p$-th percentile of the standard normal distribution can be obtained as

>qnorm(p)

EXAMPLE 4.9   Find the following percentiles of the standard normal distribution.

    a. $P(Z \leq z) = 0.937$
        In Table A.1, we find $\Phi(z) = 0.937$ at $z = 1.53$.
        Using R, it can be obtained as
        >qnorm(0.937)
    b. $P(Z \geq z) = 0.4920$
        For this problem, we need to express the probability in terms of the cdf as follows:
        $1 - P(Z \leq z) = 0.492$, thus $P(Z \leq z) = 1 - 0.492 = 0.508$
        In Table A.1, we find $\Phi(z) = 0.508$ at $z = 0.02$.
        Using R, it can be obtained as
        >qnorm(0.508)

We define z_α as the $100(1 - \alpha)$-th percentile of the standard normal distribution. In other words, the right tail area of the standard normal pdf beyond z_α is α and $1 - \Phi(z_\alpha) = \alpha$. Figure 4.14 illustrates the meaning of z_α. To compute $z_{0.33}$, for example, we solve $1 - \Phi(z_{0.33}) = 0.33$. In Table A.1, we find $\Phi(z_{0.33}) = 0.6700$ at $z_{0.33} = 0.44$. Thus, the 67th percentile is 0.44. This percentile was carefully chosen to correspond to a value given in Table A.1. Using R, the same answer can be obtained by

>qnorm(0.67)

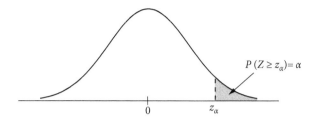

FIGURE 4.14   The shaded area is α in this normal pdf plot.

Table 4.1 provides the values of z_α for some selected values of α. These α values are frequently used for inferences to be covered in later chapters.

TABLE 4.1   The values of z_α for selected values of α

Percentile	90	95	97.5	99	99.5
α (tail area)	0.1	0.05	0.025	0.01	0.005
z_α	1.28	1.645	1.96	2.33	2.58

In Figure 4.15, $z_{0.025} = 1.96$ from Table 4.1. It can also be obtained from Table A.1. Because the standard normal pdf is symmetric about 0, the left tail area is also 0.025. Therefore, the total shaded area is 0.05.

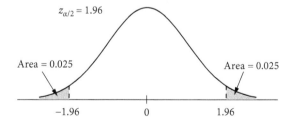

FIGURE 4.15   The total shaded area in this standard normal pdf plot is 0.05.

Of course, it would be extremely fortunate if a random variable following the normal distribution happened to have a mean of zero and a variance of 1. However, one can use the standard normal distribution to evaluate *any* normal distribution. Suppose that X has a normal distribution with mean μ and variance σ^2. Let $Z = (X - \mu)/\sigma$. Then Z has a standard normal distribution.

$$X \sim N(\mu, \sigma^2) \Rightarrow Z = \frac{X - \mu}{\sigma} \sim N(0, 1)$$

Therefore,

$$P(a \leq X \leq b) = P\left(\frac{a - \mu}{\sigma} \leq Z \leq \frac{b - \mu}{\sigma}\right) = \Phi\left(\frac{b - \mu}{\sigma}\right) - \Phi\left(\frac{a - \mu}{\sigma}\right)$$

EXAMPLE 4.10 Suppose X has a normal distribution with $\mu = 50$ and $\sigma^2 = 25$. Find $P(44 \leq X \leq 53)$.

$X \sim N(50, 25)$

$$P(44 \leq X \leq 53) = P\left(\frac{44 - 50}{5} \leq Z \leq \frac{53 - 50}{5}\right) = P(-1.2 \leq Z \leq 0.6)$$

$$= \Phi(0.6) - \Phi(-1.2) = 0.7257 - 0.1151 = 0.6106$$

Using R, it can be obtained by

>pnorm(53, 50, 5)-pnorm(44, 50, 5)

EXAMPLE 4.11 Suppose the raw scores on an SAT math test are normally distributed with a mean of 510 and a standard deviation of 110.

   a. What proportion of the students scored below 620?

$$P(X < 620) = P\left(Z < \frac{620 - 510}{110}\right) = \Phi(1) = 0.8413$$

Using R, it can be obtained by

>pnorm(620, 510, 110)

b. Find the 42nd percentile of the scores.

$$\Phi(z) = 0.42$$

From Table A.1, $z = -0.20$. Therefore,

$$z = \frac{x - 510}{110} \rightarrow x = 110z + 510 = 110(-0.2) + 510 = 488$$

Using R, it can be obtained by

>qnorm(0.42, 510, 110)

EXAMPLE 4.12 For $X \sim N(\mu, \sigma^2)$, find the following probabilities.

a. Probability that X falls within one standard deviation

$$P(\mu - \sigma \leq X \leq \mu + \sigma) = P\left(\frac{\mu - \sigma - \mu}{\sigma} \leq Z \leq \frac{\mu + \sigma - \mu}{\sigma}\right) = P(-1 \leq Z \leq 1)$$

$$= \Phi(1) - \Phi(-1) = 0.8413 - 0.1587 = 0.6826$$

b. Probability that X falls within two standard deviations

$$P(\mu - 2\sigma \leq X \leq \mu + 2\sigma) = P\left(\frac{\mu - 2\sigma - \mu}{\sigma} \leq Z \leq \frac{\mu + 2\sigma - \mu}{\sigma}\right) = P(-2 \leq Z \leq 2)$$

$$= \Phi(2) - \Phi(-2) = 0.9772 - 0.0228 = 0.9544$$

c. Probability that X falls within three standard deviations

$$P(\mu - 3\sigma \leq X \leq \mu + 3\sigma) = P\left(\frac{\mu - 3\sigma - \mu}{\sigma} \leq Z \leq \frac{\mu + 3\sigma - \mu}{\sigma}\right) = P(-3 \leq Z \leq 3)$$

$$= \Phi(3) - \Phi(-3) = 0.9987 - 0.0013 = 0.9974$$

Recall the empirical rule about a bell-shaped distribution discussed in Chapter 1. According to the empirical rule, roughly 68% of the data fall within one standard deviation from the mean, 95% of the data fall within two standard deviations, and 99.7% of the data fall within three standard deviations from the mean. These are based on the probabilities of a normal distribution as shown above. See Figure 4.16.

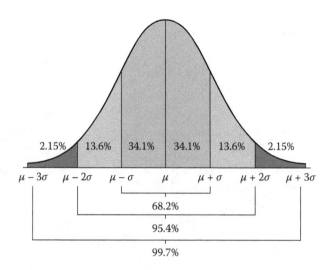

FIGURE 4.16   Illustration of the empirical rule
based on the normal distribution.

Let X be the height, in inches, of a randomly selected adult. Then X is a continuous random variable, with an associated probability density function $f(x)$. Suppose that we know $f(x)$, and we want to compute the probability that a randomly selected adult is 66 inches tall. Is the answer $P(X = 66)$? Well, no, because $P(X = 66)$ is the probability that the person is *exactly* 66 inches tall, i.e., 66.00000000 inches tall, which we know is an impossibility. It's an infinitesimal probability that an adult exists with precisely such a height, and even if one did exist, we would never be able to measure the height with infinite precision anyway. What we are really asking then is the probability that the height measurement, rounded off to the nearest inch, is 66. In other words, we want to determine

$$P(65.5 < X < 66.5)$$

which is

$$\int_{65.5}^{66.5} f(x)dx$$

If the interval of integration is sufficiently small, then we might approximate the integral as follows:

$$\int_{65.5}^{66.5} f(x)dx \approx f(66)[66.5 - 65.5]$$

This example illustrates the *continuity correction*. Given the pdf for a continuous random variable (adult height in inches) and the precision in the measurements (inches), we derive a distribution for the corresponding *discrete* random variable Y (adult height to the nearest inch). Then the probability distribution is

$$P(Y = y) = P(y - 0.5 < X < y + 0.5) = \int_{y-0.5}^{y+0.5} f(x)dx$$

The normal distribution approximates the binomial distribution when n is large and p is close to a half. The latter requirement is sensible for the reason that the binomial distribution is symmetric only when $p = 0.5$. However, the normal distribution approximates the binomial distribution for *any* (positive and less than 1) value of p, provided that n is sufficiently large.

THEOREM 4.2   The normal approximation to the binomial: Let $X \sim \text{Bin}(n, p)$, and define Z by

$$Z = \frac{X - np}{\sqrt{np(1-p)}}$$

Then Z approaches a standard normal distribution as $n \to \infty$.

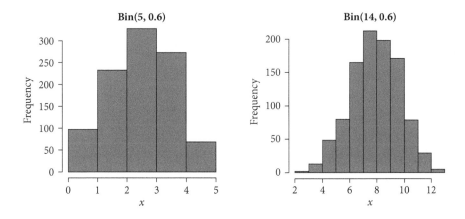

FIGURE 4.17   Shape change of the $\text{Bin}(n, p)$ distribution function as n increases.

Figure 4.17 illustrates the normal approximation to the binomial distribution. As n increases, the distribution function becomes closer to a continuous function. The normal distribution approximates the binomial distribution when both np and $n(1 - p)$ are large. Then how large is large? A rule of thumb is that the approximation is good when both np and $n(1 - p)$ exceed 10. However, sometimes the normal approximation is used when both np and $n(1 - p)$ exceed 5. When a higher precision is needed, then the approximation is used when both np and $n(1 - p)$ exceed 15.

Let $X \sim Bin(n, p)$.

When both np and $n(1 - p)$ are large (≥ 10),

$$Z = \frac{X - np}{\sqrt{np(1 - p)}} \text{ is approximately } N(0,1).$$

The method of approximation is as follows:

Step 1: Find $\mu = np$ and $\sigma^2 = np(1 - p)$.

Step 2: Use $N(\mu, \sigma^2) = N(np, np(1 - p))$.

Step 3: Remember to use continuity correction.

EXAMPLE 4.13 Let's practice continuity correction for the following problems, where $X \sim Bin(n, p)$ and $Z \sim N(0, 1)$.

a. $P(X = 3) = P(\frac{2.5-\mu}{\sigma} < Z < \frac{3.5-\mu}{\sigma})$
b. $P(X < 3) = P(Z < \frac{2.5-\mu}{\sigma})$
c. $P(X \leq 3) = P(Z < \frac{3.5-\mu}{\sigma})$
d. $P(X > 3) = P(Z > \frac{3.5-\mu}{\sigma})$
e. $P(X \geq 3) = P(Z > \frac{2.5-\mu}{\sigma})$

The above approximations are illustrated in Figure 4.18.

(a)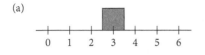

(b)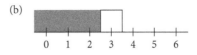

(c)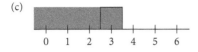

(d)

(e)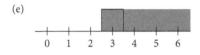

FIGURE 4.18   Illustration of the normal approximation to the binomial for Example 4.13.

EXAMPLE 4.14 Let $X \sim \text{Bin}(n, p)$ with $n = 30$, $p = 0.4$. Find $P(X = 14)$ in the following ways:

a. Find the probability using the exact distribution.

$$P(X = 14) = \binom{30}{14}(0.4)^{14}(0.6)^{16} = 0.1101$$

b. Find the probability using a normal approximation.

$\mu = np = 30(0.4) = 12$, $n(1 - p) = 30(0.6) = 18$ both exceed 10

$$\sigma = \sqrt{np(1 - p)} = \sqrt{30 \cdot 0.4 \cdot 0.6} = \sqrt{7.2} = 2.6833$$

$$P(13.5 \le X \le 14.5) = P\left(\frac{13.5 - 12}{2.6833} \le Z \le \frac{14.5 - 12}{2.6833}\right) = P(0.56 \le Z \le 0.93)$$

$$\approx \Phi(0.93) - \Phi(0.56) = 0.8238 - 0.7123 = 0.1115$$

By comparing with the answer from part (a), we see that the approximation is pretty good.

EXAMPLE 4.15 Suppose $X \sim \text{Bin}(160, 0.6)$. Then $\mu = np = 160(0.6) = 96$ and $n(1 - p) = 160(0.4) = 64$, which both exceed 10 by a comfortable margin. Therefore, we expect X to be well approximated by a normal distribution with $\mu = 96$ and $\sigma = \sqrt{np(1 - p)} = \sqrt{160(0.6)(0.4)} = \sqrt{38.4} = 6.197$. Find the following approximate probabilities.

a. $P(90 \leq X \leq 100) = P\left(\frac{89.5-96}{6.197} \leq Z \leq \frac{100.5-96}{6.197}\right) = P(-1.05 \leq Z \leq 0.73)$

$$\approx \Phi(0.73) - \Phi(-1.05) = 0.7673 - 0.1469 = 0.6204$$

b. $P(90 \leq X < 100) = P\left(\frac{89.5-96}{6.197} \leq Z \leq \frac{99.5-96}{6.197}\right) = P(-1.05 \leq Z \leq 0.56)$

$$\approx \Phi(0.56) - \Phi(-1.05) = 0.7123 - 0.1469 = 0.5654$$

The above approximations are illustrated in Figure 4.19.

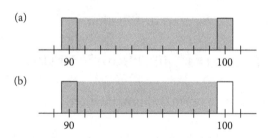

FIGURE 4.19   Illustration of the normal approximation to the binomial for Example 4.15.

EXAMPLE 4.16 In a certain country, 30% of the adult male population smoke regularly. In a random sample of 750 adults, what is the probability that (a) fewer than 200 and (b) 240 or more are smokers?

Let X be the number of smokers in the random sample. Then $X \sim \text{Bin}(750, 0.3)$. We expect $\mu = np = 750(0.3) = 225$ of the adults to be smokers, with a standard deviation of $\sigma = \sqrt{np(1 - p)} = \sqrt{750(0.3)(0.7)} = \sqrt{157.5} = 12.55$. Since $np = 225$ and $n(1 - p) = 750(0.7) = 525$ far exceed 10, X has a normal distribution, approximately.

a. $P(X < 200) = P\left(Z \leq \frac{199.5-225}{12.55}\right) \approx \Phi(-2.03) = 0.0212$

b. $P(X \geq 240) = P\left(Z \geq \frac{239.5-225}{12.55}\right) = P(Z \geq 1.16) \approx 1 - \Phi(1.16) = 1 - 0.8770 = 0.1230$

The above probabilities are illustrated in Figure 4.20.

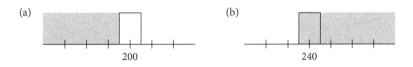

FIGURE 4.20   Illustration of the normal approximation to
the binomial for Example 4.16.

7. *The Gamma Distribution*

The gamma function is included in the pdf of the gamma distribution. For $\alpha > 0$, the gamma
function $\Gamma(\alpha)$ is defined by

$$\Gamma(\alpha) = \int_0^\infty x^{\alpha-1} e^{-x} dx \tag{4-1}$$

The gamma function has the following properties:
1.   For any $\alpha > 1$, $\Gamma(\alpha) = (\alpha - 1)\Gamma(\alpha - 1)$
2.   For any positive integer n, $\Gamma(n) = (n-1)!$
3. $\Gamma\left(\frac{1}{2}\right) = \sqrt{\pi}$

Property 1 can be obtained from (4-1) using integration by parts. Based on property 2, one
can think of the gamma function as extending the factorial function $n!$ to noninteger n. The
gamma function is needed to evaluate the gamma distribution $\text{Gamma}(\alpha, \beta)$, which is

$$f(x) = \begin{cases} \dfrac{1}{\Gamma(\alpha)\beta^\alpha} x^{\alpha-1} e^{-x/\beta}, & x \geq 0, \alpha > 0, \beta > 0 \\ \\ 0, & \text{otherwise} \end{cases}$$

with positive parameters α and β. The mean and variance of a random variable $X \sim$ Gamma(α, β) can be obtained from (4-1) as follows:

$$E(X) = \alpha\beta, \quad Var(X) = \alpha\beta^2$$

Gamma cdf and percentile can also be obtained using R. For Gamma(α, β), $P(X \leq c)$ can be obtained by

>pgamma(c, α, 1 / β)

and the $100p$-th percentile is obtained by

>qgmma(p, α, 1 / β)

The exponential distribution is a special case of the gamma distribution with $\alpha = 1$.

If $\alpha = 1$,

$$X \sim \text{Exponential}(\beta) = \text{Gamma}(1, \beta)$$

$$f(x) = \begin{cases} \dfrac{1}{\beta}e^{-x/\beta}, & x > 0, \beta > 0 \\ 0, & \text{otherwise} \end{cases}$$

$$E(X) = \beta, \quad Var(X) = \beta^2$$

and the cdf is of the form

$$F(x) = \begin{cases} 1 - e^{-x/\beta}, & x \geq 0 \\ 0, & x < 0 \end{cases}$$

Another special case that will be of major importance when we study sampling distributions in Chapter 6 is the χ^2 (chi-square) distribution with parameter $v > 0$, which is obtained from the gamma distribution with $\alpha = v/2$ and $\beta = 2$. This is called the χ^2 distribution with degrees of freedom (df) v (Gamma($v/2$, 2), denoted as χ^2_v) with the pdf of the form

$$f(x) = \begin{cases} \dfrac{1}{\Gamma\left(\dfrac{v}{2}\right) 2^{v/2}} x^{\frac{v}{2}-1} e^{-x/2}, & x > 0, v > 0 \\ \\ 0, & \text{otherwise} \end{cases}$$

The mean and variance of the χ^2 random variable are

$$E(X) = \alpha\beta = \frac{v}{2} \cdot 2 = v, \qquad \text{Var}(X) = \alpha\beta^2 = \frac{v}{2} \cdot 4 = 2v$$

The χ^2 distribution is used for inference about a variance.

8. The Beta Distribution

In some problems a random variable represents a proportion of something. An example taken from an old news item is the fraction of merchandise bar codes that indicate a price consistent with what is marked on the shelf—numerous studies at various department stores have demonstrated that the price marked on the shelf is often different from the one that is charged to the customer at checkout. Assuming that mere incompetence is behind this phenomenon (a big assumption, which is inconsistent with the observation that the customer is usually charged more than the price marked on the shelf), we may regard this fraction as a random variable; in a sufficiently busy department store, it changes from week to week with no clear trend up or down. A pdf that is often applied to random variables that are restricted to values between 0 and 1 is the beta distribution with positive parameters α and β, denoted as Beta(α, β), of the form

$$f(x) = \begin{cases} \dfrac{\Gamma(\alpha+\beta)}{\Gamma(\alpha)\Gamma(\beta)} x^{\alpha-1}(1-x)^{\beta-1}, & 0 \le x \le 1, \alpha > 0, \beta > 0 \\ \\ 0, & \text{otherwise} \end{cases}$$

For $X \sim \text{Beta}(\alpha, \beta)$, the mean and variance are

$$E(X) = \frac{\alpha}{\alpha + \beta}, \qquad Var(X) = \frac{\alpha \beta}{(\alpha + \beta)^2 (\alpha + \beta + 1)}$$

Beta cdf can also be obtained using the statistical software R. For $X \sim \text{Beta}(\alpha, \beta)$, $P(X \leq c)$ can be obtained as

>pbeta(c, α, β)

and the $100p$-th percentile can be obtained as

>qbeta(p, α, β)

The beta distribution with $\alpha = 1$ and $\beta = 1$ is the uniform$(0, 1)$ distribution.

EXAMPLE 4.17 At a gas station, the fraction of customers who purchase gasoline with the highest octane level (supreme gas) in any given week is a random variable with Beta$(1, 4)$.

a.   If the gas station serves an average of 850 customers every week, how many customers are expected to purchase supreme gas over the next week?

Let X be the fraction of the customers who purchase supreme gas. Then $X \sim \text{Beta}(1, 4)$.

$$E(X) = \frac{\alpha}{\alpha + \beta} = \frac{1}{1 + 4} = 0.2$$

The expected number of customers who purchase supreme gas over the next week is

$$850(0.2) = 170$$

b. What is the probability that more than 25% of the customers will purchase supreme gas over the next week?

$$f(x) = \frac{\Gamma(1+4)}{\Gamma(1)\Gamma(4)} x^{1-1}(1-x)^{4-1} = \frac{4!}{3!}(1-x)^3 = 4(1-x)^3, 0 < x < 1$$

$$P(X > 1/4) = \int_{1/4}^{1} 4(1-x)^3 dx = -(1-x)^4 \Big|_{1/4}^{1} = \left(\frac{3}{4}\right)^4 = \frac{81}{256} = 0.3164$$

Using R, it can be obtained as

>1-pbeta(1/4, 1, 4)

SUMMARY OF CHAPTER 4

1. Continuous Random Variables
   a. A random variable is <u>continuous</u> if it represents some measurement on a continuous scale.
   b. <u>Probability Density Function</u> (pdf) $f(x)$ has the following properties:

   i. $P(a \le X \le b) = \int_a^b f(x)\,dx$ for $a \le b$

   ii. $f(x) \ge 0$ for all x

   iii. $\int_{-\infty}^{\infty} f(x)\,dx = 1$

   c. <u>Cumulative Distribution Function</u> (cdf) of a continuous random variable X:

   $$F(x) = P(X \le x) = \int_{-\infty}^{x} f(t)\,dt$$

   d. $P(X = c) = 0$ for any number c.
   e. If a and b are real constants with $a \le b$, then

   $$P(a \le X \le b) = P(a \le X < b) = P(a < X \le b) = P(a < X < b) = F(b) - F(a).$$

   f. If a is a real constant, then

   $$P(X \ge a) = 1 - P(X \le a) = 1 - F(a).$$

g. $F'(x) = f(x)$ at every x at which $F'(x)$ exists.
h. The $100p$-th <u>percentile</u>: x such that $p = F(x)$
    Median: x such that $F(x) = 0.5$

i. Mean: $E(X) = \mu = \int_{-\infty}^{\infty} xf(x)\,dx$

    Variance: $Var(X) = \sigma^2 = E[(X - \mu)^2] = E(X^2) - \mu^2$

2. The <u>Uniform Distribution</u>: $X \sim \text{Uniform}(a,b)$

a. $f(x) = \begin{cases} \frac{1}{b-a}, & a < x < b \\ 0, & \text{otherwise} \end{cases}$

b. $E(X) = \frac{a+b}{2}, Var(X) = \frac{(a-b)^2}{12}$

3. The <u>Beta Distribution</u>: $X \sim \text{Beta}(\alpha, \beta)$

a. $f(x) = \begin{cases} \frac{\Gamma(\alpha+\beta)}{\Gamma(\alpha)\Gamma(\beta)} x^{\alpha-1}(1-x)^{\beta-1}, & 0 \le x \le 1, \alpha > 0, \beta > 0 \\ 0, & \text{otherwise} \end{cases}$

b. $E(X) = \frac{\alpha}{\alpha+\beta}, \; Var(X) = \frac{\alpha\beta}{(\alpha+\beta)^2(\alpha+\beta+1)}$

4. The <u>Gamma Distribution</u>: $X \sim \text{Gamma}(\alpha, \beta)$

a. $f(x) = \begin{cases} \frac{1}{\Gamma(\alpha)\beta^\alpha} x^{\alpha-1}e^{-x/\beta}, & x \ge 0, \alpha > 0, \beta > 0 \\ 0, & \text{otherwise} \end{cases}$

b. $E(X) = \alpha\beta, \; Var(X) = \alpha\beta^2$

5. The <u>Exponential Distribution</u> (the gamma distribution with $\alpha = 1$): $X \sim \text{Exponential}(\lambda)$

a. $f(x) = \begin{cases} \dfrac{1}{\beta}e^{-x/\beta}, & x \ge 0, \beta > 0 \\ 0, & \text{otherwise} \end{cases}$

b. $E(X) = \beta, \; Var(X) = \beta^2$

6. The Chi-Square (χ^2) Distribution with degrees of freedom v (the gamma distribution with $\alpha = v/2$ and $\beta = 2$): $X \sim \chi_v^2$

   a. $f(x) = \begin{cases} \dfrac{1}{\Gamma\left(\frac{v}{2}\right) 2^{v/2}} x^{\frac{v}{2}-1} e^{-x/2}, & x > 0, v > 0 \\ \\ 0, & \text{otherwise} \end{cases}$

   b. $E(X) = v, \ Var(X) = 2v$

7. The Normal Distribution

   a. With mean μ and variance σ^2: $X \sim N(\mu, \sigma^2)$

   $$f(x) = \frac{1}{\sqrt{2\pi}\sigma} e^{-\frac{(x-\mu)^2}{2\sigma^2}}, \ -\infty < x < \infty$$

   b. The standard normal distribution: $Z = \dfrac{X - \mu}{\sigma} \sim N(0, 1)$

   $$f(z) = \frac{1}{\sqrt{2\pi}} e^{-z^2/2}, \ -\infty < z < \infty$$

   i. $\Phi(z)$: standard normal cdf
   ii. Z is symmetric about zero: $\Phi(z) = 1 - \Phi(-z)$
   iii. Standard normal percentiles and critical values:

Percentile	90	95	97.5	99	99.5
α (tail area)	0.1	0.05	0.025	0.01	0.005
z_α	1.28	1.645	1.96	2.33	2.58

8. Normal Approximation to the Binomial Distribution

   If $X \sim \text{Bin}(n, p)$ with a large n, then $Z = \frac{X - np}{\sqrt{np(1-p)}}$ is approximately standard normal.
   a. Type of problem
      A discrete random variable X is involved. $X \sim \text{Bin}(n, p)$. Typically, n is large and a binomial table cannot be used, or $np \geq 10, n(1-p) \geq 10$
   b. Method of approximation
      i. Step 1: $\mu = np, \sigma^2 = np(1-p)$
      ii. Step 2: Use $N(\mu, \sigma^2) = N(np, np(1-p))$
      iii. Remember to do continuity correction.

EXERCISES

4.1   The lifetime of a certain brand of lightbulb has an exponential distribution with a mean of 800 hours.

    a.   Find the probability that a randomly selected lightbulb of this kind lasts 700 to 900 hours.

    b.   Find the probability that a randomly selected lightbulb of this kind lasts longer than 850 hours.

    c.   Find the 80th percentile of the lifetime of this kind of lightbulb.

4.2   The lifetime of a certain brand of lightbulb (in hours) is a random variable with pdf $f(x) = 0.005e^{-0.005x}$, $x > 0$. Find the probability that the bulb will last no more than 300 hours.

4.3   Suppose the survival time (in months) of a certain type of a fatal cancer is a random variable X with pdf $f(x) = 0.1\, e^{-0.1x}$, $x > 0$ from onset.

    a.   Find the probability that a patient will live less than 1 year from onset.

    b.   Find the probability that a patient will live 6 months to 18 months from onset.

    c.   Find the probability that a patient will live more than 2 years from onset.

4.4   In a certain crime-infested neighborhood, the time that a dwelling is burglarized is approximately exponentially distributed with a mean of 2 years.

    a.   If you move into an apartment in this neighborhood and intend to live there for five years, what is the probability that your apartment will not be burglarized during this time?

    b.   What is the probability that your apartment will be burglarized within a month after you move in?

4.5   Let X be a random variable with pdf

$$f(x) = \begin{cases} -cx, & -2 < x < 0 \\ cx, & 0 \le x < 2 \\ 0, & \text{otherwise} \end{cases}$$

where c is a constant.
a. Find the value of c.
b. Find the mean of X.
c. Find the variance of X.
d. Find $P(-1 < X < 2)$.
e. Find $P(X > 1/2)$.
f. Find the third quartile.

4.6   Let X be a random variable with pdf

$$f(x) = \begin{cases} x^2, & 0 < x < 1 \\ (7 - 3x)/4, & 1 \le x < 7/3 \\ 0, & \text{otherwise} \end{cases}$$

a. Find $P(1/2 < X < 2)$.
b. Find $P(X < 1/4)$.
c. Find the median.
d. Find $P(X > 1.5)$.

4.7   Let X be a continuous random variable with pdf

$$f(x) = \begin{cases} x^2, & 0 < x < 1 \\ x, & 1 \le x < c \\ 0, & \text{otherwise} \end{cases}$$

a. Find the value of c.
b. Find the cdf $F(x)$.
c. Find $P\left(\frac{1}{2} < X < \frac{5}{4}\right)$.
d. Find the median.

4.8 The time (in minutes) that it takes a mechanic to change oil has an exponential distribution with mean 20.
a. Find $P(X < 25)$, $P(X > 15)$, and $P(15 < X < 25)$.
b. Find the 40th percentile.

4.9 The pdf of X is $f(x) = 0.2, 1 < x < 6$.
a. Show that this is a pdf.
b. Find the cdf $F(x)$.
c. Find $P(2 < X < 5)$.
d. Find $P(X > 4)$.
e. Find $F(3)$.
f. Find the 80th percentile.

4.10 Redo Exercise 4.9 (c) through (f) using R.

4.11 Let X be a random variable with pdf $f(x) = \frac{2}{x^2}, x \geq 2$.
a. Find the cdf $F(x)$.
b. Find the median of this distribution.
c. Find $P(1 < X < 3)$.
d. Find $P(X > 3)$.
e. Find $P(3 < X < 4)$.

4.12 Let X be a random variable with cdf

$$F(x) = \begin{cases} 0, & x < 2 \\ (x-2)/2, & 2 \leq x < 4 \\ 1, & x \geq 4 \end{cases}$$

a. Find the pdf of X.
b. Find $P(2/3 < X < 3)$.
c. Find $P(X > 3.5)$.
d. Find the 60th percentile.
e. Find $P(X = 3)$.

4.13   Let X be a random variable with cdf

$$F(x) = \begin{cases} 0, & x < 0 \\ 1 - e^{-2x}, & x \geq 0 \end{cases}$$

a.   Find the pdf of X.
b.   Find $P(X < 1/2)$.
c.   Find $P(X > 3)$.
d.   Find the median.
e.   Identify the distribution of X.

4.14   Redo Exercise 4.13 (b), (c) and (d) using R.

4.15   Let X be a random variable with cdf

$$F(x) = \begin{cases} 0, & x < 2 \\ 1 - \dfrac{2}{x}, & x \geq 2 \end{cases}$$

a.   Find the pdf of X.
b.   Find the first quartile of this distribution.
c.   Find $P(0 < X < 3)$.
d.   Find $P(3 < X < 5)$.

4.16   The wait time for service at a local DMV is uniformly distributed from 1 minute to 9 minutes.

a.   Find the probability that a randomly selected person waits 1.5 to 5 minutes.
b.   Find the expected wait time.
c.   Find the standard deviation of the wait time.

4.17 The weekly demand for drinking water, in thousand gallons, from a local wholesale store, is a random variable X with pdf

$$f(x) = \begin{cases} a(x-1), & 2 \le x \le 4 \\ 0, & \text{otherwise} \end{cases}$$

where a is a constant.
   a. Find the value of a.
   b. Find the cdf of X.
   c. Find the mean of X.
   d. Compute $P(3 < X < 4)$.
   e. Find the first quartile of the distribution.

4.18 Suppose the random variable X has pdf $f(x) = (x+1)/2, -1 < x < 1$.
   a. Find E(X).
   b. Find Var(X).

4.19 Suppose the random variable X has pdf $f(x) = \frac{x}{2}, 0 < x < 2$.
   a. Find the cdf $F(x)$ and sketch its graph.
   b. Use the cdf to find $P(0.5 < X < 1.5)$ and also to find $P(X > 1.5)$.
   c. Find the median of the distribution of X.
   d. Find $E(X)$.
   e. Find $Var(X)$.

4.20 Let X be a random variable with pdf $f(x) = kx^2, 0 < x < 1$.
   a. Find the value of k.
   b. Find the cdf of X.
   c. Find $P(1/4 < X < 2/3)$.
   d. Find $E(X)$.
   e. Find $Var(X)$.
   f. Find the 70th percentile.

4.21 Let X be a random variable with pdf $f(x) = kx$, $0 < x < 4$.
    a.   Find the value of k.
    b.   Find the cdf of X.
    c.   Find $P(X < 1 \text{ or } X > 3)$.
    d.   Find $P\left(\frac{1}{2} < X \le 5\right)$.
    e.   Find $P(X > 2)$.
    f.   Find $E(X)$.
    g.   Find $Var(X)$.

4.22 The weekly demand for cereals, in thousands of boxes, from a wholesale store is a random variable X with pdf $f(x) = k(x - 2)$, $2 \le x \le 4$.
    a.   Find the value of k.
    b.   Find the cdf of X.
    c.   Find the probability that the daily demand is 2,500 to 3,500 boxes.
    d.   Find the mean of X.
    e.   Find the standard deviation of X.
    f.   At the time that 3,600 boxes were delivered to the store for a new week, there were no boxes of cereal in stock. Find the probability that the cereal boxes are sold out at the end of the week.
    g.   Find the third quartile of the distribution.

4.23 Let the random variable X have pdf $f(x) = kx - 1$, $2 < x < 4$.
    a.   Find the value of k.
    b.   Find the cdf of X.
    c.   Find $P(2.5 < X < 3)$.
    d.   Find the mean and variance of X.
    e.   Find the 90th percentile.

4.24 Let X be a random variable with pdf $f(x) = \frac{1}{2}$, $0 < x < 2$.
    a.   Find the cdf $F(x)$.
    b.   Find the mean of X.
    c.   Find the variance of X.
    d.   Find $F(1.4)$.
    e.   Find $P\left(\frac{1}{2} < X < 1\right)$.
    f.   Find $P(X > 3)$.
    g.   Find the 35th percentile.

4.25 Let X be a random variable with pdf

$$f(x) = \frac{4}{x^2}, \ x \geq a$$

    a. Find the value of the constant a.
    b. Find the cdf $F(x)$ of X.
    c. Find the median of this distribution.

4.26 Suppose the lifetime (in months) of a certain type of lightbulb is a random variable X with pdf $f(x) = \frac{2}{x^3}, \ x > 1$.
    a. Find the cdf $F(x)$.
    b. Find the probability that a lightbulb of this kind lasts longer than 2 months.
    c. Find the mean and variance of X.
    d. In a random sample of 5 such lightbulbs, what is the probability that at least 2 of them will work for more than 2 months?

4.27 Let Z be a standard normal random variable. Find the following probabilities:
    a. $P(Z < 1.53)$
    b. $P(Z < -1.5)$
    c. $P(Z > 1.32)$
    d. $P(-1.8 < Z < 1.28)$
    e. $P(Z < 0.25 \ \text{or} \ Z > 1.28)$
    f. $P(Z = 1.75)$
    g. $P(-0.75 < Z < -0.32)$

4.28 If $Z \sim N(0, 1)$, find the following probabilities:
    a. $P(Z < 1.38)$
    b. $P(Z \geq 2.14)$
    c. $P(-1.27 < Z < -0.48)$

4.29 Let Z be a standard normal random variable. Find x in the following equations.
    a. $P(Z < x) = 0.6103$
    b. $P(Z > x) = 0.7324$
    c. $P(-x < Z < x) = 0.758$
    d. $P(Z < -x) = 0.8577$
    e. $P(Z < -x \ \text{or} \ Z > x) = 0.2006$

4.30   If $Z \sim N(0, 1)$, find the following values:
    a. z such that $P(Z > z) = 0.75$
    b.   The 67th percentile of the distribution

4.31   If $Z \sim N(0, 1)$, find the following values:
    a. z such that $P(Z > z) = 0.35$
    b.   The 34th percentile of the distribution

4.32   If X is normally distributed with mean 120 and standard deviation 10, find
    a. $P(105 < X < 130)$
    b.   58th percentile

4.33   Let X have a $N(60, 900)$ distribution.
    a.   Find $P(X < 78)$.
    b.   Find the 88th percentile of the distribution.

4.34   In America, the lengths of full-term babies at birth follow a normal distribution with mean 20 inches and standard deviation 1 inch. The length of a randomly selected new born infant is measured. Find the probability that it is between 19 and 21 inches.

4.35   Suppose the weight (in pounds) of an adult male sheep in a pasture is distributed as $N(100, 225)$.
    a.   Find the probability that the weight of a sheep is more than 120 pounds.
    b.   What value of the weight separates the heaviest 10% of all the sheep in the pasture from the other 90%?

4.36   The mean and standard deviation of SAT math exam scores were 527 and 120, respectively, in 2017. Assume that the scores are normally distributed and answer the following questions.
    a.   Find the 30th percentile.
    b.   If your score is 700, what percentage of the students got higher scores than you in 2017?
    c.   What percentage of students got SAT math scores from 400 to 600?
    d.   If your score is 650, what is the percentile of your score?

4.37 Answer the questions in Exercise 4.36 by computing in R.

4.38 The compressive strength (in MPa) of bottles produced by a factory follows a normal distribution $N(\mu, \sigma^2)$. A bottle having the compressive strength less than 40 MPa is considered a defect. The process capacity index G to evaluate the process capacity is defined as

$$G = \frac{\mu - 40}{3\sigma}$$

If $G = 0.8$, find the probability that a randomly selected bottle is defective.

4.39 It is known that the IQ scores of people in the United States have a normal distribution with mean 100 and standard deviation 15.
   a. If a person is selected at random, find the probability that the person's IQ score is less than 85.
   b. Jason's IQ score is equal to the 75th percentile. What is his IQ? Round it off to the nearest integer.
   c. What proportion of this population have IQ scores above 120?

4.40 The average grade for an exam is 72, and the standard deviation is 12. If 22% of the class are given A's and the grades are curved to follow a normal distribution, what is the lowest possible score for an A?

4.41 Let X be a normal random variable with mean μ and variance σ^2. Perform the following calculations when $P(X < 7.58) = 0.95$ and $P(X < 8.84) = 0.975$.
   a. Find μ and σ^2.
   b. Find $P(1 < X < 5)$.

4.42 Find the mean μ and variance σ^2 of a random variable X when $P(X < 2) = 0.117$ and $P(2 < X < 4) = 0.674$, where $X \sim N(\mu, \sigma^2)$.

4.43 The duration of a human's pregnancy is known to have a normal distribution with mean 40 weeks and standard deviation 1 week, but it has been shortened by a week with the same standard deviation in the United States, according to a report published in a recent year.
   a. A baby was born yesterday, and his gestation period was in the 70th percentile of the $N(39, 1)$ distribution. What percentile would this be in the old days?
   b. Find the probability that the duration of a pregnancy was 37 to 40 weeks in 2018.

4.44   The lifetime of the timing belt of a certain make of cars is normally distributed with mean 125,000 miles and standard deviation 10,000 miles.
   a.   Find the probability that a timing belt lasts until the car runs 140,000 miles.
   b.   The automaker recommends that owners have the timing belt replaced when the mileage reaches 90,000 miles. What is the probability that the timing belt fails before the car reaches the manufacturer's recommended mileage?
   c.   An owner of this type of car wants to take a chance and replace the timing belt at the 1st percentile of the distribution. What should be the mileage of the car when he has the timing belt replaced?

4.45   The average grade for an examination is 70, and the standard deviation is 12. If 23% of the students in this class are given As and the grades are curved to follow a normal distribution, what is the lowest possible score for an A? Assume that the scores are integers.

4.46   Suppose X is a normal random variable with mean 7 and variance 9.
   a.   Find $P(2.5 \le X < 13)$.
   b.   Find the value c such that $P(X > c) = 0.05$.

4.47   Random variable X has a normal distribution with mean μ and standard deviation 5. The pdf $f(x)$ of X satisfies the following conditions: (A) $f(8) > f(18)$, (B) $f(2) < f(20)$. When μ is an integer, what is $P(15 \le X < 16)$?

4.48   To classify fossils of an animal, archeologists measure the length of the femur. The length of femur of a type A animal is distributed as $N(10, 0.4^2)$ and the length of femur of a type B animal is distributed as $N(12, 0.6^2)$ (in inches). The fossil is classified as type A if the length is less than d inches and classified as type B if the length is larger than or equal to d inches. Find the value of d that the probability of correctly classifying a type A animal equals the probability of correctly classifying a type B animal.

4.49   If X has a binomial distribution with $n = 150$ and the success probability $p = 0.4$, find the following probabilities approximately:
   a. $P(48 \le X < 66)$
   b. $P(X > 69)$
   c. $P(48 < X \le 66)$
   d. $P(X \ge 65)$
   e. $P(X < 60)$

4.50   Find the probabilities in Exercise 4.49 using R in the following ways:
   a.   Find the exact probabilities for (a) through (e).
   b.   Find the probabilities for (a) through (e) using normal approximation to the binomial distribution.

4.51   In the United States, the mean and standard deviation of adult women's heights are 65 inches (5 feet 5 inches) and 3.5 inches, respectively. Suppose the American adult women's heights have a normal distribution.
   a.   If a woman is selected at random in the United States, find the probability that she is taller than 5 feet 8 inches.
   b.   Find the 72nd percentile of the distribution of heights of American women.
   c.   If 100 women are selected at random in the United States, find an approximate probability that exactly 20 of them are taller than 5 feet 8 inches.

4.52   A fair die is tossed 180 times. Find the approximate probability that the number 6 is obtained more than 40 times.

4.53   In the United States, 45% of the population has type O blood. If you randomly select 50 people in the nation, what is the approximate probability that more than half will have type O blood?

4.54   Let X have a binomial distribution with $n = 400$ and the success probability $p = 0.5$. Find $P(185 \leq X \leq 210)$ approximately.

4.55   A fair coin is tossed 200 times.
   a.   Find an approximate probability of getting fewer than 90 heads.
   b.   Find an approximate probability of getting exactly 100 heads.

4.56   Forty percent of adult males recovered from flu within five days last winter. If 100 adult males are known to have caught flu last winter, what is the approximate probability that more than half of them recovered within five days?

4.57   The flight experiences (in hours) of pilots in an airline company have approximately a normal distribution with mean 8,500 hours and standard deviation 400 hours.
   a.   Find the probability that a pilot in this company has flight hours more than 8,700.
   b.   Suppose 50 pilots are randomly chosen from this company. Find an approximate probability that 20 of them have flight hours over 8,700.

4.58 Suppose the annual amount of rainfall (in million tons) accumulated in a lake follows a gamma distribution with $\alpha = 3$ and $\beta = 5$.
   a. Find the expected annual amount of rainfall accumulated in this lake.
   b. Find the standard deviation of the annual amount of rainfall accumulated in this lake.

4.59 Let X be a random variable representing the load of a web server. Suppose X is distributed as gamma with $\alpha = 1$ and $\beta = 5$.
   a. Find $P(X < 2)$.
   b. Find $P(1 < X < 3)$.
   c. Find the mean and variance of X.

4.60 Answer the following questions regarding Exercise 4.59 using R.
   a. Find the probabilities in (a) and (b).
   b. Find the 80th percentile.

4.61 Let X be a random variable from the Beta$(4, 1)$ distribution.
   a. What is the pdf of the distribution?
   b. Find the cdf of the distribution.
   c. Find $P(0.25 < X < 0.75)$.
   d. Find the mean and variance of X.
   e. Find the median of X.

4.62 Answer the following questions using R.
   a. Find the probability in Exercise 4.61 (c).
   b. Find the median in Exercise 4.61 (e).

4.63 The proportion of defective electronic items in a shipment has the Beta$(2, 3)$ distribution.
   a. What is the pdf of the distribution?
   b. Find the cdf of the distribution.
   c. Find the expected proportion of defective items.
   d. Find the probability that at least 20% of the items in the shipment are defective.
   e. Find the probability that the shipment has 10% to 25% defective items.

4.64 Answer the following questions regarding the Beta$(1, 1)$ distribution.
   a. What is the pdf of this distribution?
   b. Write another name for this distribution.
   c. Find the cdf.

4.65   What are the outcomes of the following R commands?
   a.   pnorm(2)+pnorm(-2,0,1)
   b.   punif(pnorm(1,1,4),-1,1)
   c.   qgamma(pgamma(3,2,4),2,4)
   d.   qexp(2,3)

4.66   What are the outcomes of the following R commands?
   a.   pbeta(0.6,1,1)-punif(1,0,2)
   b.   pgamma(2,1,2)-pexp(2,2)
   c.   pnorm(1,1,4)+qunif(0.6,1,2)
   d.   qbinom(0.5,12,0.5)

Multiple Random Variables

The theoretical development since Chapter 3 has primarily concerned the probability distribution of a single random variable. However, many of the problems discussed in Chapter 2 had two or more distinct aspects of interest; it would be awkward to attempt to describe their sample spaces using a single random variable. In this chapter, we develop the theory for distributions of multiple random variables, called *joint* probability distributions.

1. Discrete Distributions

EXAMPLE 5.1   A survey reports the number of sons and daughters of 100 married couples. The results are summarized in the following table.

	B_0	B_1	B_2	Total
G_0	6	8	8	22
G_1	9	17	10	36
G_2	15	15	12	42
Total	30	40	30	100

Here, G_0 = no daughters, G_1 = 1 daughter, G_2 = 2 or more daughters, B_0 = no sons, B_1 = 1 son, B_2 = 2 or more sons.

An outcome in this sample space consists of the number of sons and daughters of randomly selected couples in this group. We could map this sample space onto a finite set of integers and therefore employ a single random variable, but it is more natural to define two random variables for this problem. Let X_1 be the number of daughters of a randomly chosen couple, and let X_2 be the number of sons. Then we can define the joint probability distribution for this pair of discrete random variables as a straightforward extension of the probability distribution of a single random variable, namely:

$$f(x_1, x_2) = P(X_1 = x_1, X_2 = x_2)$$

The probability that a randomly selected couple has no daughters and two sons is

$$P(G_0 \cap B_2) = P(X_1 = 0, X_2 = 2) = f(0, 2) = \frac{8}{100} = 0.08$$

Some events are a little simpler to express in the notation of random variables; e.g.,

$$P(X_1 + X_2 \geq 2) = P(G_2 \cup B_2 \cup (G_1 \cap B_1)) = P(G_2 \cup B_2) + P(G_1 \cap B_1)$$
$$= P(G_2) + P(B_2) - P(G_2 \cap B_2) + P(G_1 \cap B_1)$$
$$= \frac{42 + 30 - 12 + 17}{100} = \frac{77}{100} = 0.77$$

Once again, the requirement that the sample space has probability 1 is equivalent to the condition

$$\sum_{\text{all } x_1} \sum_{\text{all } x_2} f(x_1, x_2) = 1$$

Joint probability distribution

$f(x_1, x_2) = P(X_1 = x_1, X_2 = x_2)$	$f(x_1, x_2) \geq 0$ for all x_1, x_2
$P(X_1, X_2 \in A) = \displaystyle\sum_{(x_1, x_2) \in A} \sum f(x_1, x_2)$	$\displaystyle\sum_{\text{all } x_1} \sum_{\text{all } x_2} f(x_1, x_2) = 1$

From the joint probability distribution, one can extract the probability distribution for any single random variable. Such a probability distribution is called a *marginal distribution*, and to keep the notation clear, we use a subscript to indicate which random variable is being retained. For example, the marginal distribution of X_1 is

$$f_1(x_1) = P(X_1 = x_1)$$

The meaning of $f_1(x_1)$ is that it gives the probability of different values of X_1 *regardless* of the value of X_2. For example,

$$f_1(1) = P(X_1 = 1) = P(G_1) = \frac{36}{100} = 0.36$$

Notice that the values of the *marginal* distributions $f_1(x_1)$ and $f_2(x_2)$ are obtained from the row and column totals, respectively, which are contained in the margins of the table in Example 5.1. We obtain the marginal distribution $f_2(x_2)$, for example, by summing the joint distribution $f(x_1, x_2)$ over all possible values of x_1, and vice versa, to obtain $f_1(x_1)$; i.e.,

$$f_1(x_1) = \sum_{\text{all } x_2} f(x_1, x_2), \quad f_2(x_2) = \sum_{\text{all } x_1} f(x_1, x_2)$$

Thus, we have

$$f_1(x_1) = \begin{cases} 0.22 & x_1 = 0 \\ 0.36, & x_1 = 1 \\ 0.42, & x_1 = 2 \end{cases} \qquad f_2(x_2) = \begin{cases} 0.3 & x_2 = 0 \\ 0.4, & x_2 = 1 \\ 0.3, & x_2 = 2 \end{cases}$$

Now it is easy to generalize to the case of n discrete random variables $X_1, X_2, \cdots, X_n$. The joint distribution function $f(x_1, x_2, \cdots, x_n)$ is defined as follows:

$$f(x_1, x_2, \cdots, x_n) = P(X_1 = x_1, X_2 = x_2, \cdots, X_n = x_n)$$

The total probability is then

$$\sum_{\text{all } x_1} \cdots \sum_{\text{all } x_n} f(x_1, \cdots, x_n) = 1$$

The notation introduced in Example 5.1 for marginal distributions extends easily to arbitrary numbers of random variables; for example,

$$f_2(x_2) = \sum_{\text{all } x_1} \sum_{\text{all } x_3} \cdots \sum_{\text{all } x_n} f(x_1, x_2, \cdots, x_n)$$

More complex marginal distributions are possible when there are more than 2 random variables. For example, we can obtain, from the joint distribution for 4 random variables, X_1, X_2, X_3, X_4, a marginal distribution which is itself a joint distribution for 2 random variables,

$$f_{2,4}(x_2, x_4) = \sum_{\text{all } x_1} \sum_{\text{all } x_3} f(x_1, x_2, x_3, x_4)$$

2. *Continuous Distributions*

For the case of multiple *continuous* random variables, the notation is the same as in the discrete case (just as it has been all along); however, the values of the function $f(x_1, \cdots, x_n)$ now give probability densities. For $n = 2$, the meaning of the *joint pdf* $f(x_1, x_2)$ is as follows: For any real numbers a, b, c, d with $a \le b$ and $c \le d$,

$$P(a \le X_1 \le b, c \le X_2 \le d) = \int_c^d \int_a^b f(x_1, x_2) \, dx_1 dx_2$$

Figure 5.1 illustrates this probability.

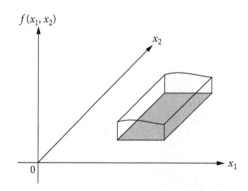

FIGURE 5.1   Joint probability.

For n continuous random variables $X_1, \cdots, X_n$ and constants $a_1, \cdots, a_n$ and $b_1, \cdots, b_n$, the joint pdf is defined as

$$P(a_1 \le X_1 \le b_1, \cdots, a_n \le X_n \le b_n) = \int_{a_n}^{b_n} \cdots \int_{a_1}^{b_1} f(x_1, \cdots, x_n) \, dx_1 \cdots dx_n$$

It is evident that the probability of any event is determined from an integral of the joint pdf over a region of the (x_1, x_2) plane. The region does not have to be rectangular, as it is in the equation above; more generally, if A is any region of the (x_1, x_2) plane, then

$$P((X_1, X_2) \in A) = \iint\limits_{A} f(x_1, x_2)\, dx_1\, dx_2$$

In direct analogy with the discrete case, we obtain the marginal pdf $f_1(x_1)$ by integrating the joint pdf over all possible values of x_2 (and similarly for $f_2(x_2)$),

$$f_1(x_1) = \int_{-\infty}^{\infty} f(x_1, x_2)\, dx_2, \qquad f_2(x_2) = \int_{-\infty}^{\infty} f(x_1, x_2)\, dx_1$$

Again, as a straightforward extension of the case for $n = 1$, there are two mathematical constraints on the joint pdf $f(x_1, x_2)$, namely nonnegativity and the probability axiom $P(S) = 1$,

$$f(x_1, x_2) \ge 0, \qquad -\infty < x_1 < \infty, \qquad -\infty < x_2 < \infty,$$

$$\int_{-\infty}^{\infty}\int_{-\infty}^{\infty} f(x_1, x_2)\, dx_1\, dx_2 = 1$$

Extension of the definitions of joint and marginal pdf's to n continuous random variables is straightforward.

Joint pdf: $f(x_1, x_2)$

$$P((X_1, X_2) \in A) = \iint\limits_{A} f(x_1, x_2)\, dx_1\, dx_2$$

$$P(a \le X_1 \le b, c \le X_2 \le d) = \int_{c}^{d}\int_{a}^{b} f(x_1, x_2)\, dx_1\, dx_2$$

$f(x_1, x_2) \ge 0$ for all x_1, x_2

$$\int_{-\infty}^{\infty}\int_{-\infty}^{\infty} f(x_1, x_2)\, dx_1\, dx_2 = 1$$

Marginal pdf:

$$f_1(x_1) = \int_{-\infty}^{\infty} f(x_1, x_2)\, dx_2, \qquad f_2(x_2) = \int_{-\infty}^{\infty} f(x_1, x_2)\, dx_1$$

EXAMPLE 5.2   Consider the joint pdf :

$$f(x, y) = \begin{cases} \dfrac{3}{8}(x + y^2), & 0 \le x \le 2, \ 0 \le y \le 1 \\ 0, & \text{otherwise} \end{cases}$$

$$f_1(x) = \int_{-\infty}^{\infty} f(x, y)dy = \int_0^1 \frac{3}{8}(x + y^2)dy = \frac{3}{8}\left[xy + \frac{y^3}{3} \right]_{y=0}^1 = \frac{3}{8}\left(x + \frac{1}{3} \right) = \frac{1}{8}(3x + 1)$$

$$f_1(x) = \begin{cases} \dfrac{1}{8}(3x + 1), & 0 \le x \le 2 \\ 0, & \text{otherwise} \end{cases}$$

$$f_2(y) = \int_{-\infty}^{\infty} f(x, y)dx = \int_0^2 \frac{3}{8}(x + y^2)dx = \frac{3}{8}\left[\frac{x^2}{2} + xy^2 \right]_{x=0}^2 = \frac{3}{8}(2 + 2y^2) = \frac{3}{4}(y^2 + 1)$$

$$f_2(y) = \begin{cases} \dfrac{3}{4}(y^2 + 1), & 0 \le y \le 1 \\ 0, & \text{otherwise} \end{cases}$$

$$P\left(\frac{1}{2} \le Y \le \frac{3}{4} \right) = \int_{1/2}^{3/4} f_2(y)dy = \frac{3}{4}\int_{1/2}^{3/4} (y^2 + 1)dy = \frac{3}{4}\left[\frac{y^3}{3} + y \right]_{\frac{1}{2}}^{\frac{3}{4}} = \left[\frac{y^3}{4} + \frac{3y}{4} \right]_{\frac{1}{2}}^{\frac{3}{4}}$$

$$= \left(\frac{27}{256} + \frac{9}{16} \right) - \left(\frac{1}{32} + \frac{3}{8} \right) = \frac{67}{256} = 0.2617$$

Another useful function to generalize to the case of n random variables is the cumulative distribution function. The random variables $X_1, \cdots, X_n$ have a *joint cdf*, defined as follows:

$$F(x_1, \cdots, x_n) = P(X_1 \le x_1, \cdots, X_n \le x_n)$$

The *marginal cdf* for X_i is denoted as follows:

$$F_i(x_i) = P(X_i \le x_i), \quad i = 1, \cdots, n$$

and we can, of course, determine marginal cdf's of any combination of random variables, if the need arises.

3. *Independent Random Variables*

DEFINITION 5.1 Two random variables X_1 and X_2 are independent if for all possible values of x_1 and x_2,

$$f(x_1, x_2) = f_1(x_1)f_2(x_2)$$

where $f(x_1, x_2)$ is the joint distribution function (or joint pdf) of X_1 and X_2.

EXAMPLE 5.3 Consider a joint distribution with random variables X_1 and X_2, given by the following table.

		x_2		
$f(x_1, x_2)$		0	10	20
	5	0.22	0.10	0.10
x_1	10	0.10	0.08	0.12
	20	0.05	0.05	0.18

Are X_1 and X_2 independent?

We require, for their independence, that $f(x_1, x_2) = f_1(x_1)f_2(x_2)$ for all possible values of (x_1, x_2). For example,

$$f_1(10)f_2(20) = (0.3) \cdot (0.4) = 0.12 = f(10, 20)$$

However,

$$f_1(10)f_2(10) = (0.3) \cdot (0.23) = 0.069 \neq 0.08 = f(10, 10)$$

Thus, X_1 and X_2 are not independent.

Note the distinction between the definition of independence introduced in Chapter 2 and the one used here. The former definition refers to individual events; e.g., in the example above, the events $X_1 = 10$ and $X_2 = 20$ are independent. However, for X_1 and X_2 to be independent, we require that the events $X_1 = x_1$ and $X_2 = x_2$ be independent for all possible x_1 and x_2.

An equivalent definition of independence is the following: X_1 and X_2 are *independent* if and only if, for all constants a, b, c, d with $a \le b$ and $c \le d$,

$$P(a \le X_1 \le b, c \le X_2 \le d) = P(a \le X_1 \le b)P(c \le X_2 \le d)$$

One can also show that $X_1, \cdots, X_n$ are independent if and only if

$$F(x_1, \cdots, x_n) = F_1(x_1) \cdots F_n(x_n)$$

EXAMPLE 5.4   Suppose that the lifetime of two lightbulbs are independent of each other and that the lifetime of the first bulb X and the lifetime of the second bulb Y have exponential distributions with mean λ and β, respectively. Then the joint pdf of X and Y is

$$f(x, y) = f_1(x)f_2(y) = \begin{cases} \dfrac{1}{\lambda}e^{-x/\lambda}\dfrac{1}{\beta}e^{-y/\beta} = \dfrac{1}{\lambda\beta}e^{-\frac{x}{\lambda}-\frac{y}{\beta}}, & x \ge 0, \ y \ge 0 \\ 0, & \text{otherwise} \end{cases}$$

Let $\lambda = 800$ hours and $\beta = 1,000$ hours. Then the probability that both lightbulbs last longer than 1200 hours is

$$P(X \ge 1200, Y \ge 1200) = P(X \ge 1200)P(Y \ge 1200) = e^{-\frac{1200}{\lambda}}e^{-\frac{1200}{\beta}}$$
$$= e^{-\frac{1200}{800}}e^{-\frac{1200}{1000}} = e^{-1.5}e^{-1.2} = e^{-2.7} = 0.0672$$

4. *Conditional Distributions*

The conditional probability distribution (or conditional pdf) of X_2 given X_1 is defined as

$$f_2(x_2|x_1) = \frac{f(x_1, x_2)}{f_1(x_1)}, \quad f_1(x_1) > 0$$

If X_1 and X_2 are discrete, then $f_2(x_2|x_1)$ is the conditional probability given as follows:

$$f_2(x_2|x_1) = P(X_2 = x_2|X_1 = x_1)$$

For continuous random variables, $f_2(x_2|x_1)$ should be read as the conditional pdf of $X_2 = x_2$ given $X_1 = x_1$. Notice that both $f(x_1, x_2)$ and $f_2(x_2|x_1)$ appear to be functions of the two variables x_1 and x_2, yet the difference in the notation, and thus the meaning of the functions, should be clear: $f(x_1, x_2)$, being the joint pdf for X_1 and X_2, is a function of two variables, whereas $f_2(x_2|x_1)$ is a function of only one variable, x_2, with x_1 acting as a constant.

EXAMPLE 5.5   Consider the joint pdf of Example 5.2,

$$f(x, y) = \begin{cases} \dfrac{3}{8}(x + y^2), & 0 \le x \le 2, \ 0 \le y \le 1 \\ 0, & \text{otherwise} \end{cases}$$

where we calculated the marginal pdf of X to be

$$f_1(x) = \begin{cases} \dfrac{1}{8}(3x + 1), & 0 \le x \le 2 \\ 0, & \text{otherwise} \end{cases}$$

What is $P(Y < 1/2 | X = 4/3)$?

Let's first compute the conditional pdf of Y given X. Since X is given, it must be between 0 and 1 because all other values of X are impossible. Given that $0 \le X \le 2$, we have

$$f_2(y|x) = \frac{f(x, y)}{f_1(x)} = \frac{\frac{3}{8}(x + y^2)}{\frac{1}{8}(3x + 1)} = \frac{3(x + y^2)}{3x + 1}, \qquad 0 \le y \le 1$$

Therefore, for $0 \le y \le 1$,

$$f_2(y|4/3) = \frac{3\left(\frac{4}{3} + y^2\right)}{3\left(\frac{4}{3}\right) + 1} = \frac{3}{5}\left(\frac{4}{3} + y^2\right) = \frac{3}{5}y^2 + \frac{4}{5}$$

so that

$$P(Y < 1/2 | X = 4/3) = \int_{-\infty}^{1/2} f_2(y|4/3)\,dy = \int_0^{1/2}\left(\frac{3}{5}y^2 + \frac{4}{5}\right)dy = \left[\frac{y^3}{5} + \frac{4y}{5}\right]_0^{1/2} = \frac{1}{40} + \frac{2}{5}$$

$$= \frac{17}{40} = 0.425$$

EXAMPLE 5.6 Consider the following joint distribution for the discrete random variables X_1 and X_2.

$f(x_1, x_2)$		x_2		
		0	1	2	Total
x_1	0	0.15	0.30	0.10	0.55
	1	0.25	0.15	0.05	0.45
Total		0.40	0.45	0.15	1

The row and column sums indicate the values of the marginal probability distributions $f_1(x_1)$ and $f_2(x_2)$, respectively, so that we can compute the following conditional probabilities (for example):

$$f_1(0|0) = \frac{f(0,0)}{f_2(0)} = \frac{0.15}{0.40} = \frac{3}{8}, \qquad f_1(1|0) = \frac{f(1,0)}{f_2(0)} = \frac{0.25}{0.40} = \frac{5}{8},$$

$$f_2(0|0) = \frac{f(0,0)}{f_1(0)} = \frac{0.15}{0.55} = \frac{3}{11}, \qquad f_2(1|0) = \frac{f(0,1)}{f_1(0)} = \frac{0.30}{0.55} = \frac{6}{11},$$

$$f_2(2|0) = \frac{f(0,2)}{f_1(0)} = \frac{0.10}{0.55} = \frac{2}{11}$$

Note that X_1 and X_2 are independent if $f_1(x_1|x_2) = f_1(x_1)$ for all x_1 and x_2. Since $f_1(1|0) = 5/8 \neq 0.45 = f_1(1)$, the conditional probability distribution of X_1 given X_2 is different from the marginal probability distribution of X_1. It follows that X_1 and X_2 are not independent. Equivalently, $f(1,0) = 0.25 \neq 0.18 = (0.45)(0.4) = f_1(1)f_2(0)$.

5. *Expectations*

If $h(x)$ is any function of a real number x and X is any random variable, then $h(X)$ is also a random variable, and its expectation is

$$E[h(X)] = \begin{cases} \displaystyle\sum_{\text{all } x} h(x)f(x) & \text{(for a discrete distribution)} \\ \displaystyle\int_{-\infty}^{\infty} h(x)f(x)dx & \text{(for a continuous distribution)} \end{cases}$$

EXAMPLE 5.7   Define the following:

$$h(x) = 10 + 2x + x^2$$

Given the probability distribution of X, we can compute the expected value of $h(X)$ as follows:

x	$f(x)$	$h(x)$	$h(x)f(x)$
2	0.5	18	9.0
3	0.3	25	7.5
4	0.2	34	6.8
Total	1	77	23.3

$$E[h(X)] = \sum_{x=2}^{4} h(x)f(x) = 18(0.5) + 25(0.3) + 34(0.2) = 23.3$$

THEOREM 5.1   Let X be a random variable, and let Y be a random variable related to X as follows: $Y = aX + b$, where a and b are constants. Then

a. $E(Y) = aE(X) + b$

b. $Var(Y) = a^2 Var(X)$

PROOF

a. $E(Y) = E(aX + b) = \int_{-\infty}^{\infty} (ax + b)f(x)dx = a\int_{-\infty}^{\infty} xf(x)dx + b = aE(X) + b$

b. $Var(Y) = E(Y^2) - [E(Y)]^2 = E((aX + b)^2) - [E(aX + b)]^2$

$= E(a^2 X^2 + 2abX + b^2) - [aE(X) + b]^2$

$= a^2 E(X^2) + 2abE(X) + b^2 - [a^2 [E(X)]^2 + 2abE(X) + b^2]$

$= a^2 E(X^2) - a^2 [E(X)]^2 = a^2 \{E(X^2) - [E(X)]^2\}$

$= a^2 Var(X)$

EXAMPLE 5.8   Let X be a random variable with mean μ and standard deviation σ, and define the random variable $Z = (X - \mu)/\sigma$. Show that Z has a mean of zero and a standard deviation of 1.

$$Z = \frac{X - \mu}{\sigma} \Rightarrow Z = \frac{1}{\sigma}X - \frac{\mu}{\sigma}$$

Applying Theorem 5.1 (a) to evaluate $E(Z)$,

$$E(Z) = E\left(\frac{1}{\sigma}X - \frac{\mu}{\sigma}\right) = \frac{1}{\sigma}E(X) - \frac{\mu}{\sigma} = \frac{\mu}{\sigma} - \frac{\mu}{\sigma} = 0$$

Applying Theorem 5.1 (b) to evaluate $Var(Z)$,

$$Var(Z) = Var\left(\frac{1}{\sigma}X - \frac{\mu}{\sigma}\right) = \frac{1}{\sigma^2}Var(X) = \frac{\sigma^2}{\sigma^2} = 1$$

Therefore, the standard deviation of Z is 1.

The definition of expectation extends easily to joint distributions. Consider the case of n random variables $X_1, \cdots, X_n$ having a joint distribution $f(x_1, \cdots, x_n)$ and an arbitrary function $h(x_1, \cdots, x_n)$. Then

$$E(h(X_1, \cdots, X_n)) = \begin{cases} \displaystyle\sum_{x_1} \cdots \sum_{x_n} h(x_1, \cdots, x_n) f(x_1, \cdots, x_n) & \text{discrete} \\[2em] \displaystyle\int_{-\infty}^{\infty} \cdots \int_{-\infty}^{\infty} h(x_1, \cdots, x_n) f(x_1 \cdots x_n) dx_1 \cdots dx_n & \text{continuous} \end{cases}$$

EXAMPLE 5.9   Consider the case of two continuous random variables X_1 and X_2, and let $h(x_1, x_2) = x_1$. Then $E(h(X_1, X_2)) = E(X_1)$. Likewise,

$$E(h(X_1, X_2)) = \int_{-\infty}^{\infty}\int_{-\infty}^{\infty} h(x_1, x_2) f(x_1, x_2) dx_2 dx_1 = \int_{-\infty}^{\infty}\int_{-\infty}^{\infty} x_1 f(x_1, x_2) dx_2 dx_1$$

$$= \int_{-\infty}^{\infty} x_1 \int_{-\infty}^{\infty} f(x_1, x_2) dx_2 dx_1 = \int_{-\infty}^{\infty} x_1 f_1(x_1) dx_1 = E(X_1)$$

All this expression tells us is that we obtain the pdf of any single random variable by integrating out all of the others. Thus, $E(X_1)$ can be computed either from the joint pdf $f(x_1, x_2)$ or the marginal pdf $f_1(x_1)$; the latter calculation is certainly faster if $f_1(x_1)$ was previously evaluated. Of course, this result extends to an arbitrary number of random variables. Note that we are not making any profound statements here; we are simply using a notation to say the same thing in different ways.

Let $n = 2$, and $h(x_1, x_2) = (x_1 - \mu_1)(x_2 - \mu_2)$, where $\mu_1 = E(X_1)$ and $\mu_2 = E(X_2)$. The expectation of this function is an important quantity, known as the *covariance* of X_1 and X_2.

Covariance of X_1 and X_2: $Cov(X_1, X_2) = E((X_1 - \mu_1)(X_2 - \mu_2))$

THEOREM 5.2   An equivalent formula of the covariance can be obtained as follows:

$$Cov(X_1, X_2) = E(X_1 X_2) - \mu_1 \mu_2$$

PROOF

$$\begin{aligned} \text{Cov}(X_1, X_2) &= E[(X_1 - \mu_1)(X_2 - \mu_2)] = E(X_1 X_2 - \mu_1 X_2 - \mu_2 X_1 + \mu_1 \mu_2) \\ &= E(X_1 X_2) - \mu_1 E(X_2) - \mu_2 E(X_1) + \mu_1 \mu_2 = E(X_1 X_2) - \mu_1 \mu_2 - \mu_1 \mu_2 + \mu_1 \mu_2 \\ &= E(X_1 X_2) - \mu_1 \mu_2 \end{aligned}$$

THEOREM 5.3   Suppose X_1 and X_2 are independent random variables. Then
$Cov(X_1, X_2) = 0$.

PROOF

$$Cov(X_1, X_2) = E(X_1 X_2) - \mu_1 \mu_2 = E(X_1)E(X_2) - \mu_1 \mu_2 = 0$$

As discussed in Chapter 1, covariance is not an efficient measure of the relationship between two variables. The covariance measures the direction of the relationship between two variables, and the correlation coefficient measures the strength of it. The covariance is used to show if there is any pattern to the way two variables move together. The correlation coefficient measures how strong the linear relationship between two variables is.

DEFINITION 5.2 Correlation coefficient of two random variables X_1 and X_2 is defined as follows:

$$\rho = Cor(X_1, X_2) = \frac{Cov(X_1, X_2)}{\sigma_{X_1} \sigma_{X_2}}.$$

The correlation coefficient has the following condition:

$$-1 \leq \rho \leq 1$$

EXAMPLE 5.10 Let's find the correlation coefficient of X and Y in Example 5.2.

$$f(x, y) = \begin{cases} \dfrac{3}{8}(x + y^2), & 0 \leq x \leq 2, \ 0 \leq y \leq 1 \\ 0, & \text{otherwise} \end{cases}$$

$$f_1(x) = \begin{cases} \dfrac{1}{8}(3x + 1), & 0 \leq x \leq 2 \\ 0, & \text{otherwise} \end{cases}$$

and

$$f_2(y) = \begin{cases} \dfrac{3}{4}(y^2 + 1), & 0 \leq y \leq 1 \\ 0, & \text{otherwise} \end{cases}$$

from Example 5.2.

$$E(X) = \int_{-\infty}^{\infty} x f_1(x) dx = \int_0^2 \frac{1}{8}(3x^2 + x) dx = \frac{1}{8}\left[x^3 + \frac{x^2}{2}\right]_0^2 = \frac{1}{8}(8 + 2) = \frac{5}{4}$$

$$E(X^2) = \int_{-\infty}^{\infty} x^2 f_1(x) dx = \int_0^2 \frac{1}{8}(3x^3 + x^2) dx = \frac{1}{8}\left[\frac{3x^4}{4} + \frac{x^3}{3}\right]_0^2 = \frac{1}{8}\left(12 + \frac{8}{3}\right) = \frac{11}{6}$$

$$Var(X) = E(X^2) - [E(X)]^2 = \frac{11}{6} - \left(\frac{5}{4}\right)^2 = \frac{13}{48}$$

$$E(Y) = \int_{-\infty}^{\infty} y f_2(y) dy = \int_0^1 \frac{3}{4}(y^3 + y) dy = \frac{3}{4}\left[\frac{y^4}{4} + \frac{y^2}{2}\right]_0^1 = \frac{3}{4}\left(\frac{1}{4} + \frac{1}{2}\right) = \frac{9}{16}$$

$$E(Y^2) = \int_{-\infty}^{\infty} y^2 f_2(y) dy = \int_0^1 \frac{3}{4}(y^4 + y^2) dx = \frac{3}{4}\left[\frac{y^5}{5} + \frac{y^3}{3}\right]_0^1 = \frac{3}{4}\left(\frac{1}{5} + \frac{1}{3}\right) = \frac{2}{5}$$

$$Var(Y) = E(Y^2) - [E(Y)]^2 = \frac{2}{5} - \left(\frac{9}{16}\right)^2 = \frac{107}{1280}$$

$$E(XY) = \int_{-\infty}^{\infty} xy f(x, y) dx dy = \int_0^1 \int_0^2 \frac{3}{8}(x^2 y + xy^3) dx dy = \int_0^1 \left[\frac{3}{8}\left(\frac{x^3 y}{3} + \frac{x^2 y^3}{2}\right)\right]_0^2 dy$$

$$= \int_0^1 \left(y + \frac{3y^3}{4}\right) dy = \left(\frac{y^2}{2} + \frac{3y^4}{16}\right)\Big|_0^1 = \frac{1}{2} + \frac{3}{16} = \frac{11}{16}$$

$$Cov(X, Y) = E(XY) - E(X)E(Y) = \frac{11}{16} - \left(\frac{5}{4}\right)\left(\frac{9}{16}\right) = -\frac{1}{64}$$

$$\rho = \frac{Cov(X, Y)}{\sigma_X \sigma_Y} = \frac{-\frac{1}{64}}{\sqrt{\left(\frac{13}{48}\right)\left(\frac{107}{1280}\right)}} = -0.104$$

DEFINITION 5.3 Let $X_1, \cdots, X_n$ be n random variables and $a_1, \cdots, a_n$ be constants. Then

$$Y = a_1 X_1 + \cdots + a_n X_n = \sum_{i=1}^{n} a_i X_i$$

is called a linear combination of the X_i. A few special cases of linear combination are given below.

$$\text{If } a_1 = \cdots = a_n = 1, \text{ then } Y = \sum_{i=1}^{n} X_i$$

$$\text{If } a_1 = \cdots = a_n = \frac{1}{n}, \text{ then } Y = \bar{X}$$

THEOREM 5.4   Let $X_1, \cdots, X_n$ be n random variables and $a_1, \cdots, a_n$ be constants. Let X_i have mean μ_i and variance σ_i^2 for $i = 1, 2, \cdots, n$. Then

a. $E(a_1 X_1 + \cdots + a_n X_n) = a_1 E(X_1) + \cdots + a_n E(X_n)$
b.   If $X_1, \cdots, X_n$ are independent, then

$$Var(a_1 X_1 + \cdots + a_n X_n) = a_1^2 Var(X_1) + \cdots + a_n^2 Var(X_n)$$

Let X_i have mean μ_i and variance σ_i^2 for $i = 1, 2, \cdots, n$. Then
1. $E(a_1 X_1 + \cdots + a_n X_n) = a_1 E(X_1) + \cdots + a_n E(X_n)$ $\left(\mu_Y = \sum_{i=1}^{n} a_i \mu_i\right)$
2. If $X_1, \cdots, X_n$ are independent, then

$$Var(a_1 X_1 + \cdots + a_n X_n) = a_1^2 Var(X_1) + \cdots + a_n^2 Var(X_n) \left(\sigma_Y^2 = \sum_{i=1}^{n} a_i^2 \sigma_i^2\right)$$

EXAMPLE 5.11 The prices of three different brands of toothpaste at a particular store are $2.40, $2.70, and $2.90 per pack. Let X_i be the total amount, in packs, of brand i purchased on a particular day. Assume that X_1, X_2, and X_3 are independent, and that $\mu_1 = 500$, $\mu_2 = 250$, $\mu_3 = 200$, $\sigma_1 = 50$, $\sigma_2 = 40$, and $\sigma_3 = 25$. Here, $\mu_i = E(X_i)$ and $\sigma_i^2 = Var(X_i)$ for $i = 1, 2, 3$. Compute the expected revenue from the sale of the toothpaste and its standard deviation.

Let Y be the total revenue. Then $Y = 2.4X_1 + 2.7X_2 + 2.9X_3$. Applying Theorem 5.4,

$$E(Y) = 2.4\mu_1 + 2.7\mu_2 + 2.9\mu_3 = 2.4(500) + 2.7(250) + 2.9(200) = \$2{,}455$$

$$Var(Y) = (2.4)^2\sigma_1^2 + (2.7)^2\sigma_2^2 + (2.9)^2\sigma_3^2 = \$31{,}320.25$$

$$\sigma_Y = \sqrt{31{,}320.25} = \$176.98$$

In case $n = 2$, if $a_1 = 1$ and $a_2 = -1$, then $Y = a_1X_1 + a_2X_2 = X_1 - X_2$, and thus

$$E(X_1 - X_2) = \mu_1 - \mu_2$$

If X_1 and X_2 are independent, then

$$Var(X_1 - X_2) = \sigma_1^2 + \sigma_2^2$$

EXAMPLE 5.12 The tar contents of the cigarettes from two different brands are known to be different. Let X_1 and X_2 be the tar contents of the cigarettes from brand 1 and brand 2, respectively, and suppose that $\mu_1 = 15$ mg, $\mu_2 = 17$ mg, $\sigma_1 = 1.1$ mg, and $\sigma_2 = 1.4$ mg. What is the expected difference in the tar contents between these two cigarettes and the standard deviation if X_1 and X_2 are independent?

$$E(X_1 - X_2) = \mu_1 - \mu_2 = 15 - 17 = -2 \text{ mg}$$

$$Var(X_1 - X_2) = \sigma_1^2 + (-1)^2\sigma_2^2 = \sigma_1^2 + \sigma_2^2 = (1.1)^2 + (1.4)^2 = 3.17 \text{ mg}$$

$$\sigma_{X_1 - X_2} = \sqrt{3.17} = 1.78 \text{ mg}$$

THEOREM 5.5   If $X_1, \cdots, X_n$ are independent, $X_i \sim N(\mu_i, \sigma_i^2)$ for $i = 1, 2, \cdots, n$, then

$$Y = a_1X_1 + \ldots + a_nX_n \sim N(a_1\mu_1 + \ldots + a_n\mu_n, a_1^2\sigma_1^2 + \ldots + a_n^2\sigma_n^2)$$

Therefore,

$$X_1 - X_2 \sim N(\mu_1 - \mu_2, \sigma_1^2 + \sigma_2^2)$$

EXAMPLE 5.13 From Example 5.11, suppose X_1, X_2, and X_3 are normal. Then for $Y = 2.4X_1 + 2.7X_2 + 2.9X_3$, $Y \sim N(2.4\mu_1 + 2.7\mu_2 + 2.9\mu_3, (2.4)^2\sigma_1^2 + (2.7)^2\sigma_2^2 + (2.9)^2\sigma_3^2)$.

SUMMARY OF CHAPTER 5

1. Joint Distributions

	Discrete distribution	Continuous distribution
Joint distribution of X_1 and X_2	$f(x_1, x_2) = P(X_1 = x_1, X_2 = x_2)$ $f(x_1, x_2) \geq 0$ $\sum_{x_1}\sum_{x_2} f(x_1, x_2) = 1$	$P(a \leq X_1 \leq b, c \leq X_2 \leq d)$ $= \int_c^d \int_a^b f(x_1, x_2)\, dx_1 dx_2$ $f(x_1, x_2) \geq 0$ $\int_{-\infty}^{\infty}\int_{-\infty}^{\infty} f(x_1, x_2)dx_1 dx_2 = 1$
Joint cdf of X_1 and X_2	$F(x_1, x_2) = P(X_1 \leq x_1, X_2 \leq x_2)$ $= \sum_{u \leq x_1}\sum_{v \leq x_2} f(u, v)$	$F(x_1, x_2) = P(X_1 \leq x_1, X_2 \leq x_2)$ $= \int_{-\infty}^{x_2}\int_{-\infty}^{x_1} f(u, v)\, dudv$
Marginal distribution	$f_1(x_1) = \sum_{x_2} f(x_1, x_2)$ $f_2(x_2) = \sum_{x_1} f(x_1, x_2)$	$f_1(x_1) = \int_{-\infty}^{\infty} f(x_1, x_2)dx_2$ $f_2(x_2) = \int_{-\infty}^{\infty} f(x_1, x_2)dx_1$
Conditional distribution	$f_1(x_1|x_2) = \dfrac{f(x_1, x_2)}{f_2(x_2)}$, $f_2(x_2) > 0$ $f_2(x_2|x_1) = \dfrac{f(x_1, x_2)}{f_1(x_1)}$, $f_1(x_1) > 0$	

2. Random variables X_1 and X_2 are <u>independent</u> if and only if $f(x_1, x_2) = f_1(x_1)f_2(x_2)$, or equivalently $F(x_1, x_2) = F_1(x_1)F_2(x_2)$ or $P(X_1 \in A, X_2 \in B) = P(X_1 \in A)P(X_2 \in B)$.

3. Properties of Expectation

a. $E[h(X)] = \begin{cases} \sum_{\text{all } x} h(x)f(x) & \text{(for a discrete distribution)} \\ \int_{-\infty}^{\infty} h(x)f(x)dx & \text{(for a continuous distribution)} \end{cases}$

b. For constants a and b, $E(aX + b) = aE(X) + b$ and $Var(aX + b) = a^2 Var(X)$.

c. $E[h(X_1, \cdots, X_n)] = \begin{cases} \sum_{x_1} \cdots \sum_{x_n} h(x_1, \cdots, x_n) f(x_1, \cdots, x_n) & \text{discrete} \\ \int_{-\infty}^{\infty} \cdots \int_{-\infty}^{\infty} h(x_1, \cdots, x_n) f(x_1, \cdots, x_n) dx_1 \cdots dx_n & \text{continuous} \end{cases}$

d. $Cov(X_1, X_2) = E[(X_1 - \mu_1)(X_2 - \mu_2)]$

e. Correlation coefficient: $\rho = Cor(X_1, X_2) = \frac{Cov(X_1, X_2)}{\sigma_{X_1} \sigma_{X_2}}$

4. Distribution of a Linear Combination of Random Variables

a. $E(a_1 X_1 + \cdots + a_n X_n) = a_1 E(X_1) + \cdots + a_n E(X_n)$

b. If $X_1, X_2, \cdots, X_n$ are independent,
$Var(a_1 X_1 + \cdots + a_n X_n) = a_1^2 Var(X_1) + \cdots + a_n^2 Var(X_n)$

c. If $X_1, X_2, \cdots, X_n$ are independent with $X_i \sim N(\mu_i, \sigma_i^2)$ for $i = 1, 2, \cdots, n$, then
$Y = a_1 X_1 + \ldots + a_n X_n \sim N(a_1\mu_1 + \ldots + a_n\mu_n, a_1^2\sigma_1^2 + \ldots + a_n^2\sigma_n^2)$

EXERCISES

5.1   Random variables X_1 and X_2 have the following distribution.

	$f(x_1, x_2)$		x_2 0	x_2 10	x_2 20
		5	0.22	0.10	0.10
x_1		10	0.10	0.08	0.12
		20	0.05	0.05	0.18

Answer the following questions.

a. Find $f(10, 10)$.

b. Find $P(X_1 \geq 10, X_2 \leq 10)$.

c. Find $f_1(10)$.

d. Find $f_2(20)$.

e. Find the marginal distributions of X_1 and X_2.

f. Find $P(X_1 \geq 10)$.

g. Find $P(X_2 \leq 10)$.

5.2   Let the random variables X and Y have joint distribution

$$f(x, y) = \frac{2x + y}{30}, \quad x = 1, 2; \ y = 1, 2, 3$$

a. Find the marginal distributions $f_1(x)$ of X and $f_2(y)$ of Y.
b. Are X and Y independent? Justify your answer.

5.3   Suppose the random variables X and Y have joint distribution as follows:

$$f(x, y) = \frac{1}{12}, \quad x = 1, 2, 3; \ y = 1, 2, 3, 4$$

a. Find the marginal distributions $f_1(x)$ of X and $f_2(y)$ of Y.
b. Show that X and Y are independent.

5.4   Suppose the random variables X and Y have joint distribution as follows:

$$f(x, y) = \frac{xy}{36}, \quad x = 1, 2, 3; \ y = 1, 2, 3$$

a. Find $f_1(x)$.
b. Find $f_2(y)$.
c. Find $f_1(x|y)$.
d. Find $f_2(y|x)$.
e. Find $P(X \leq 2)$.
f. Find $P(Y > 1)$.
g. Find $P(X \leq 2|Y > 1)$.
h. Are X and Y independent?

5.5   For the distribution given in Example 5.1, perform the following calculations.
a. Find $f_1(x_1)$.
b. Find $f_2(x_2)$.
c. Find $f_1(x_1|x_2)$.
d. Find $f_2(x_2|x_1)$.
e. Find $P(X_1 \leq 1)$.
f. Find $P(X_2 > 0)$.
g. Find $P(X_1 \leq 1, X_2 < 2)$.
h. Find $P(X_1 + X_2 \leq 1)$.

5.6   Suppose the random variables X, Y, and Z have joint distribution as follows:

$$f(x, y, z) = \frac{xy^2z}{180}, \quad x = 1, 2, 3; \; y = 1, 2; \; z = 1, 2, 3$$

    a. Find the two-dimensional marginal distributions $f_{1,2}(x, y)$, $f_{1,3}(x, z)$ and $f_{2,3}(y, z)$.

    b. Find the marginal distributions $f_1(x)$, $f_2(y)$, and $f_3(z)$.

    c. Find $P(Y = 2 | X = 1, Z = 3)$.

    d. Find $P(X \geq 2, Y = 2 | Z = 2)$.

    e. Are X, Y, and Z independent?

5.7   Let X denote the sum of the points in two tosses of a fair die.

    a. Find the probability distribution and events corresponding to the values of X.

    b. Obtain the cdf $F(x)$ of X.

    c. Find $P(3 < X \leq 6)$.

5.8   If X and Y are independent exponential random variables with pdf $f(x) = e^{-x}$, $x > 0$, find $P(X > 2, Y > 1)$.

5.9   Let the random variables X and Y have joint cdf as follows:

$$F(x, y) = \begin{cases} 1 - e^{-2x} - e^{-3y} + e^{-2x-3y}, & x > 0, \; y > 0 \\ 0, & \text{otherwise} \end{cases}$$

    a. Find the joint pdf $f(x, y)$ of X and Y.

    b. Are X and Y independent?

    c. Find $f_1(x|y)$.

    d. Find $P(X \leq 2 \text{ and } Y \geq 3)$.

    e. Find the means of the two variables.

    f. Find the variances of the two variables.

    g. Find the correlation coefficient of X and Y.

5.10   Let the random variables X and Y have joint pdf given as

$$f(x, y) = 12x(1-x)y, \quad 0 < x < 1, 0 < y < 1$$

    a.   Find the marginal pdf $f_1(x)$ of X.
    b.   Find the marginal pdf $f_2(y)$ of Y.
    c.   Find the conditional pdf $f_1(x|y)$.
    d.   Are X and Y independent?
    e.   Find $P(X \leq 0.5|Y = 0.5)$.

5.11   The joint pdf of X and Y is given as

$$f(x, y) = \frac{2(x + 2y)}{3}, \quad 0 < x < a, \ 0 < y < 1,$$

where a is a constant.
    a.   Find the value of a.
    b.   Using the value of a obtained in part (a), determine the marginal pdf's of X and Y.
    c.   Determine the conditional pdf $f_1(x|y)$.
    d.   Find $P(X \leq 1/2|Y = 1/2)$.

5.12   Suppose the random variables X and Y have joint pdf as follows:

$$f(x, y) = 15xy^2, \quad 0 < y < x < 1$$

    a.   Find the marginal pdf $f_1(x)$ of X.
    b.   Find the conditional pdf $f_2(y\,|\,x)$.
    c.   Find $P(Y > 1/3\,|\,X = x)$ for any $1/3 < x < 1$.
    d.   Are X and Y independent? Justify your answer.

5.13   Let the random variables X and Y have joint pdf as follows:

$$f(x, y) = \frac{4}{7}\left(x^2 + \frac{xy}{3}\right), \quad 0 < x < 1, 0 < y < 3$$

a. Find the marginal densities of X and Y.
b. Find the cdf of X and cdf of Y.
c. Find $P(Y < 2)$.
d. Find $P(X > \frac{1}{2}, Y < 1)$.
e. Find $P(X > \frac{1}{2})$ and $P(Y < 1)$.
f. Are X and Y independent?
g. Determine the conditional pdf of Y given $X = x$.
h. Find $P(1 < Y < 2 | X = \frac{1}{2})$.

5.14   Let the joint pdf of X and Y be $f(x, y) = 12e^{-4x-3y}$, $x > 0$, $y > 0$.
a. Find the marginal pdf's of X and Y.
b. Are X and Y independent?
c. Find the conditional pdf $f_1(x|y)$.
d. Find the marginal cdf's of X and Y.
e. Find $P(1 < Y < 3)$.
f. Find $P(1 < Y < 3 | X = 3)$.
g. Find $P(X > 2, 1 < Y < 3)$.
h. Find $E(X)$ and $E(Y)$.
i. Find $Var(X)$ and $Var(Y)$.

5.15   - Random variables X and Y have the following joint probability distribution.

		x		
$f(x, y)$		1	2	3
y	1	0.1	0.3	0.2
	2	0.2	0.15	0.05

a. Find $P(X + Y > 3)$.
b. Find the marginal probability distributions $f_1(x)$ and $f_2(y)$.
c. Find $f_1(x|y = 2)$.
d. Are X and Y independent?
e. Find $E(X)$ and $E(Y)$.
f. Find $Var(X)$ and $Var(Y)$.
g. Find the correlation coefficient of X and Y.

5.16 Let random variables X and Y have the joint distribution given in the following table:

$f(x, y)$		y		
		0	1	2
	0	1/12	1/12	1/6
x	1	0	1/3	1/4
	2	0	0	1/12

Answer the following questions:
a. Find the marginal probability distributions $f_1(x)$ and $f_2(y)$.
b. Find $E(X)$ and $E(Y)$.
c. Find $Var(X)$ and $Var(Y)$.
d. Find the correlation coefficient of X and Y.

5.17 Random variables X and Y have the following joint probability distribution.

$f(x, y)$		x		
		1	2	3
	1	0.1	0.1	0.15
y	2	0.05	0.05	0.2
	3	0.15	0.1	0.1

a. Find $P(X + Y \leq 4)$.
b. Find the marginal probability distributions $f_1(x)$ and $f_2(y)$.
c. Find $P(X < 2 | Y = 2)$.
d. Are X and Y independent?
e. Find $E(X)$ and $E(Y)$.
f. Find $Var(X)$ and $Var(Y)$.
g. Find the correlation coefficient of X and Y.

5.18 Let the random variables X and Y have joint pdf as follows:

$$f(x, y) = cx + y, \quad 0 < x < \frac{1}{2}, 0 < y < 1$$

a. Find the value of constant c.
b. Find the marginal densities of X and Y.
c. Are the random variables X and Y independent?
d. Find the conditional pdf's $f_1(x|y)$ and $f_2(y|x)$.
e. Find $E(X)$ and $E(Y)$.
f. Find $Var(X)$ and $Var(Y)$.
g. Find $Cov(X, Y)$.
h. Find the correlation coefficient of X and Y.

5.19   Suppose the random variables X and Y have joint pdf $f(x, y) = 6y, 0 < y < x < 1$.
   a.   Find the marginal pdf of X and marginal pdf of Y.
   b.   Find the conditional pdf of X given $Y = y$.
   c.   Find $P(X > \frac{1}{2}|Y = \frac{1}{3})$.
   d.   Find $E(X)$ and $E(Y)$.
   e.   Find $Var(X)$ and $Var(Y)$.
   f.   Find $Cov(X, Y)$.
   g.   Find the correlation coefficient of X and Y.

5.20   Suppose the random variables X and Y have joint pdf $f(x, y) = \frac{1}{2}, 0 < y < x < 2$.
   a.   Find the marginal pdf of X and marginal pdf of Y.
   b.   Find the conditional pdf of Y given $X = x$.
   c.   Find the conditional pdf of X given $Y = y$.
   d.   Find $P(X > \frac{3}{4}|Y = \frac{1}{3})$.
   e.   Find $E(X)$ and $E(Y)$.
   f.   Find $E(Y|X = x)$ and $E(X|Y = y)$.
   g.   Find $Var(X)$ and $Var(Y)$.
   h.   Find $Cov(X, Y)$.
   i.   Find the correlation coefficient of X and Y.

5.21   Let the random variables X and Y have joint pdf as follows:

$$f(x, y) = \frac{3}{4}(x^2 + 3y^2), \quad 0 < x < 1, 0 < y < 1$$

   a.   Find the marginal densities of X and Y.
   b.   Find the cdf of X and cdf of Y.
   c.   Determine the conditional pdf's $f_1(x|y)$ and $f_2(y|x)$.
   d.   Find $E(X)$ and $E(Y)$.
   e.   Find $Var(X)$ and $Var(Y)$.
   f.   Find $Cov(X, Y)$.
   g.   Find $P(Y < 1/3|X = 1/3)$.
   h.   Find $E(Y|X = \frac{1}{2})$.

5.22   Let X have the following distribution.

x	0	1	2	3
$f(x)$	0.1	0.2	0.4	0.3

   a.   Find $E(X)$.
   b.   Find the standard deviation of X.
   c.   Find $E(X^2 - 2)$.

5.23   Let X be a continuous random variable with pdf

$$f(x) = 3x^2, \ 0 < x < 1.$$

   a.   Find $E(X)$.
   b.   Find $E(4X + 5X^2)$.

5.24   Let X and Y be independent random variables representing the lifetime (in 100 hours) of type A and type B lightbulbs, respectively. Both variables have exponential distributions, and the mean of X is 2 and the mean of Y is 3.
   a.   Find the joint pdf $f(x, y)$ of X and Y.
   b.   Find the conditional pdf $f_2(y|x)$ of Y given $X = x$.
   c.   Find the probability that a type A bulb lasts at least 300 hours and a type B bulb lasts at least 400 hours.
   d.   Given that a type B bulb fails at 300 hours, find the probability that a type A bulb lasts longer than 300 hours.
   e.   What is the expected total lifetime of two type A bulbs and one type B bulb?
   f.   What is the variance of the total lifetime of two type A bulbs and one type B bulb?

5.25   Suppose X and Y are independent random variables such that $E(X) = 4, Var(X) = 9, E(Y) = 5, Var(Y) = 25.$ Find $E(U)$ and $Var(U)$ where $U = 3X - Y + 2.$

5.26   Suppose X and Y are independent random variables such that $E(X) = 5, \ Var(X) = 8, E(Y) = 3, Var(Y) = 5.$ Find $E(V)$ and $Var(V)$ where $V = 2X - 3Y - 1.$

5.27   Let X_1 and X_2 be independent normal random variables, distributed as $N(\mu_1, \sigma^2)$ and $N(\mu_2, \sigma^2)$, respectively. Consider a random variable $Y = 3X_1 - 2X_2$.
   a.   Find $E(Y)$.
   b.   Find $Var(Y)$.
   c.   Find the distribution of Y.

5.28 Let X_1, X_2, X_3 be three independent normal random variables with expected values μ_1, μ_2, μ_3 and variances $\sigma_1^2, \sigma_2^2, \sigma_3^2$, respectively.

a. If $\mu_1 = \mu_2 = \mu_3 = 50$ and $\sigma_1^2 = \sigma_2^2 = \sigma_3^2 = 27$, find $P(\bar{X} < 45)$.

b. If $\mu_1 = \mu_2 = \mu_3 = 60$ and $\sigma_1^2 = \sigma_2^2 = \sigma_3^2 = 27$, find $P(X_1 + X_2 + X_3 > 170)$.

c. If $\mu_1 = \mu_2 = \mu_3 = 60$ and $\sigma_1^2 = \sigma_2^2 = \sigma_3^2 = 4$, find $P(-20 < 5X_1 - 3X_2 - 2X_3 < 20)$.

d. If $\mu_1 = 40$, $\mu_2 = 50$, $\mu_3 = 60$ and $\sigma_1^2 = \sigma_2^2 = \sigma_3^2 = 27$, find $P(130 < X_1 + X_2 + X_3 < 160)$.

e. If $\mu_1 = 40$, $\mu_2 = 50$, $\mu_3 = 60$ and $\sigma_1^2 = 4, \sigma_2^2 = 9, \sigma_3^2 = 16$, find $P(-10 < 4X_1 - 2X_2 - X_3 < 10)$.

5.29 Let $X_1, \cdots, X_4$ be independent normal random variables and X_i be distributed as $N(\mu_i, \sigma_i^2)$ for $i = 1, \cdots, 4$.

a. Find $P(35 < X_1 + \cdots + X_4 < 45)$ when $\mu_1 = \cdots = \mu_4 = 10$ and $\sigma_1^2 = \cdots = \sigma_4^2 = 2$.

b. Find $P(\bar{X} < 11)$ when $\mu_1 = \cdots = \mu_4 = 10$ and $\sigma_1^2 = \cdots = \sigma_4^2 = 2$.

c. Find $P(4X_1 + X_2 - X_3 - 4X_4 < 3)$ when $\mu_1 = \cdots = \mu_4 = 10$ and $\sigma_1^2 = \cdots = \sigma_4^2 = 2$.

d. Find $P(35 < X_1 + \cdots + X_4 < 45)$ when $\mu_1 = \mu_2 = 8$, $\mu_3 = \mu_4 = 12$, $\sigma_1^2 = \sigma_2^2 = 1$ and $\sigma_3^2 = \sigma_4^2 = 2$.

5.30 In a certain city, the mean price of a two-liter bottle of soda is $1.50 and the standard deviation is $0.20 at various grocery stores. The mean price of one gallon of milk is $3.80 and the standard deviation is $0.30. Ten tourists came to the city together. They separately went to randomly chosen grocery stores, and each person bought one item. Six of them bought two-liter bottles of soda, and each of the remaining four bought one gallon of milk.

a. What is the expected total amount of money that the 10 people spent?

b. If the prices of individual items are independent and normally distributed, what is the probability that the total amount of money the 10 people spent is at least $25?

Sampling Distributions

1. Populations and Samples

As a subject, *statistics* provides methods for designing the process of data collection, summarizing and interpreting the data, and drawing conclusions. The vast collection of all potential observations of a particular type is called the *population*. If a population is large, it is impractical to observe it entirely. Thus, it is necessary to use a *sample* of the population to infer a quantity that characterizes the entire population.

> **Population:** The vast collection of all potential observations of a particular type;
> e.g., enrollment figures, height, and so on.
> **Sample:** The data set that is acquired through the process of observation.

EXAMPLE 6.1

a. To predict the result of a presidential election, it is not realistic to ask every voter's opinion due to time and financial constraint. A pollster selects a sample carefully and makes an inference using the answers from the sample. Figure 6.1 illustrates this.

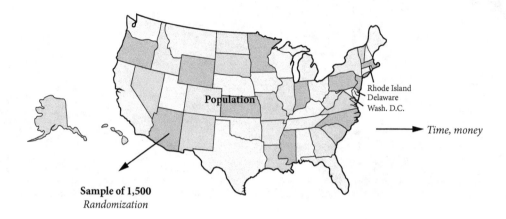

Rhode Island
Delaware
Wash. D.C.

Time, money

Population

Sample of 1,500
Randomization

FIGURE 6.1   Illustration of random sampling.

b. In an automobile collision study, an exhaustive data set is not available. Instead, a small sample is selected from each brand.

In probability, properties of the population are assumed to be known. Questions regarding a sample taken from the population are posed and answered. For example, the probability of obtaining a head in the toss of a fair coin is 1/2; the probability of obtaining a 1 in the toss of a fair die is 1/6. In tossing 6 fair coins, we expect $E(X) = 6(1/2) = 3$ heads, where X denotes the number of heads.

In statistics, we use the known characteristics of a sample to draw conclusions about the population. The major objectives of statistics are (1) to make inferences about a population from an analysis of information contained in the sample data, and (2) to design the process and the extent of the sampling so that the observations form a basis for drawing valid inferences.

> The random variables $X_1, \cdots, X_n$ are a <u>random sample</u> of size n (iid) if:
> 1. $X_1, \cdots, X_n$ are independent, and
> 2. every X_i has the same distribution.

In other words, $X_1, \cdots, X_n$ are iid, *independent and identically distributed.*

Statistical inference is directed toward drawing some type of conclusion about one or more population quantities, such as the mean μ and variance σ^2. The population quantity is called a *parameter*. A function of one or more random variables that is not dependent on any unknown parameter is called a *statistic*. For example, $\bar{X}$, $\sum_{i=1}^{n} X_i$, and S^2 are statistics. Notice that any statistic is itself a random variable; thus, it has a probability distribution. The next few theorems concern the probability distribution of the sample mean $\bar{X}$.

2. Distribution of the Sample Mean When σ Is Known

THEOREM 6.1   Let $X_1, \cdots, X_n$ be a random sample from a distribution with mean μ and standard distribution σ. Then

   a. $E(\bar{X}) = \mu$

   b. $Var(\bar{X}) = \dfrac{\sigma^2}{n}, \quad \sigma_{\bar{X}} = \dfrac{\sigma}{\sqrt{n}}$

Consequently,

$$E\left(\sum_{i=1}^{n} X_i\right) = n\mu, \quad \text{and} \quad Var\left(\sum_{i=1}^{n} X_i\right) = n\sigma^2, \quad \sigma_{\Sigma X_i} = \sqrt{n}\sigma$$

PROOF

   a.   Since $X_1, \cdots, X_n$ are identically distributed,

$$E(\bar{X}) = E\left(\sum_{i=1}^{n} \frac{X_i}{n}\right) = \frac{1}{n}\sum_{i=1}^{n} E(X_i) = \frac{1}{n}(n\mu) = \mu$$

   b.   By independence,

$$Var(\bar{X}) = Var\left(\sum_{i=1}^{n} \frac{X_i}{n}\right) = \frac{1}{n^2} Var\left(\sum_{i=1}^{n} X_i\right) = \frac{1}{n^2}\sum_{i=1}^{n} Var(X_i) = \frac{1}{n^2}(n\sigma^2) = \frac{\sigma^2}{n}$$

The value of $\bar{X}$ should be thought of as an estimate of μ. Then, $\sigma_{\bar{X}}$ tells us how close to μ we can reasonably expect $\bar{X}$ to be. As the value of $\sigma_{\bar{X}}$ decreases as the sample size grows while $E(\bar{X})$ stays the same, the distribution of $\bar{X}$ tends to become more concentrated around μ. We are more certain about the estimate of μ if more information (larger data) is given.

EXAMPLE 6.2   The average grade of a nationwide examination is $\mu = 70$ and the standard deviation is $\sigma = 14$. Let $X_1, X_2, \cdots, X_{36}$ be a random sample of size 36, where each X_i is the score of a randomly selected student. Then the expected value of the sample mean score and the expected total of the scores for the 36 students are

$$E(\bar{X}) = \mu = 70 \ \text{ and } \ E\left(\sum_{i=1}^{n} X_i\right) = n\mu = 36(70) = 2{,}520$$

respectively, and the standard deviations of $\bar{X}$ and $\sum_{i=1}^{n} X_i$ are

$$\sigma_{\bar{X}} = \frac{\sigma}{\sqrt{n}} = \frac{14}{6} = 2.33 \ \text{ and } \ \sigma_{\Sigma X_i} = \sqrt{n}\sigma = 6 \cdot 14 = 84$$

respectively.

Theorem 6.1 tells us about the expectation and variance of $\bar{X}$, but to make useful inferences we still need to know its probability distribution. The next theorem says that if we sample from a normal distribution, then the sample mean is normally distributed. This is a special case of Theorem 5.5.

THEOREM 6.2   If $X_1, \cdots, X_n$ is a random sample from a *normal population*, then $\bar{X}$ is distributed as normal with mean μ and variance σ^2 / n.

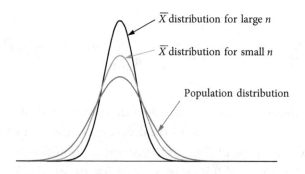

FIGURE 6.2   Change in the distribution of $\bar{X}$ for a normal random sample as n varies.

The standard deviation $\sigma_{\bar{X}}$ of $\bar{X}$ decreases as the sample size n increases even though $E(\bar{X})$ remains the same. The distribution of $\bar{X}$ tends to become more concentrated around $\bar{X}$ as n increases. More information from larger data makes us more certain about the estimation of μ. Figure 6.2 illustrates this phenomenon.

EXAMPLE 6.3   The time that it takes a randomly selected sports car from the same brand to accelerate from 0 to 60 miles per hour (mph) is normally distributed with mean 6.2 seconds and standard deviation 0.15 seconds. Suppose 6 cars are selected at random, and let $X_1, \cdots, X_6$ denote their times to reach 60 mph.

a. What is the probability that the total time $T = \sum_{i=1}^{6} X_i$ is between 36.5 and 37.5 seconds?

$$E(T) = E\left(\sum_{i=1}^{n} X_i\right) = n\mu = 6(6.2) = 37.2 \text{ and } \sigma_T = \sigma_{\sum_{i=1}^{n} X_i} = \sqrt{n}\sigma = 0.15\sqrt{6} = 0.367$$

Hence, $T \sim N(37.2, (0.367)^2)$.

$$P(36.5 < T < 37.5) = P\left(\frac{36.5 - 37.2}{0.367} < Z < \frac{37.5 - 37.2}{0.367}\right) = P(-1.91 < Z < 0.82)$$

$$= \Phi(0.82) - \Phi(-1.91) = 0.7939 - 0.0281 = 0.7658$$

b. What is the probability that the sample average time $\bar{X}$ is at most 6.0 seconds?

$$E(\bar{X}) = \mu = 6.2 \text{ and } \sigma_{\bar{X}} = \frac{\sigma}{\sqrt{n}} = \frac{0.15}{\sqrt{6}} = 0.0612$$

Hence, $\bar{X} \sim N(6.2, (0.0612)^2)$.

$$P(\bar{X} < 6.0) = P\left(Z < \frac{6.0 - 6.2}{0.0612}\right) = \Phi(-3.27) = 0.0005$$

3. *Central Limit Theorem*

The following theorem is a major result in statistics. It says that for a sufficiently large sample, the sample mean is approximately normal, no matter how the population is actually distributed!

THEOREM 6.3   Central limit theorem (CLT):

Let $X_1, \cdots, X_n$ be a random sample from *any* distribution with mean μ and variance σ^2. Then if n is large,

$$Z = \frac{\bar{X} - \mu}{\sigma / \sqrt{n}} \text{ is approximately } N(0,1).$$

The rule of thumb for applying CLT is $n \geq 30$.

EXAMPLE 6.4   The normal approximation to the binomial distribution is a special case of CLT. Let $X_1, \cdots, X_n$ be a random sample of Bernoulli trials with success probability p. Then $E(X_i) = p$, $Var(X_i) = p(1-p)$, and $Y = \sum_{i=1}^{n} X_i$ is distributed as $\text{Bin}(n, p)$.

$$Z = \frac{\bar{X} - \mu}{\sigma / \sqrt{n}} = \frac{\bar{X} - p}{\sqrt{p(1-p)} / \sqrt{n}} = \frac{n(\bar{X} - p)}{\sqrt{np(1-p)}} = \frac{n\bar{X} - np}{\sqrt{np(1-p)}} = \frac{Y - np}{\sqrt{np(1-p)}}$$

is approximately $N(0,1)$ by CLT if n is large.

EXAMPLE 6.5 Let $X_1, \cdots, X_n$ be a random sample from a skewed population distribution with $\mu = 2$, $\sigma = 1.41$. For the sample sizes $n = 3$ or $n = 10$, the distribution of $\overline{X}$ is changed, as in Figure 6.3.

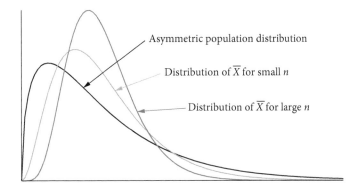

Asymmetric population distribution

Distribution of $\overline{X}$ for small n

Distribution of $\overline{X}$ for large n

FIGURE 6.3   Change in the distribution of $\overline{X}$ for a skewed population distribution as n varies.

EXAMPLE 6.6 The scores of a national mathematics examination are normally distributed with mean 84 and standard deviation 14. Let $X_1, \cdots, X_n$ be a random sample from this distribution. Find an approximate probability that the mean score of the sample is between 82.6 and 86.8 in the following cases.

a. When the sample size is 49.

$$E(\overline{X}) = \mu = 84, \quad \sigma_{\overline{x}} = \frac{\sigma}{\sqrt{n}} = \frac{14}{\sqrt{49}} = 2$$

$$Z = \frac{\overline{X} - \mu}{\sigma/\sqrt{n}} = \frac{\overline{X} - 84}{2} \text{ is approximately } N(0,1)$$

$$P(82.6 < \overline{X} < 86.8) = P\left(\frac{82.6 - 84}{2} < Z < \frac{86.8 - 84}{2}\right) = P(-0.7 < Z < 1.4)$$

$$\approx \Phi(1.4) - \Phi(-0.7) = 0.9192 - 0.2420 = 0.6772$$

b. When the sample size is 100.

$$E(\overline{X}) = \mu = 84, \quad \sigma_{\overline{x}} = \frac{\sigma}{\sqrt{n}} = \frac{14}{\sqrt{100}} = 1.4$$

$$Z = \frac{\bar{X} - \mu}{\sigma / \sqrt{n}} = \frac{\bar{X} - 84}{1.4} \text{ is approximately } N(0,1)$$

$$P(82.6 < \bar{X} < 86.8) = P\left(\frac{82.6 - 84}{1.4} < Z < \frac{86.8 - 84}{1.4}\right) = P(-1 < Z < 2)$$

$$\approx \Phi(2) - \Phi(-1) = 0.9772 - 0.1587 = 0.8185$$

Note that the probability is much larger when the sample size is 100. This is because the distribution of $\bar{X}$ is more concentrated around the mean.

EXAMPLE 6.7   The weights of seventh-grade male students of a certain school have mean 109 pounds and standard deviation 12. Let $X_1, X_2, \cdots, X_{64}$ be a random sample from this population. Find an approximate probability that the sample mean exceeds 112 pounds.

$$Z = \frac{\bar{X} - \mu}{\sigma / \sqrt{n}} = \frac{\bar{X} - 109}{12 / \sqrt{64}} = \frac{\bar{X} - 109}{1.5} \text{ is approximately } N(0,1)$$

$$P(\bar{X} > 112) = P\left(Z > \frac{112 - 109}{1.5}\right) = P(Z > 2) \approx \Phi(-2) = 0.0228$$

4. Distribution of the Sample Mean for a Normal Population When σ Is Unknown

In most practical cases the variance σ^2 is unknown. Thus, to get any idea of the variability of $\bar{X}$, it is necessary to estimate this population variance using the sample variance S^2. But using S^2 in place of σ^2 changes the distribution, particularly at small n.

THEOREM 6.4   If $X_1, \cdots, X_n$ is a random sample from $N(\mu, \sigma^2)$ and $S^2 = \dfrac{\sum_{i=1}^{n}(X_i - \bar{X})^2}{n-1}$, then

$$T = \frac{\bar{X} - \mu}{S / \sqrt{n}}$$

has a t distribution (student's t distribution) with degrees of freedom (df) $n - 1$, which is denoted as t_{n-1}.

Like the standard normal distribution $N(0, 1)$, the t distribution is bell shaped and centered around zero, but its graph is more spread out than $N(0, 1)$. The tails of the distribution become thinner, approaching $N(0, 1)$ as the df increases (not surprisingly, as S^2 is a closer estimate of σ^2 as $n \to \infty$). Again, the rule of thumb is that $N(0, 1)$ is a good estimate of the t distribution when $n \geq 30$. Figure 6.4 illustrates the shape change of the pdf of the t distribution as the df varies.

t distribution with df v:
1. Bell shaped, symmetric about 0.
2. More spread out than $N(0, 1)$ curve.
3. As v increases, the tails become thinner.
4. As $v \to \infty$, $t_v \to N(0,1)$. (Approximation is good when $v \geq 30$)

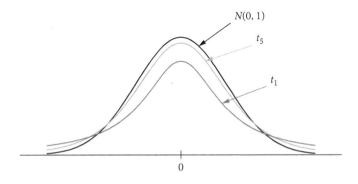

FIGURE 6.4   Change in the density function of a t distribution as df v varies.

EXAMPLE 6.8 The average gasoline consumption of an engine for half-hour runs is 2.4 gallons. Let $X_1, \cdots, X_9$ be gallons used from 9 half-hour random test runs. Then

$$T = \frac{\bar{X} - \mu}{S / \sqrt{n}} = \frac{\bar{X} - 2.4}{S / \sqrt{9}} \sim t_{n-1} = t_8$$

Let $t_{\alpha,v}$ denote the value of t_v such that the area under the curve to the right of $t_{\alpha,v}$ (right tail area) is α. In other words, $t_{\alpha,v}$, called a t critical value, is the $100(1 - \alpha)$-th percentile of the t distribution with v degrees of freedom. Figure 6.5 illustrates this. Because the t distribution is symmetric about 0, the area under the curve to the left of $-t_{\alpha,v}$ (left tail area) is also α. Table A.4 gives the values of $t_{\alpha,v}$ for various α and v.

The t critical value can also be obtained using R. In R, pt is the function for the cdf, and qt is the function for the quantile ($100p$-th percentile) of the t distribution. Thus, $P(T < c)$, where T is a t random variable with v degrees of freedom, is obtained as follows:

>pt(c, v)

and the $100p$-th percentile is obtained as

>qt(p, v)

Therefore, $t_{\alpha, v}$ can be obtained as

>qt($1 - \alpha$, v)

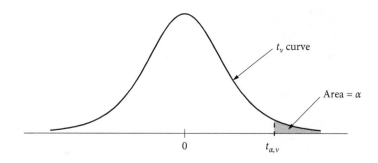

FIGURE 6.5 Right tail area corresponding to a t critical value.

EXAMPLE 6.9   The times between the call for an ambulance and a patient's arrival at the hospital are 27, 15, 20, 32, 18, and 26 minutes. The ambulance service claims that its average is 20 minutes. In view of the data, is this claim reasonable?

We have $n = 6$, $\bar{x} = 23$ minutes, and $s = 6.39$ minutes. The question can be restated as follows: Did we, by pure chance, happen to get a batch of long times in this sample? Or do these data provide strong evidence of a mean that is substantially longer than 20 minutes?

Let's assume $\mu = 20$ minutes and compute the probability of obtaining a sample mean of 23 minutes *or more* by blind chance. As σ is unknown and n is small, it is appropriate to apply the t distribution with degrees of freedom $v = n - 1 = 5$. Then

$$t = \frac{\bar{x} - \mu}{s / \sqrt{n}} = \frac{23 - 20}{6.39 / \sqrt{6}} = 1.15,$$

so that $P(\bar{X} > 23) = P(T > 1.15)$. From Table A.4, we find $t_{0.1,\,5} = 1.476$; i.e., $P(T > 1.476) = 0.1$. Therefore, $P(T > 1.15) > 0.1$. In other words, there is some (greater than 10%) chance of finding $\bar{X} > 23$ by chance, and on this basis it is not reasonable to reject the ambulance company's claim that $\mu = 20$ minutes. In R, the answer to

>pt(1.476, 5)

is 0.9; thus, the right tail area of 0.1 can be obtained using

>1-pt(1.476, 5)

The answer to

>qt(0.9, 5)

is 1.476. $P(T > 1.15)$ can be obtained using

>1-pt(1.15, 5)

and the answer is 0.151.

5. *Sampling Distribution of the Variance*

Now that we have brought up the issue of using S^2 to estimate σ^2, it is important to know the probability distribution of S^2.

THEOREM 6.5   Let $X_1, \cdots, X_n$ be a random sample from $N(\mu, \sigma^2)$. Then

$$\frac{(n-1)S^2}{\sigma^2} = \frac{\sum_{i=1}^{n}(X_i - \bar{X})^2}{\sigma^2}$$

has a chi-square (χ^2) distribution with degrees of freedom $n-1$, which is denoted as χ^2_{n-1}.

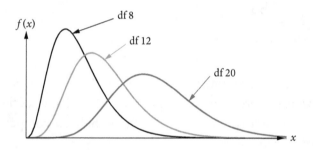

FIGURE 6.6   Change in the density function of a χ^2_v distribution as df v varies.

Let $\chi^2_{\alpha, v}$ denote the value of χ^2_v such that the area under the curve to the right of $\chi^2_{\alpha, v}$ is α. In other words, $\chi^2_{\alpha, v}$, called a χ^2 critical value, is the $100(1-\alpha)$-th percentile of the χ^2 distribution with v degrees of freedom. Figure 6.6 shows shapes of the chi-square density functions with different degrees of freedom. Table A.5 gives the values of $\chi^2_{\alpha, v}$ for various α and v.

Because the distribution is not symmetric, the left tail area cannot be obtained using $-\chi^2_{\alpha, v}$. Instead of using the negative critical value, $\chi^2_{1-\alpha, v}$ is used to find the left tail area. The area under the curve to the left of $\chi^2_{1-\alpha, v}$ (left tail area) is α. Figure 6.7 shows the right and left tail areas.

The χ^2 critical value can also be obtained using R. In R, pchisq is the function for the cdf, and qchisq is the function for the quantile of the χ^2 distribution. Thus, $P(X < c)$, where X is a χ^2 random variable with v degrees of freedom, is obtained as follows:

```
>pchisq(c, v)
```

and the $100p$-th percentile is obtained as

```
>qchisq(p, v)
```

Therefore, $\chi^2_{\alpha,v}$ can be obtained as

```
>qchisq(1-α, v)
```

and $\chi^2_{1-\alpha,v}$ can be obtained as

```
>qchisq(α, v)
```

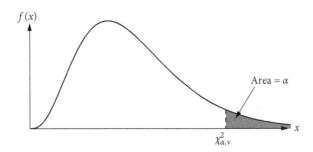

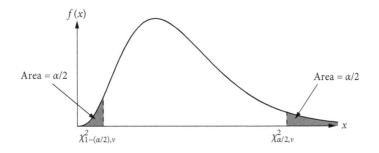

FIGURE 6.7   Right and left tail areas corresponding to χ^2 critical values.

The χ^2 distribution with degrees of freedom v, introduced in Chapter 4, is a skewed distribution on the domain 0 to ∞. As v increases, the distribution becomes more symmetric about its mean v. This phenomenon was illustrated in Figure 6.3 in Section 6.3.

EXAMPLE 6.10 The claim that $\sigma^2 = 21.3$ for a normal population is rejected if s^2 for a sample of $n = 15$ exceeds 39.74. What's the probability that the claim is rejected even though $\sigma^2 = 21.3$?

Here, we are computing the probability of an erroneous inference. To do so, we compute $P(S^2 > 39.74)$ assuming $\sigma^2 = 21.3$.

$$\chi^2 = \frac{(n-1)s^2}{\sigma^2} = \frac{14(39.74)}{21.3} = 26.12$$

so that $P(S^2 > 39.74) = P(\chi^2 > 26.12)$. Table A.5 indicates $\chi^2_{0.025,\, n-1} = \chi^2_{0.025,\, 14} = 26.12$. Thus, $P(S^2 > 39.74) = 0.025$. In R, the same answer is obtained by

>1-pchisq(26.12, 14)

Another important distribution is the F distribution. The F distribution arises naturally as the distribution of a ratio of variances.

THEOREM 6.6   Let $X_1, \cdots, X_m$ be a random sample from $N(\mu_1, \sigma^2)$ with sample variance S_1^2, and let $Y_1, \cdots, Y_n$ be a random sample from $N(\mu_2, \sigma^2)$ with sample variance S_2^2. Assume the X_i and Y_j are independent of one another. Then

$$F = \frac{S_1^2}{S_2^2}$$

is distributed as F with degrees of freedom $m - 1$ and $n - 1$, which is denoted as $F_{m-1,\, n-1}$.

Let $F_{\alpha,\, v_1,\, v_2}$ denote the value of $F_{v_1,\, v_2}$ such that the area under the curve to the right of $F_{\alpha,\, v_1,\, v_2}$ is α. In other words, $F_{\alpha,\, v_1,\, v_2}$, called an F critical value, is the $100(1 - \alpha)$-th percentile of the F distribution with v_1 and v_2 degrees of freedom. Table A.6 gives the values of $F_{\alpha,\, v_1,\, v_2}$ for selected α, v_1, and v_2. Like the χ^2 distribution, the F distribution is skewed to the right. The area under the curve to the left of $F_{1-\alpha,\, v_1,\, v_2}$ (left tail area) is α.

In R, pf is the function for the cdf, and qf is the function for the quantile of the F distribution. Thus, $P(X < c)$, where X is an F random variable with v_1 and v_2 degrees of freedom, is obtained as follows:

>pf(c, v_1, v_2)

and the $100p$-th percentile is obtained as

>qf(p, v_1, v_2)

Therefore, F_{α, v_1, v_2} can be obtained as

>qf($1-\alpha$, v_1, v_2)

and $F_{1-\alpha, v_1, v_2}$ can be obtained as

>qf(α, v_1, v_2)

In Theorem 6.6, the ratio of the sample variances from two independent samples with the same population variance has the F distribution with $m-1$ and $n-1$ degrees of freedom. We can see that if $F \sim F_{m-1, n-1}$, then $\frac{1}{F} \sim F_{n-1, m-1}$ by swapping the numerator and the denominator.

EXAMPLE 6.11 Let $X_1, X_2, \cdots, X_{10}$ be a random sample from $N(\mu_1, \sigma^2)$, independent with $Y_1, Y_2, \cdots, Y_{13}$, which is a random sample from $N(\mu_2, \sigma^2)$. Let S_X^2 and S_Y^2 be their sample variances, respectively. Find $P(S_X^2 > 2.8S_Y^2)$.

$$\frac{S_X^2}{S_Y^2} \sim F_{v_1, v_2}, \quad v_1 = n_1 - 1 = 9, \quad v_2 = n_2 - 1 = 12$$

$$F_{0.05, 9, 12} = 2.80 \text{ from Table A.6. Thus,}$$

$$P(S_X^2 > 2.8S_Y^2) = P\left(\frac{S_X^2}{S_Y^2} > 2.8\right) = P(F_{9, 12} > 2.8) = 0.05$$

This result can also be obtained using R as

>1-pf(2.8, 9, 12)

Figure 6.8 illustrates the right and left tail areas of the F distribution.

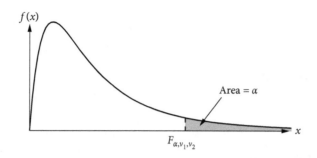

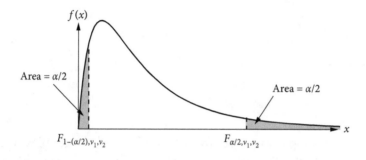

FIGURE 6.8 Right and left tail areas corresponding to F critical values.

Recall that $\frac{S_X^2}{S_Y^2} \sim F_{v_1, v_2}$ implies $\frac{S_Y^2}{S_X^2} \sim F_{v_2, v_1}$. Table A.6 provides the F critical values for small values of the right tail area α. To find the critical value for a large α, we use the following relationship:

$$\alpha = P\left(\frac{S_X^2}{S_Y^2} > F_{\alpha, \, v_1, v_2}\right) = P\left(\frac{S_Y^2}{S_X^2} < F_{1-\alpha, \, v_2, v_1}\right) = P\left(\frac{S_X^2}{S_Y^2} > \frac{1}{F_{1-\alpha, \, v_2, v_1}}\right)$$

EXAMPLE 6.12 The F critical value $F_{0.95, 7, 10}$ can be found using the above relationship as

$$F_{0.95, 7, 10} = \frac{1}{F_{0.05, 10, 7}} = \frac{1}{3.64} = 0.27$$

SUMMARY OF CHAPTER 6

1. Random variables $X_1, X_2, \cdots, X_n$ are a <u>random sample</u> if:
   a. they are independent, and
   b. every X_i has the same probability distribution.

2. Let $X_1, X_2, \cdots, X_n$ be a random sample from a distribution with mean μ and variance σ^2. Then

   a. $E(\bar{X}) = \mu$

   b. $Var(\bar{X}) = \dfrac{\sigma^2}{n}, \sigma_{\bar{x}} = \dfrac{\sigma}{\sqrt{n}}$

   c. $E(\sum_{i=1}^{n} X_i) = n\mu$

   d. $Var(\sum_{i=1}^{n} X_i) = n\sigma^2, \ \sigma_{\sum X_i} = \sqrt{n}\sigma$

3. Let $X_1, X_2, \cdots, X_n$ be a random sample from $N(\mu, \sigma^2)$. Then
   a. $\bar{X} \sim N\left(\mu, \frac{\sigma^2}{n}\right)$
   b. $\sum_{i=1}^{n} X_i \sim N(n\mu, n\sigma^2)$

4. Let $X_1, X_2, \cdots, X_n$ be a random sample from <u>any</u> distribution with mean μ and variance σ^2. Then for large n ($n \geq 30$)
   a. $\bar{X}$ is approximately $N\left(\mu, \frac{\sigma^2}{n}\right)$
   b. $\sum_{i=1}^{n} X_i$ is approximately $N(n\mu, n\sigma^2)$
   c. Central Limit Theorem (CLT):
      $Z = \frac{\bar{X} - \mu}{\sigma/\sqrt{n}}$ is approximately $N(0,1)$

5. For unknown σ, if $X_1, X_2, \cdots, X_n$ is a random sample from $N(\mu, \sigma^2)$ and $S^2 = \frac{\sum_{i=1}^{n}(X_i - \bar{X})^2}{n-1}$, then $T = \frac{\bar{X} - \mu}{S/\sqrt{n}} \sim t_{n-1}$.

6. Let $X_1, \cdots, X_n$ be a random sample from $N(\mu, \sigma^2)$. Then $\frac{(n-1)S^2}{\sigma^2} \sim \chi_{n-1}^2$.

7. Let $X_1, \cdots, X_m$ be a random sample from $N(\mu_1, \sigma^2)$ with sample variance S_1^2, and $Y_1, \cdots, Y_n$ a random sample from $N(\mu_2, \sigma^2)$ with sample variance S_2^2. Assume the X_i and Y_j are independent of one another. Then $\frac{S_1^2}{S_2^2} \sim F_{m-1, n-1}$.

EXERCISES

6.1   The population standard deviation is $\sigma = 9$. For a random sample, the standard deviation of the sample mean is 1.5. What is the sample size?

6.2   It is known that IQ scores have a normal distribution with a mean of 100 and a standard deviation of 15.

    a.   A random sample of 36 students is selected. What is the probability that the sample mean IQ score of these 36 students is between 95 and 110?

    b.   A random sample of 100 students is selected. What is the probability that the sample mean IQ score of these 100 students is between 95 and 110?

6.3   The length of a TV show program has a normal distribution with mean 50 minutes and standard deviation 2 minutes. Find the probability that the mean length of randomly selected 9 shows is between 49 and 51 minutes.

6.4   The weight of an adult bottlenose dolphin was found to follow a normal distribution with a mean of 550 pounds and a standard deviation of 50 pounds.

    a.   What percentage of adult bottlenose dolphins weigh from 400 to 600 pounds?

    b.   If $\bar{X}$ represents the mean weight of a random sample of 9 adult bottlenose dolphins, what is $P(500 < \bar{X} < 580)$?

    c.   In a random sample of 9 adult bottlenose dolphins, what is the probability that 5 of them are heavier than 560 pounds?

6.5   The amount of time that a ten-year-old boy plays video games in a week is normally distributed with a mean of 10 hours and a standard deviation of 4 hours.

    a.   Suppose 9 ten-year-old boys are randomly chosen. What is the probability that the sample mean time for playing video games per week is 8 to 12 hours?

    b.   Suppose a boy is considered addicted if he plays video games for more than 16 hours a week. If 12 ten-year-old boys are to be chosen at random, find the probability that at least 1 is addicted to video games.

6.6   *Roystonea regia* is a large palm tree. The height of a fully grown *Roystonea regia* is normally distributed with a mean of 82 feet and a standard deviation of 10 feet. What is the probability that the mean of a random sample of 16 *Roystonea regia* palm trees is 80 to 85 feet?

6.7   Let $X_1, X_2, \cdots, X_9$ be a random sample from $N(0, 3^2)$ and $Y_1, Y_2, \cdots, Y_{16}$ a random sample from $N(2, 2^2)$. Find a constant c satisfying $P(\bar{X} \geq 1) = P(\bar{Y} \leq c)$.

6.8   There are 60 students in an introductory statistics class. The amount of time needed for the instructor to grade a randomly chosen final exam paper is a random variable with a mean of 5 minutes and a standard deviation of 5 minutes.

a.   If grading times are independent, what is the probability that the instructor can finish grading in 5 hours?

b.   What is the probability that the instructor cannot finish grading in 6 hours and 30 minutes?

6.9   A sample X_1, X_2, $\cdots$, X_{25} is randomly chosen from a population of $N(20, 36)$ and another sample Y_1, Y_2, $\cdots$, Y_{36} is randomly chosen from a population of $N(30, \sigma^2)$. If

$$P(\bar{X} < 22.4) + P(\bar{Y} < 28) = 1,$$

find $P(\bar{Y} < 31)$.

6.10   The distribution of hourly rates of registered nurses in a large city has mean $31 and standard deviation $5.

a.   Find an approximate distribution of the sample mean based on a random sample of 100 registered nurses in the city.

b.   Find an approximate probability that the average hourly rate of the 100 registered nurses sampled exceeds $31.5.

6.11   The average age of the residents in a city is 37, and the standard deviation is 18 years. The distribution of ages is known to be normal. Suppose a group of 20 people is formed to represent all age groups. The average age of this group is 45. What is the chance that the average age of a randomly selected group of 20 people from this population is at least 45 years old?

6.12   Assume the population has a distribution with a mean of 100 and a standard deviation of 10. For a random sample of size 50, find the following.

a. $P(99 < \bar{X} < 102)$

b. $P(\bar{X} > 97)$

c.   70th percentile of $\bar{X}$

6.13   Let $X_1, X_2, \cdots, X_{64}$ be a random sample from a Poisson distribution with a mean of 4. Find an approximate probability that the sample mean $\bar{X}$ is greater than 3.5.

6.14   A fair four-sided die with four equilateral triangle-shaped faces is tossed 200 times. Each of the die's four faces shows a different number from 1 to 4.
   a.   Find the expected value of the sample mean of the values obtained in these 200 tosses.
   b.   Find the standard deviation of the number obtained in 1 toss.
   c.   Find the standard deviation of the sample mean obtained in these 200 tosses.
   d.   Find an approximate probability that the sample mean of the 200 numbers obtained is smaller than 2.7.

6.15   Let a random variable X from a population have a mean of 70 and a standard deviation of 15. A random sample of 64 is selected from that population.
   a.   Find an approximate distribution of the sample mean of the 64 observations.
   b.   Use the answer to part (a) to find an approximate probability that the sample mean will be greater than 75.

6.16   Heights of men in America have a normal distribution with a mean of 69.5 inches and a standard deviation of 3 inches. Perform the following calculations.
   a.   In a random sample of 20 adult men in the United Sates, find $P(68 < \bar{X} < 70)$.
   b.   Let $\bar{X}$ represent the mean height of a random sample of n American adults. Find n if $P(68.52 < \bar{X} < 70.48) = 0.95$.
   c.   If 100 American men are chosen at random, find an approximate probability that at least 25 of them are shorter than 68 inches.

6.17   Eighteen percent of Caucasians have Rh-negative blood types. In a random sample of 64 Caucasians, find an approximate probability that the sample proportion with Rh-negative blood types will be greater than 20%.

6.18   Let $X_1, X_2, \cdots, X_{100}$ be a random sample from a distribution with pdf

$$f(x) = \begin{cases} \dfrac{3x^2}{2} + x, & 0 \le x \le 1 \\[2mm] 0, & \text{otherwise} \end{cases}$$

   a.   Find the mean of X_1.
   b.   Find the variance of X_1.

c. Use the central limit theorem to find an approximate probability of $P(0.7 < \bar{X} < 0.75)$.

6.19 Let $X_1, X_2, \cdots, X_{64}$ be a random sample from a distribution with pdf

$$f(x) = \begin{cases} 2(1-x), & 0 \le x \le 1 \\ 0, & \text{otherwise} \end{cases}$$

a. Find the mean of X_1.
b. Find the variance of X_1.
c. Use the central limit theorem to find an approximate probability of $P(0.33 < \bar{X} < 0.34)$.

6.20 Let $X_1, X_2, \cdots, X_{400}$ be a random sample from Uniform$(0, 4)$. Find an approximate probability that $\bar{X}$ is less than 1.9.

6.21 Let $X_1, X_2, \cdots, X_{100}$ be a random sample from a distribution with pdf

$$f(x) = \begin{cases} \dfrac{4}{3} - x^2, & 0 \le x \le 1 \\ 0, & \text{otherwise} \end{cases}$$

a. Find the mean of X_1.
b. Find the variance of X_1.
c. Use the central limit theorem to find an aproximate probability of $P(0.4 < \bar{X} < 0.5)$.

6.22 A random sample of size 64 is taken from an infinite population with a mean of 60 and a standard deviation of 4. With what probability can we assert that the sample mean is within 59 and 61, if we use:
a. the central limit theorem?
b. Chebyshev's inequality?

6.23 The amount of cola in a 355 ml bottle from a certain company is a random variable with a mean of 355 ml and a standard deviation of 2 ml. For a sample of size 32, perform the following calculations.
a. Find an approximate probability that the sample mean is less than 354.8 ml.
b. Suppose the amount of cola is distributed as $N(355, 4)$. Find an approximate probability that 10 of the bottles in the sample contain less than 354.8 ml of cola.

6.24 If the variance of a normal population is 4, what is the probability that the variance of a random sample of size 10 exceeds 6.526?
    a. Find the probability using the distribution table.
    b. Find the probability using R.

6.25 If the variance of a normal population is 3, what is the 95th percentile of the variance of a random sample of size 15?
    a. Find the percentile using the distribution table.
    b. Find the percentile using R.

6.26 Two independent random samples are obtained from normal populations with equal variance. If the sample sizes are 8 and 12, respectively, perform the following calculations using the distribution table.

    a. Find $P\left(\frac{S_1^2}{S_2^2} < 1.7298\right)$.

    b. Find $P\left(0.3726 < \frac{S_1^2}{S_2^2} < 1.7298\right)$.

6.27 Find the probabilities in Exercise 6.26 using R.

6.28 Random samples are obtained from two normal populations with equal variance of $\sigma^2 = 12$. If the sample sizes are 20 and 18, respectively, find the 90th percentile of the ratio of the sample variances $\frac{S_1^2}{S_2^2}$ as instructed below.
    a. Find the percentile using the distribution table.
    b. Find the percentile using R.

Introduction to Point Estimation and Testing

7

1. Point Estimation

A *point estimator* is a statistic intended for estimating a parameter, and a *point estimate* is an observed value of the estimator. An estimator is a function of the sample, while an estimate is the realized value of an estimator that is obtained when a sample is actually taken. For estimating a parameter θ, normally we denote the estimator as $\hat{\theta}$.

EXAMPLE 7.1

a. Let X_1, X_2, X_3 be a random sample from a population. Consider $\bar{X}$ as an estimator of μ. If the values of the sample are $x_1 = 7.1$, $x_2 = 5.5$, and $x_3 = 6.6$, then

$$\bar{x} = \frac{7.1 + 5.5 + 6.6}{3} = 6.40$$

Thus, 6.40 is the estimate of μ.

b. Consider

$$\hat{\sigma}^2 = S^2 = \frac{\sum_{i=1}^{n}(X_i - \bar{X})^2}{n-1} = \frac{\sum X_i^2 - \left(\sum X_i\right)^2 / n}{n-1}$$

as an estimator of σ^2. Then the estimate is

$$\frac{\sum x_i^2 - \left(\sum x_i\right)^2 / n}{n-1} = \frac{\sum x_i^2 - n(\bar{x})^2}{n-1} = \frac{(7.1^2 + 5.5^2 + 6.6^2) - 3(6.40)^2}{3-1} = 0.67$$

The standard deviation of an estimator is called a *standard error* (s.e.). For example, for an estimator $\bar{X}$ of μ, the standard error of $\bar{X}$ is $\sigma / \sqrt{n}$. An estimator $\hat{\theta}$ is an *unbiased estimator* of θ if $E(\hat{\theta}) = \theta$ for all θ. If $\hat{\theta}$ is not unbiased (biased), then $E(\hat{\theta}) - \theta$ is called the *bias*.

EXAMPLE 7.2   Suppose $X \sim \text{Bin}(n, p)$ with $n = 25$. Then

$$E(X) = np, \quad E\left(\frac{X}{n}\right) = \frac{1}{n}E(X) = \frac{1}{n} \cdot np = p$$

Therefore, $\hat{p} = X / n$ is an unbiased estimator of p. If $x = 15$, then $x / n = 15 / 25 = 3 / 5$ is the estimate. The standard error of $\hat{p}$ is

$$\sigma_{\hat{p}} = \sqrt{Var\left(\frac{X}{n}\right)} = \sqrt{\frac{Var(X)}{n^2}} = \sqrt{\frac{np(1-p)}{n^2}} = \sqrt{p(1-p)/n}$$

Since the value of p is not known, we need to estimate the standard error. The estimator of p can be substituted into the form of the standard error. An estimator of the standard error is

$$\hat{\sigma}_{\hat{p}} = \sqrt{\hat{p}(1-\hat{p})/n}$$

Thus, the estimate of the standard error is

$$\sqrt{(0.6)(0.4)/25} = 0.098$$

EXAMPLE 7.3   In a presidential election poll conducted the day before an election, a sample of 1,500 voters were asked for their intended vote. From this sample, 45% preferred the candidate from a conservative party. Assume everyone in the sample expressed a preference. The unbiased estimate of this ratio is 0.45, and the estimate of the standard error is $\sqrt{(0.45)(0.55)/1500} = 0.013$.

THEOREM 7.1 Let $X_1, \cdots, X_n$ be a random sample from a distribution with mean μ and variance σ^2. Then $\hat{\mu} = \bar{X}$ is an unbiased estimator of μ and $\hat{\sigma}^2 = S^2 = \Sigma(X_i - \bar{X})^2 / (n-1)$ is an unbiased estimator of σ^2.

PROOF

For the proof that $E(\bar{X}) = \mu$, see Theorem 6.1 in Chapter 6.

Since $Var(Y) = E(Y^2) - [E(Y)]^2$, $E(Y^2) = Var(Y) + [E(Y)]^2$. Thus,

$$E(S^2) = E\left\{\frac{1}{n-1}\left[\sum_{i=1}^{n}X_i^2 - \frac{\left(\sum_{i=1}^{n}X_i\right)^2}{n}\right]\right\} = \frac{1}{n-1}\left\{\sum_{i=1}^{n}E(X_i^2) - \frac{1}{n}E\left[\left(\sum_{i=1}^{n}X_i\right)^2\right]\right\}$$

$$= \frac{1}{n-1}\left\{\sum_{i=1}^{n}(\sigma^2 + \mu^2) - \frac{1}{n}\left[Var\left(\sum_{i=1}^{n}X_i\right) + \left[E\left(\sum_{i=1}^{n}X_i\right)\right]^2\right]\right\}$$

$$= \frac{1}{n-1}\left\{n(\sigma^2 + \mu^2) - \frac{1}{n}[n\sigma^2 + (n\mu)^2]\right\} = \frac{1}{n-1}(n\sigma^2 - \sigma^2) = \sigma^2$$

2. Tests of Hypotheses

A *statistical hypothesis* is a statement about a population parameter. For example, a new im-proved laundry soap is claimed to have better cleaning power than the old formula. This claim is tested statistically by examining whether the data support the claim.

A. THE NULL AND THE ALTERNATIVE HYPOTHESES

As an example, suppose the cure rate for a given disease using a standard medication is known to be 35%. The cure rate of a new drug is claimed to be better. A sample of 18 patients is gath-ered. Let X denote the number of patients who are cured by the new drug. Is there substantial evidence that the new drug has a higher cure rate than the standard medication? In hypothesis testing, we formulate a hypothesis to be tested as a single value for a parameter. We usually hypothesize the opposite of what we hope to prove. In this problem, we hypothesize that the cure rate of the new drug is no better than the cure rate of the old one. We call a hypothesis like this the *null hypothesis* and denote it H_0. The claim that the new drug has a higher cure rate is called the *alternative hypothesis* and is denoted as H_1. If the cure rate is denoted as p, then the null hypothesis in this problem is

H_0: The new drug is not better than the old one ($p \leq 0.35$)

and the alternative hypothesis is

H_1: The new drug is better than the old one ($p > 0.35$)

If H_0 is true, we expect less than or equal to 6 cures out of 18 patients, and if H_1 is true, we expect more than 6 cures.

> **Choice of H_0 and H_1**
> When our goal is to establish an assertion, the *negation* of the assertion is H_0, and the *assertion* itself is H_1.

A decision rule is set up for a conclusion. A decision rule that tells us when to reject H_0 and when not to reject H_0 is called a *test*. A *test statistic* is a statistic whose value serves to determine the action. A *rejection region* or *critical region* is the set of values of a test statistic for which H_0 is to be rejected. In the above example, the test of the null hypothesis is performed by rejecting H_0 if $X > c$ and by accepting H_0 if $X \leq c$ for a real constant c. Here X is the test statistic, and the set $\{X > c\}$ is the rejection region. The basis of choosing a particular rejection region lies in an understanding of the errors that one might be faced with in drawing a conclusion. There are two kinds of errors. A *type I error* consists of rejecting H_0 when H_0 is true, and a *type II error* consists of failure to reject H_0 when H_1 is true. Figure 7.1 illustrates the two types of error.

> <u>Test:</u> A decision rule that tells us when to reject H_0 and when not to reject H_0.
> <u>Test statistic:</u> A statistic whose value serves to determine the action.
> <u>Rejection region:</u> The set of values of a test statistic for which H_0 is to be rejected.

B. THE TWO TYPES OF ERRORS

	Unknown True State of Nature	
Test Concludes	H_0 is True	H_0 is False
Do not Reject H_0	Correct	Wrong (Type II error)
Reject H_0	Wrong (Type I error)	Correct

FIGURE 7.1   The two types of errors.

Two types of errors:
Type I error: The rejection of H_0 when H_0 is true.
Type II error: The failure to reject H_0 when H_1 is true.

EXAMPLE 7.4   The two opposing hypotheses for the previous example are $H_0 : p \leq 0.35$ versus $H_1 : p > 0.35$. Suppose the rejection region is chosen to be $X \geq 10$ (see Figure 7.2). Determine the type of error that can occur and calculate the error probability when:
a. $p = 0.3$
b. $p = 0.6$

If $p = 0.3$, then H_0 is true. A possible error is the type I error, which is a rejection of H_0. Therefore,

$$P(\text{type I error given } p = 0.3) = P(X \geq 10 \text{ given } p = 0.3)$$

The random variable X is binomial with $n = 18$ and $p = 0.3$. Thus,

$$P(X \geq 10) = 1 - P(X \leq 9) = 1 - 0.979 = 0.021$$

Therefore, the probability of a type I error given $p = 0.3$ is 0.021.

If $p = 0.6$, then H_1 is true. A possible error is the type II error, which is failure to reject H_0. Therefore,

$$P(\text{type II error given } p = 0.6) = P(X \le 9 \text{ given } p = 0.6) = 0.263$$

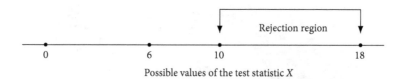

FIGURE 7.2   Rejection region for Example 7.4.

The probability of a type I error varies with the value of p as follows:

	p in H_0		
Type I error probability	0.2	0.3	0.35
$P(X \ge 10)$	0.001	0.021	0.060

The maximum type I error probability of a test is called the *level of significance* or *significance level* and is denoted as α. In the above example, $\alpha = 0.060$. The type II error probability is denoted as $\beta = P(\text{type II error})$. The power is defined as $1 - \beta$. The power of this test is $1 - 0.263 = 0.737$ when $p = 0.6$.

Level of significance (significance level; α): The maximum type I error probability of a test.

Type II error probability: $\beta = P(\text{type II error})$.

The test in Example 7.4 of $H_0 : p \le 0.35$ versus $H_1 : p > 0.35$ can be written as $H_0 : p = 0.35$ versus $H_1 : p > 0.35$ because the type I error probability normally reaches the maximum at the boundary value 0.35 of H_0 and H_1 and the test is conducted at the significance level of α. In Example 7.5 and hypothesis testing problems throughout this book, we will set up the hypotheses in one of the two ways.

EXAMPLE 7.5   The lifetime of a certain type of car battery is known to be normally distributed with mean $\mu = 5$ years and standard deviation $\sigma = 0.9$ years. A new kind of battery is designed to increase the average lifetime. The hypotheses are $H_0: \mu = 5$ and $H_1: \mu > 5$. Assume that $X_1, X_2, \cdots, X_{25}$ is a random sample from the distribution of new batteries. Assume that the variance remains unchanged in the new distribution. Then the distribution of $\bar{X}$ is $N\left(\mu, \left(\sigma/\sqrt{n}\right)^2\right) = N(\mu, 0.18^2)$. Suppose the rejection region is chosen to be $\{\bar{X} \geq 5.42\}$. Then the probability of a type I error is

$$
\begin{aligned}
\alpha &= P(\text{type I error}) \\
&= P(\text{reject } H_0 \text{ when it is true}) \\
&= P(\bar{X} \geq 5.42 \text{ when } \bar{X} \sim N(5, 0.18^2)) \\
&= P\left(Z \geq \tfrac{5.42-5}{0.18}\right) = P(Z \geq 2.33) \\
&= \Phi(-2.33) = 0.01
\end{aligned}
$$

The probability of a type II error when $\mu = 5.2$ is

$$
\begin{aligned}
\beta(5.2) &= P(\text{type II error when } \mu = 5.2) \\
&= P(\text{do not reject } H_0 \text{ when } \mu = 5.2) \\
&= P(\bar{X} < 5.42 \text{ when } \bar{X} \sim N(5.2, 0.18^2)) \\
&= P\left(Z < \tfrac{5.42-5.2}{0.18}\right) = P(Z < 1.22) = \Phi(1.22) = 0.8888
\end{aligned}
$$

and the power is $1 - 0.8888 = 0.1112$.

The probability of a type II error when $\mu = 5.5$ is

$$
\beta(5.5) = P\left(Z < \tfrac{5.42-5.5}{0.18}\right) = \Phi(-0.44) = 0.33
$$

and the power is $1 - 0.33 = 0.67$.

Figure 7.3 illustrates the above calculations.

For a fixed n, decreasing the size of the rejection region occurs simultaneously with smaller α and larger β.

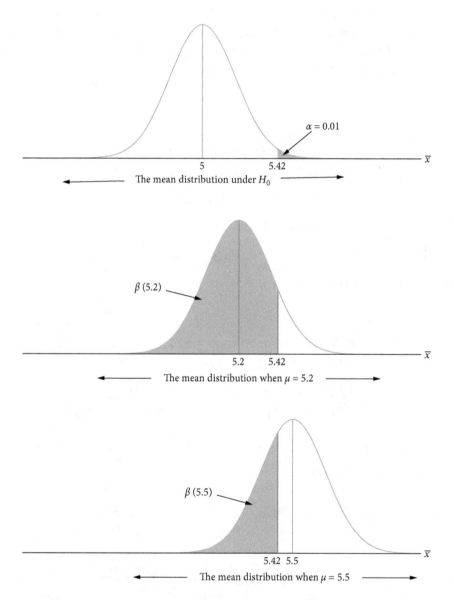

FIGURE 7.3   Type I error probability and type II error probability.

> Suppose an experiment and n are fixed. Then decreasing the size of the rejection region $\Rightarrow$ smaller α $\Rightarrow$ larger β.

In general, hypothesis testing may be described as a five-step procedure:

- Step 1: Set up H_0 and H_1.

- Step 2: Choose the value of α.

- Step 3: Choose the test statistic and rejection region.

- Step 4: Substitute the values in the test statistic and draw a conclusion.

- Step 5: Find the p-value.

Assuming that H_0 is correct, the actual value of the test statistic is some distance away from its expectation. What is the probability that, if the experiment were repeated, the value of the test statistic would be even *farther* away? This quantity is defined to be the p-value. A small value for the p-value implies that it would be very unlikely to obtain a value of the test statistic such as the one we observed if H_0 actually were true. The smaller the value of p, therefore, the more contradictory the sample results are to H_0 (i.e., the stronger the evidence is for H_1). We reject H_0 if $p \le \alpha$, and we do not reject H_0 if $p > \alpha$.

The p-value is useful because it provides more information than simply a yes/no conclusion to the hypothesis test. The selection of the significance level is fairly subjective, and its choice affects the decision rule. Two researchers examining the same data may draw different conclusions by applying different significance levels. For example, suppose researcher A prefers $\alpha = 0.01$, while researcher B prefers $\alpha = 0.05$. If the p-value for the test is 0.03, then A does not reject H_0 and B rejects H_0.

> The p-value is the probability of obtaining a value for the test statistic that is more extreme than the value actually observed. (Probability is calculated under H_0.)

Meaning of the p-value:
Reject H_0 if $p \le \alpha$
Do not reject H_0 if $p > \alpha$

SUMMARY OF CHAPTER 7

1. Point Estimator: A statistic intended for estimating a parameter.
2. An estimator $\hat{\theta}$ is an <u>unbiased estimator</u> of θ if $E(\hat{\theta}) = \theta$ for all θ.
    a. <u>Bias</u>: $E(\hat{\theta}) - \theta$
    b. $\bar{X}$ is an unbiased estimator of μ.
    c. $S^2 = \Sigma(X_i - \bar{X})^2/(n-1)$ is an unbiased estimator of σ^2
3. Choice of H_0 and H_1: When the goal is to establish an assertion, the <u>negation</u> of the assertion is H_0, and the <u>assertion</u> itself is H_1.
4. Two Types of Errors
    a. Type I error: Rejection of H_0 when H_0 is true
    b. Type II error: Failure to reject H_0 when H_1 is true
5. Significance Level α: Maximum type I error probability of a test
6. Hypothesis Testing Procedure
    a. Set up H_0 and H_1
    b. Choose the value of α
    c. Choose the test statistic and rejection region
    d. Substitute the values in the test statistic and draw a conclusion
    e. Find the p-value
7. Meaning of the p-value
    a. Reject H_0 if $p \le \alpha$
    b. Do not reject H_0 if $p > \alpha$

EXERCISES

7.1   The professor of a large calculus class randomly selected 6 students and asked the amount of time (in hours) spent for his course per week. The data are given below.

10 8 9 7 11 13

    a.   Estimate the mean of the time spent in a week for this course by the students who are taking this course.

    b.   Estimate the standard deviation of the time spent in a week for this course by the students who are taking this course.

    c.   Estimate the standard error of the estimated time spent in a week for this course by the students who are taking this course.

7.2   A survey was conducted regarding the president's handling of issues on foreign policy. Fifty-five out of 100 people who participated in this survey support the president's handling of the issue.

    a.   Estimate the population proportion of Americans supporting the president in this matter.

    b.   Estimate the standard error of the estimated proportion of Americans supporting the president in this matter.

7.3   Match each term in the left column with a related term in the right column.

normal approximation	a. central limit theorem
random sample	b. p-value
significance level	c. iid
false rejection of H_0	d. type I error

7.4   Match each item in the left column with the correct item in the right column.

p-value = 0.02	a. do not reject H_0 at $\alpha = 0.1$
p-value = 0.07	b. reject H_0 at $\alpha = 0.1$ but not at 0.05
p-value = 0.3	c. reject H_0 at $\alpha = 0.05$ but not at 0.01
p-value = 0.006	d. reject H_0 at $\alpha = 0.01$

7.5   In a scientific study, a statistical test resulted in a p-value of 0.035. Match each α value in the left column with the correct decision in the right column. One entry in the right column may be used more than once.

Significance level	Decision
0.01	a. do not reject H_0
0.05	b. reject H_0
0.10	

7.6   In a scientific study, a statistical test yielded a p-value of 0.067. Which of the following is the correct decision?
a.   Reject H_0 at $\alpha = 0.05$ and reject it for $\alpha = 0.1$.
b.   Reject H_0 at $\alpha = 0.05$ but not for $\alpha = 0.01$.
c.   Reject H_0 at $\alpha = 0.1$ but not for $\alpha = 0.05$.
d.   Do not reject H_0 at $\alpha = 0.05$ and do not reject it for $\alpha = 0.1$.

7.7   An information technology company claims that a certain model of laser printers have mean output capacity of 400 pages. A consumer report firm wants to show that the actual mean is smaller than the company claims. What should be the null and alternative hypotheses?

7.8   A printer company claims that the mean warm-up time of a certain brand of printer is 15 seconds. An engineer of another company is conducting a statistical test to show this is an underestimate.
a.   State the testing hypotheses.
b.   The test yielded a p-value of 0.035. What would be the decision of the test if $\alpha = 0.05$?
c.   Suppose a further study establishes that the true mean warm-up time is 14 seconds. Did the engineer make the correct decision? If not, what type of error did he or she make?

7.9   The speeds of two types of drones in reaching a target from the same distance are compared. The owner of drone A claims that it reaches the target faster than drone B from the other type. The speeds are measured 5 times from the same distance. Let X be the number of times that drone A reached the target faster than drone B. When the rejection region is $\{X = 5\}$, answer the following questions:

a.   If the probability that drone A is faster than drone B is p, find the distribution of X.

b.   State the testing hypotheses.

c.   Find the probability of a type I error.

d.   When the probability that drone A is faster in reaching the target than drone B is 90%, find the power of the test.

7.10   In Exercise 7.9, suppose the rejection region is $\{X = 3, 4, 5\}$, and answer the following questions:

a.   Find the probability of a type I error.

b.   When the probability that drone A is faster in reaching the target than drone B is 60%, find the power of the test.

7.11   The survival rate of a cancer using an existing medication is known to be 30%. A pharmaceutical company claims that the survival rate of a new drug is higher. The new drug is given to 15 patients to test for this claim. Let X be the number of cures out of the 15 patients. Suppose the rejection region is $\{X \geq 8\}$.

a.   State the testing hypotheses.

b.   Determine the type of error that can occur when the true survival rate is 25%. Find the error probability.

c.   Determine the type of error that can occur when the true survival rate is 30%. Find the error probability.

d.   Determine the type of error that can occur when the true survival rate is 40%. Find the error probability.

e.   What is the level of significance?

7.12 The incubation period of Middle East respiratory syndrome (MERS) is known to have a normal distribution with a mean of 8 days and a standard deviation of 3 days. Suppose a group of researchers claimed that the true mean is shorter than 8 days. A test is conducted using a random sample of 20 patients.

a. Formulate hypotheses for this test.

b. Consider a rejection region $\{\bar{X} \leq 6\}$. Suppose the test failed to reject H_o. If the true mean incubation period is 5 days, what is the type II error probability of the test using $\sigma = 3$?

7.13 The lifetime of certain type of car engine is normally distributed with a mean of 200,000 miles and a standard deviation of 30,000 miles. An automaker claims that the new year model has an engine with a longer average lifetime. A sample of 16 cars with this type of engine from the new model is obtained for a test. Consider a rejection region $\{\bar{X} \geq 215,000\}$.

a. What hypotheses should be tested?

b. Find the probability of a type I error.

c. Suppose that a further study establishes that, in fact, the average lifetime of the new engine is 210,000 miles. Find the probability of a type II error.

Inferences Based on One Sample

1. Inferences Concerning a Population Mean

This section is concerned with the problem of estimating the mean of a population. The procedure for constructing the estimate is elementary: Estimate μ as the sample mean $\bar{X}$. However, how do we determine the accuracy of this estimate? We do not know how far $\bar{x}$ is from the real value μ because we do not know what μ is in the first place (otherwise we wouldn't need to sample!).

The theorems of the previous chapter tell us the probability distribution of $\bar{X}$ in reasonably general circumstances. The probability distribution can be used to construct a *confidence interval* for μ; i.e., an error bar for the estimate that is very likely to be correct. The estimation of the mean is also relevant to hypothesis testing, since the sample mean, as an estimate of the mean, is then compared to some value that serves as the null hypothesis.

A. NORMAL POPULATION WITH KNOWN σ

a) Confidence interval

Consider a random sample $X_1, \cdots, X_n$ from $N(\mu, \sigma^2)$, where σ is known. The sample mean $\bar{X}$ is distributed as $N(\mu, \sigma^2/n)$. Therefore,

$$Z = \frac{\bar{X} - \mu}{\sigma/\sqrt{n}} \sim N(0, 1)$$

Since z_α is the $100(1-\alpha)$-th percentile of the standard normal distribution, $z_{0.025} = 1.96$. In other words, the area under the graph of the standard normal distribution to the right of $z = 1.96$ is 0.025 (see Figure 8.1). Equivalently,

$$P(-1.96 < Z < 1.96) = 0.95$$

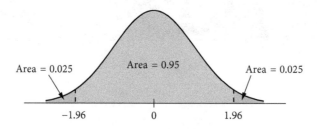

FIGURE 8.1   The $z_{0.025}$ tail area and a 95% confidence interval.

Let's express the event $-1.96 < Z < 1.96$ as an estimate of μ with a certain error bar. Since $(\bar{X} - \mu)/(\sigma/\sqrt{n}) \sim N(0, 1)$, we have

$$P\left(-1.96 < \frac{\bar{X} - \mu}{\sigma/\sqrt{n}} < 1.96\right) = 0.95$$

Multiplying through by $\sigma/\sqrt{n}$ to all sides, we find the equivalent probability

$$P\left(-1.96\frac{\sigma}{\sqrt{n}} < \bar{X} - \mu < 1.96\frac{\sigma}{\sqrt{n}}\right) = 0.95$$

This probability can be rewritten in terms of μ as follows:

$$P\left(\bar{X} - 1.96\frac{\sigma}{\sqrt{n}} < \mu < \bar{X} + 1.96\frac{\sigma}{\sqrt{n}}\right) = 0.95$$

The probability is 95% that μ lies somewhere in the interval

$$\left(\bar{X} - 1.96\frac{\sigma}{\sqrt{n}}, \bar{X} + 1.96\frac{\sigma}{\sqrt{n}}\right)$$

Another way of expressing this conclusion is that a 95% confidence interval (CI) of μ is

$$\bar{X} \pm 1.96 \frac{\sigma}{\sqrt{n}}$$

We have succeeded in constructing a *95% confidence interval* for μ.

$$X_1, \cdots, X_n \sim N(\mu, \sigma^2), \sigma \text{ is known} \Rightarrow \bar{X} \sim N(\mu, \tfrac{\sigma^2}{n})$$

$$Z = \frac{\bar{X} - \mu}{\sigma/\sqrt{n}} \sim N(0,1)$$

95% CI for μ of $N(\mu, \sigma^2)$ when σ is known: $(\bar{x} - 1.96\tfrac{\sigma}{\sqrt{n}}, \bar{x} + 1.96\tfrac{\sigma}{\sqrt{n}})$

Figure 8.2 illustrates the interpretation of the 95% CI. If we construct 95% CIs with different samples, then in the long run, 95% of the CIs will contain μ.

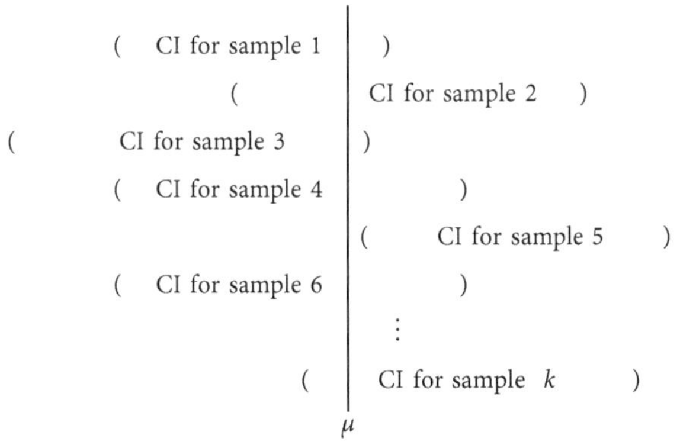

FIGURE 8.2   Interpretation of a 95% confidence interval.

It is easy to generalize the confidence interval. Note that to make a 95% confidence interval, we had to calculate $z_{0.025}$. Thus, to make a $100(1 - \alpha)\%$ confidence interval, we have to know $z_{\alpha/2}$.

$$A\ 100(1-\alpha)\%\ \text{CI for } \mu \text{ of } N(\mu, \sigma^2) \text{ when } \sigma \text{ is known:} \left(\bar{x} - z_{\alpha/2}\frac{\sigma}{\sqrt{n}}, \bar{x} + z_{\alpha/2}\frac{\sigma}{\sqrt{n}} \right)$$

Values of $z_{\alpha/2}$ that are used frequently are given in Table 8.1.

TABLE 8.1   Frequently used values of $z_{\alpha/2}$

$1-\alpha$	0.80	0.90	0.95	0.99
$z_{\alpha/2}$	1.28	1.645	1.96	2.58

EXAMPLE 8.1   For a random sample $X_1, \cdots, X_n$ of size 25 from $N(\mu, 4)$, suppose $\bar{x}$ is 19.5. Find a 90% confidence interval for μ.

Since $z_{\alpha/2} = z_{0.05} = 1.645,$

$$\bar{x} \pm z_{0.05}\frac{\sigma}{\sqrt{n}} = 19.5 \pm 1.645\frac{2}{\sqrt{25}} = 19.5 \pm 0.66$$

Therefore, a 90% confidence interval is (18.84, 20.16).

b) Sample size for estimation of μ

Sample size needs to be determined for an experiment in which the goal is to make an inference about a population from a sample. The maximum error of the estimate is used in the sample size calculation. The maximum error of the estimate $E = |\bar{x} - \mu|$ for a $100(1-\alpha)\%$ CI is given as $z_{\alpha/2}\sigma/\sqrt{n}$, as shown in Figure 8.3. It decreases as n increases. Often the values of α and E are given as accuracy requirements, and then it is necessary to determine the sample size needed to satisfy these requirements. Solving for n, we find

$$n = \left(\frac{z_{\alpha/2}\sigma}{E} \right)^2$$

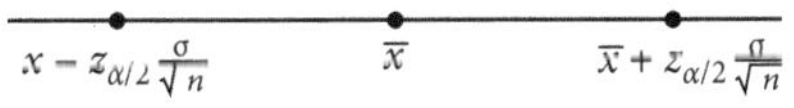

$$\overline{x} - z_{\alpha/2}\frac{\sigma}{\sqrt{n}} \qquad \overline{x} \qquad \overline{x} + z_{\alpha/2}\frac{\sigma}{\sqrt{n}}$$

FIGURE 8.3   Illustration of the maximum error of the estimate for a $100(1-\alpha)\%$ CI for μ.

The maximum error of estimate: $E = |\overline{x} - \mu|$

$E = z_{\alpha/2}\frac{\sigma}{\sqrt{n}}$: decreases as n increases

The sample size needed to ensure a $100(1-\alpha)\%$ CI with a maximum error of estimate E :

$$n = \left(\frac{z_{\alpha/2}\sigma}{E}\right)^2$$

EXAMPLE 8.2   A research worker wants to know the average drying time of a certain type of paint under a specific condition, and she wants to be able to assert with 95% confidence that the mean of her sample is off by at most 30 seconds. If $\sigma = 1.5$ minutes, how large a sample will she have to take?

$$n = \left(\frac{z_{\alpha/2}\sigma}{E}\right)^2 = \left(\frac{z_{0.025}\sigma}{E}\right)^2 = \left(\frac{1.96 \cdot 1.5}{0.5}\right)^2 = 34.6$$

Because the sample size needs to be an integer, she needs a sample of size at least 35.

c) Testing hypothesis

In statistical testing, a *one-sided test* and a *two-sided test* are alternative ways of computing the significance of a parameter using a test statistic. A one-sided test is used if the alternative hypothesis is in one direction, while a two-sided test is used if the alternative hypothesis is in either direction. The critical value at α is used for a one-sided test, while the critical value at $\alpha/2$ is used for a two-sided test. For a test of the mean of a normal population when σ is known, z_α is used for a one-sided test and $z_{\alpha/2}$ is used for a two-sided test, as shown in Figure 8.4.

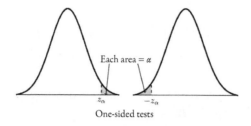

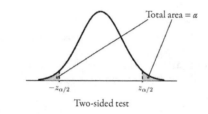

FIGURE 8.4   One-sided and two-sided tests.
Shaded area: rejection region.

For a normal population with known σ, we follow the procedure given in Table 8.2 for testing about μ.

TABLE 8.2   Test about μ for a normal population with known σ

	Case (a)	Case (b)	Case (c)
Step 1	$H_0: \mu = \mu_0, H_1: \mu \neq \mu_0$ (2-sided alternative)	$H_0: \mu = \mu_0, H_1: \mu > \mu_0$ (1-sided alternative)	$H_0: \mu = \mu_0, H_1: \mu < \mu_0$ (1-sided alternative)
Step 2	$\alpha = ?$	$\alpha = ?$	$\alpha = ?$
Step 3	$z = \dfrac{\bar{x} - \mu_0}{\sigma/\sqrt{n}}$ Rejection region: $\|z\| \geq z_{\alpha/2}$	$z = \dfrac{\bar{x} - \mu_0}{\sigma/\sqrt{n}}$ Rejection region: $z \geq z_\alpha$	$z = \dfrac{\bar{x} - \mu_0}{\sigma/\sqrt{n}}$ Rejection region: $z \leq -z_\alpha$
Step 4	Substitute $\bar{x}, \mu_0, \sigma,$ and n Calculate z Decision	Substitute $\bar{x}, \mu_0, \sigma,$ and n Calculate z Decision	Substitute $\bar{x}, \mu_0, \sigma,$ and n Calculate z Decision
Step 5	$p = 2P(Z \geq \|z\|)$	$p = P(Z \geq z)$	$p = P(Z \leq z)$

EXAMPLE 8.3   A manufacturer claims that the output for a certain electric circuit is 130 V. A sample of $n = 9$ independent readings on the voltage for this circuit, when tested, yields $\bar{x} = 131.4$ V. It is assumed that the population has a normal distribution with $\sigma = 1.5$ V. Do the data contradict the manufacturer's claim at $\alpha = 0.01$?

$H_0: \mu = \mu_0 = 130$, $H_1: \mu \neq 130$

$\alpha = 0.01$

The test statistic is

$$Z = \frac{\bar{X} - \mu_0}{\sigma/\sqrt{n}}$$

The rejection region is $|z| \geq z_{\alpha/2} = z_{0.005} = 2.58$ (see Figure 8.5).

$$z = \frac{131.4 - 130}{1.5/\sqrt{9}} = 2.8$$

Since $|2.8| = 2.8 > 2.58$, we reject H_0.

$p\text{-value} = P(|Z| > 2.8) = 2P(Z > 2.8) = 2\left[1 - \Phi(2.8)\right] = 2(1 - 0.9974) = 0.0052 < 0.01$

Based on the test, we conclude that the data contradict the manufacturer's claim.

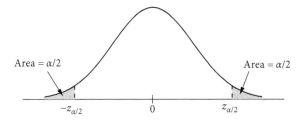

FIGURE 8.5   Rejection region for the test for Example 8.3.

d) Relationship between CI and test

The confidence interval and hypothesis test have a close relationship. The confidence interval is the same as the acceptance region in a two-sided test. In a two-sided test, the parameter is included in the confidence interval if and only if H_0 is not rejected. The parameter is not included in the confidence interval if H_0 is rejected.

EXAMPLE 8.4 In Example 8.3, consider a 99% confidence interval for μ. Since

$$\bar{x} \pm z_{\alpha/2} \frac{\sigma}{\sqrt{n}} = 131.4 \pm 2.58 \frac{1.5}{\sqrt{9}} = 131.4 \pm 1.29$$

a 99% confidence interval for μ is $(130.11, 132.69)$. Here, $\mu = 130$ is not included in the confidence interval. Therefore, we reject H_0. This result coincides with the result in Example 8.3.

CI is the same as the acceptance region in a two-sided test.
The parameter is included in the CI $\Leftrightarrow$ Do not reject H_0 in a two-sided test.
The parameter is not included in the CI $\Leftrightarrow$ Reject H_0 in a two-sided test.

e) Probability of a type II error and sample size determination

For a one-sided test, the rejection region $z \geq z_\alpha$ is equivalent to

$$\frac{\bar{X} - \mu_0}{\sigma/\sqrt{n}} \geq z_\alpha \text{ and } \bar{X} \geq \mu_0 + z_\alpha \frac{\sigma}{\sqrt{n}}$$

Thus, the probability of a type II error is

$$\beta(\mu') = P(H_0 \text{ is not rejected when } \mu = \mu')$$

$$= P\left(\bar{X} < \mu_0 + z_\alpha \frac{\sigma}{\sqrt{n}} \text{ when } \mu = \mu' \right)$$

$$= P\left(\frac{\bar{X} - \mu'}{\sigma/\sqrt{n}} < \frac{\mu_0 + z_\alpha \frac{\sigma}{\sqrt{n}} - \mu'}{\sigma/\sqrt{n}} \middle| \mu = \mu' \right)$$

$$= P\left(Z < z_\alpha + \frac{\mu_0 - \mu'}{\sigma/\sqrt{n}} \right)$$

$$= \Phi\left(z_\alpha + \frac{\mu_0 - \mu'}{\sigma/\sqrt{n}} \right)$$

$$= \beta$$

Since

$$-z_\beta = z_\alpha + \frac{\mu_0 - \mu'}{\sigma/\sqrt{n}},$$

$$z_\alpha + z_\beta = -(\mu_0 - \mu')\frac{\sqrt{n}}{\sigma}$$

Therefore, the size of the sample is

$$n = \left(\frac{\sigma(z_\alpha + z_\beta)}{\mu_0 - \mu'}\right)^2$$

This is valid for either upper- or lower-tailed test. The probability of a type II error and the sample size for a two-sided test can be derived similarly. Table 8.3 shows the type II error probability, and Table 8.4 shows required sample size for each of the alternative hypotheses.

TABLE 8.3   Calculation of the probability of a type II error for μ

Alternative Hypothesis	$\beta(\mu')$ for a Level α Test
$H_1: \mu > \mu_0$	$\Phi\left(z_\alpha + \frac{\mu_0 - \mu'}{\sigma/\sqrt{n}}\right)$
$H_1: \mu < \mu_0$	$1 - \Phi\left(-z_\alpha + \frac{\mu_0 - \mu'}{\sigma/\sqrt{n}}\right)$
$H_1: \mu \neq \mu_0$	$\Phi\left(z_{\alpha/2} + \frac{\mu_0 - \mu'}{\sigma/\sqrt{n}}\right) - \Phi\left(-z_{\alpha/2} + \frac{\mu_0 - \mu'}{\sigma/\sqrt{n}}\right)$

TABLE 8.4   Sample size determination for μ

Alternative Hypothesis	Required Sample Size
One-sided	$n = \left(\frac{\sigma(z_\alpha + z_\beta)}{\mu_0 - \mu'}\right)^2$
Two-sided	$n = \left(\frac{\sigma(z_{\alpha/2} + z_\beta)}{\mu_0 - \mu'}\right)^2$

EXAMPLE 8.5 Let μ denote the true average duration of treating a disease by a certain therapy. Consider testing H_0: $\mu = 20$ days versus H_1: $\mu > 20$ days based on a sample of size $n = 16$ from a normal population distribution with $\sigma = 2$. A test with $\alpha = 0.01$ requires $z_\alpha = z_{0.01} = 2.33$. The probability of a type II error when $\mu = 21$ is

$$\beta(21) = \Phi\left(2.33 + \frac{20-21}{2/\sqrt{16}}\right) = \Phi(0.33) = 0.6293$$

Let's find the required sample size when the probability of a type II error is 0.1. Since $z_{0.1} = 1.28$, the requirement that the level 0.01 test also has $\beta(21) = 0.1$ is

$$n = \left(\frac{\sigma(z_\alpha + z_\beta)}{\mu_0 - \mu'}\right)^2 = \left(\frac{2(2.33 + 1.28)}{20-21}\right)^2 = (-7.22)^2 = 52.13$$

Therefore, n should be at least 53.

B. LARGE SAMPLE WITH UNKNOWN σ

a) Confidence interval

Let $X_1, \cdots, X_n$ be a random sample from a population with mean μ and variance σ^2. By the central limit theorem, $Z = (\bar{X} - \mu)/(\sigma/\sqrt{n})$ is approximately $N(0, 1)$ when n is large. Therefore,

$$P\left(-z_{\alpha/2} < \frac{\bar{X} - \mu}{\sigma/\sqrt{n}} < z_{\alpha/2}\right) \approx 1 - \alpha$$

However, as σ^2 is usually unknown, we replace σ^2 with s^2 in constructing a confidence interval as

$$\left(\bar{X} - z_{\alpha/2}\frac{s}{\sqrt{n}}, \ \bar{X} + z_{\alpha/2}\frac{s}{\sqrt{n}}\right)$$

when n is large. The sample size of $n \geq 30$ is considered large in this case.

If n is large ($n \geq 30$), a $100(1 - \alpha)\%$ CI for μ is

$$\left(\bar{x} - z_{\alpha/2} \frac{s}{\sqrt{n}}, \bar{x} + z_{\alpha/2} \frac{s}{\sqrt{n}} \right)$$

EXAMPLE 8.6 To estimate the average speed of cars on a specific highway, an investigator collected speed data from a random sample of 75 cars driving on the highway. The sample mean and sample standard deviation are 58 miles per hour and 15 miles per hour, respectively.

a. Construct a 90% confidence interval for the mean speed.

$$n = 75, \bar{x} = 58, s = 15$$

Since $z_{\alpha/2} = z_{0.05} = 1.645$, $z_{\alpha/2} \frac{s}{\sqrt{n}} = 1.645 \frac{15}{\sqrt{75}} = 2.85$ and the confidence interval is

$$\left(\bar{x} - z_{\alpha/2} \frac{s}{\sqrt{n}}, \bar{x} + z_{\alpha/2} \frac{s}{\sqrt{n}} \right) = (58 - 2.85, 58 + 2.85) = (55.15, 60.85)$$

b. Construct an 80% confidence interval for the mean speed.

Since $z_{\alpha/2} = z_{0.1} = 1.28$, $z_{\alpha/2} \frac{s}{\sqrt{n}} = 1.28 \frac{15}{\sqrt{75}} = 2.22$ and the confidence interval is

$$\left(\bar{x} - z_{\alpha/2} \frac{s}{\sqrt{n}}, \bar{x} + z_{\alpha/2} \frac{s}{\sqrt{n}} \right) = (58 - 2.22, 58 + 2.22) = (55.78, 60.22)$$

b) Testing hypothesis

For a large sample, we follow the procedure given in Table 8.5 for testing about μ.

TABLE 8.5.   Test about μ for a large sample

	Case (a)	Case (b)	Case (c)
Step 1	$H_0: \mu = \mu_0, H_1: \mu \neq \mu_0$ (2-sided alternative)	$H_0: \mu = \mu_0, H_1: \mu > \mu_0$ (1-sided alternative)	$H_0: \mu = \mu_0, H_1: \mu < \mu_0$ (1-sided alternative)
Step 2	$\alpha = ?$	$\alpha = ?$	$\alpha = ?$
Step 3	$z = \dfrac{\bar{x} - \mu_0}{s/\sqrt{n}}$ Rejection region: $\|z\| \geq z_{\alpha/2}$	$z = \dfrac{\bar{x} - \mu_0}{s/\sqrt{n}}$ Rejection region: $z \geq z_{\alpha}$	$z = \dfrac{\bar{x} - \mu_0}{s/\sqrt{n}}$ Rejection region: $z \leq -z_{\alpha}$
Step 4	Substitute $\bar{x}, \mu_0, s$, and n Calculate z Decision	Substitute $\bar{x}, \mu_0, s$, and n Calculate z Decision	Substitute $\bar{x}, \mu_0, s$, and n Calculate z Decision
Step 5	$p = 2P(Z \geq \|z\|)$	$p = P(Z \geq z)$	$p = P(Z \leq z)$

EXAMPLE 8.7   From extensive records, it is known that the duration of treating a disease by a standard therapy has a mean of 15 days. It is claimed that a new therapy can reduce the treatment time. To test this claim, the new therapy is to be tried on 70 patients, and their times to recovery are to be recorded. The sample mean and sample standard deviation are 14.6 days and 3.0 days, respectively. Conduct a hypothesis test using $\alpha = 0.05$.

$H_0: \mu = 15, H_1: \mu < 15$

$\alpha = 0.05$

The test statistic is

$$Z = \frac{\bar{X} - \mu_0}{S/\sqrt{n}}$$

The rejection region is $z \leq -z_{\alpha} = -z_{0.05} = -1.645$.

$$z = \frac{14.6 - 15}{3/\sqrt{70}} = -1.12 > -1.645$$

We do not reject H_0. We do not have sufficient evidence to support the claim that the new therapy reduces treatment time.

p-value $= P(Z < -1.12) = \Phi(-1.12) = 0.1314$

C. SMALL SAMPLE, NORMAL POPULATION WITH UNKNOWN σ

a) Confidence interval

Consider a random sample $X_1, \cdots, X_n$ from $N(\mu, \sigma^2)$, where σ is unknown. In Chapter 6, we saw that $T = (\bar{X} - \mu)/(S/\sqrt{n})$ follows the t distribution with $n - 1$ degrees of freedom. Let $t_{\alpha, v}$ be the $100(1 - \alpha)$-th percentile of the t distribution with v degrees of freedom. Then

$$P(-t_{\alpha/2, n-1} < T < t_{\alpha/2, n-1}) = 1 - \alpha$$

We use this fact to construct a confidence interval for a small sample from a normal population with unknown σ as follows:

$$\left(\bar{x} - t_{\alpha/2, n-1} \frac{s}{\sqrt{n}}, \bar{x} + t_{\alpha/2, n-1} \frac{s}{\sqrt{n}} \right)$$

Let $\bar{x}$ be the sample mean and s the sample standard deviation computed from a small ($n < 30$) random sample from a normal population with mean μ. Then a $100(1 - \alpha)\%$ confidence interval for μ is

$$\left(\bar{x} - t_{\alpha/2, n-1} \frac{s}{\sqrt{n}}, \ \bar{x} + t_{\alpha/2, n-1} \frac{s}{\sqrt{n}} \right)$$

EXAMPLE 8.8 The weights, in pounds, of two-month-old babies in a sample of 15 are the following:

8.9, 8.6, 8.0, 8.3, 8.8, 8.6, 8.1, 7.2, 8.0, 8.6, 9.1, 9.0, 9.1, 8.3, 7.9

The sample mean and sample standard deviation are $\bar{x} = 8.433$ and $s = 0.533$, respectively. Assuming that the data were sampled from a normal population distribution, a 95% confidence interval for μ is

$$\bar{x} \pm t_{\alpha/2,\, n-1} \frac{s}{\sqrt{n}} = \bar{x} \pm t_{0.025,\, 14} \frac{s}{\sqrt{n}} = 8.433 \pm 2.145 \frac{0.533}{\sqrt{15}} = 8.433 \pm 0.295$$

$$= (8.138, 8.728)$$

b) Testing hypothesis

For a small sample from a normal population, we follow the procedure given in Table 8.6 for testing about μ.

TABLE 8.6   Small sample hypothesis testing for μ

	Case (a)	Case (b)	Case (c)
Step 1	$H_0: \mu = \mu_0, H_1: \mu \neq \mu_0$ (2-sided alternative)	$H_0: \mu = \mu_0, H_1: \mu > \mu_0$ (1-sided alternative)	$H_0: \mu = \mu_0, H_1: \mu < \mu_0$ (1-sided alternative)
Step 2	$\alpha = ?$	$\alpha = ?$	$\alpha = ?$
Step 3	$t = \dfrac{\bar{x} - \mu_0}{s/\sqrt{n}}$ Rejection region: $\lvert t \rvert \geq t_{\alpha/2,\, n-1}$	$t = \dfrac{\bar{x} - \mu_0}{s/\sqrt{n}}$ Rejection region: $t \geq t_{\alpha, n-1}$	$t = \dfrac{\bar{x} - \mu_0}{s/\sqrt{n}}$ Rejection region: $t \leq -t_{\alpha, n-1}$
Step 4	Substitute $\bar{x}, \mu_0, s$, and n Calculate t Decision	Substitute $\bar{x}, \mu_0, s$, and n Calculate t Decision	Substitute $\bar{x}, \mu_0, s$, and n Calculate t Decision
Step 5	$p = 2P(T \geq \lvert t \rvert)$	$p = P(T \geq t)$	$p = P(T \leq t)$

Note: Sample size $n < 30$ (assumption: population is normally distributed).

EXAMPLE 8.9   The intelligence quotients (IQs) of 5 students from one area of a city are given as follows:

98, 117, 102, 111, 109

Assuming the IQ score is normally distributed, do the data suggest that the population mean IQ exceeds 100? Carry out a test using a significance level of 0.05.

$H_0: \mu = 100, H_1: \mu > 100$

$\alpha = 0.05$

The test statistic is

$$T = \frac{\bar{X} - \mu_0}{S/\sqrt{n}}$$

$$\bar{x} = 107.4, \quad s^2 = \frac{\sum x_i^2 - (\sum x_i)^2/n}{n-1} = 56.3, \quad s = \sqrt{56.3} = 7.5$$

The rejection region is $t \geq t_{\alpha, n-1} = t_{0.05, 4} = 2.132$.

$$t = \frac{107.4 - 100}{7.5/\sqrt{5}} = 2.206 > 2.132$$

We reject H_0 and conclude that there is sufficient evidence that the population IQ exceeds 100.

From Table A.4, $t_{0.05, 4} = 2.132 < 2.206 < 2.776 = t_{0.025, 4}$.

Thus, $0.025 < p$-value < 0.05.

Alternatively, the p-value can be obtained using R as follows:

>1-pt(2.206, 4)

0.046176

This shows that the p-value is 0.046. Note that the p-value is obtained as $1 - P(T < 2.206)$, where T is distributed as t with 4 degrees of freedom because it is the right-tail probability.

The test in Example 8.9 can be done using R as follows:

>x=c(98,117,102,111,109)

>t.test(x, mu=100, alt="greater")

The output contains the value of the t statistic and the p-value. For a one-sided test with the alternative hypothesis of the other direction, alt="less" is used. For a two-sided test, you enter alt="two.sided". However, you do not need to specify the alternative hypothesis for a two-sided test, because the default is a two-sided test.

The confidence interval obtained in Example 8.8 can be obtained using R as follows:

> x=c(8.9,8.6,8.0,8.3,8.8,8.6,8.1,7.2,8.0,8.6,9.1,9.0,9.1,8.3,7.9)

>t.test(x)

The output contains the two-sided test result with the null hypothesis of H_0: $\mu = 0$ with the level of $\alpha = 0.05$ and a 95% confidence interval. For a different level of confidence, the value of α should be specified. A 90% confidence interval, for example, can be obtained by

> t.test(x,conf.level=0.90)

To help identify an appropriate method for inferences concerning a population mean, a flowchart is given in Figure 8.6.

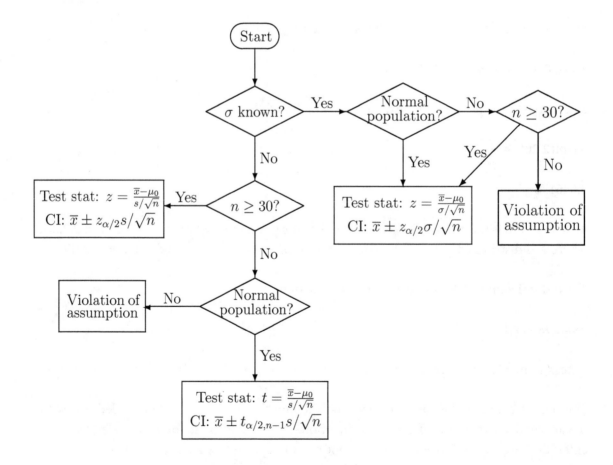

FIGURE 8.6   Flowchart for searching an appropriate method for inferences concerning a population mean.

2. *Inferences Concerning a Population Proportion*

a) Confidence interval

Suppose X is the number of successes and p is the success rate in n independent Bernoulli trials. Since X is distributed as $Bin(n, p)$, $E(X) = np$, $Var(X) = np(1 - p)$, and $\sigma_X = \sqrt{np(1 - p)}$.

Consider an unbiased estimator $\hat{p} = X/n$ of p. Then

$$Z = \frac{X - np}{\sqrt{np(1 - p)}}$$

is approximately standard normal for a large n. Also,

$$Z = \frac{X - np}{\sqrt{np(1 - p)}} = \frac{\hat{p} - p}{\sqrt{p(1 - p)/n}}$$

and $E(\hat{p}) = E(X)/n = p$ and $\sigma_{\hat{p}} = \sqrt{p(1 - p)/n}$. We estimate $\sigma_{\hat{p}}$ as $\sqrt{\hat{p}(1 - \hat{p})/n}$.

> A $100(1 - \alpha)\%$ confidence interval for p for a large n is
>
> $$(\hat{p} - z_{\alpha/2}\sqrt{\hat{p}(1 - \hat{p})/n}, \ \hat{p} + z_{\alpha/2}\sqrt{\hat{p}(1 - \hat{p})/n})$$
>
> It is good for $n\hat{p} \geq 10$ and $n(1 - \hat{p}) \geq 10$.

EXAMPLE 8.10 Out of a random sample of 1,500 people in a presidential election poll, 820 voters are found to support a certain candidate. Find a 95% confidence interval for the support rate of this candidate.

Since $\hat{p} = x/n = 820/1500 = 0.547$,

$$z_{\alpha/2}\sqrt{\hat{p}(1 - \hat{p})/n} = 1.96\sqrt{(0.547)(0.453)/1500} = 0.025$$

A 95% confidence interval for p is

$$\left(\hat{p} - z_{\alpha/2}\sqrt{\hat{p}(1-\hat{p})/n}, \, \hat{p} + z_{\alpha/2}\sqrt{\hat{p}(1-\hat{p})/n}\right) = (0.547 - 0.025, 0.547 + 0.025)$$

$$= (0.522, 0.572)$$

b) Sample size estimation

The maximum error of the estimate E for a $100(1-\alpha)\%$ confidence interval is $z_{\alpha/2}\sqrt{\hat{p}(1-\hat{p})/n}$. Thus, $E = z_{\alpha/2}\sqrt{p(1-p)/n}$ and $E^2 = z_{\alpha/2}^2 p(1-p)/n$ (see Figure 8.7). Therefore, the required sample size is

$$n = p(1-p)\left(\frac{z_{\alpha/2}}{E}\right)^2$$

$$\hat{p} - z_{\alpha/2}\sqrt{\frac{\hat{p}(1-\hat{p})}{n}} \qquad \hat{p} \qquad \hat{p} + z_{\alpha/2}\sqrt{\frac{\hat{p}(1-\hat{p})}{n}}$$

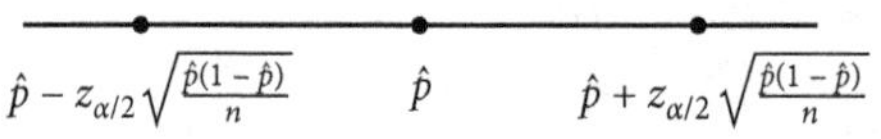

FIGURE 8.7   Illustration of the maximum error of the estimate for a $100(1-\alpha)\%$ CI for p.

If we have prior knowledge of p, then the required sample size is $n = p(1-p)(z_{\alpha/2}/E)^2$. Because the sample size is estimated before the experiment starts, the estimate of p may not be available. If we do not have prior knowledge of p, we estimate the maximum possible sample size for all possible values of p. The value of $p(1-p)$ is maximized when $p = 0.5$, and thus $p(1-p) = 0.25$. Therefore, if $n = (1/4)(z_{\alpha/2}/E)^2$, then the error is at most E regardless of the value of p. This is a conservative approach to finding the required sample size.

The maximum error of Estimate: $E = \left|\frac{x}{n} - p\right|$

$$E = z_{\alpha/2}\sqrt{\frac{\hat{p}(1-\hat{p})}{n}}$$

Sample size needed to attain maximum error of estimate E:

$$n = p(1-p)\left(\frac{z_{\alpha/2}}{E}\right)^2 \text{ with prior knowledge of } p$$

$$n = \frac{1}{4}\left(\frac{z_{\alpha/2}}{E}\right)^2 \text{ without prior knowledge of } p$$

EXAMPLE 8.11 A public health survey is to be designed to estimate the proportion p of a population having a certain disease. How many persons should be examined if the investigators wish to be 99% certain that the error of estimation is below 0.04 when:

a.   there is no knowledge about the value of p?

$$n = \frac{1}{4}\left(\frac{z_{\alpha/2}}{E}\right)^2 = \frac{1}{4}\left(\frac{z_{0.005}}{E}\right)^2 = \frac{1}{4}\left(\frac{2.58}{0.04}\right)^2 = 1040.1$$

Thus, the sample size should be at least 1,041.

b. p is known to be 0.2?

$$n = p(1-p)\left(\frac{z_{\alpha/2}}{E}\right)^2 = (0.2)(0.8)\left(\frac{2.58}{0.04}\right)^2 = 665.6$$

Thus, the sample size should be at least 666.

c) Testing hypothesis

For a large sample, we follow the procedure given in Table 8.7 for testing about p. Notice that in the standard deviation $\sqrt{p(1-p)/n}$ of $\bar{x}$, $\hat{p}$ is substituted for p in the formula of the confidence interval, but p_0 is substituted in the test statistic. This is because we assume that the test statistic follows the distribution under H_0 in testing hypothesis.

TABLE 8.7   Test for a population proportion (large sample)

	Case (a)	Case (b)	Case (c)
Step 1	$H_0: p = p_0, H_1: p \neq p_0$ (2-sided alternative)	$H_0: p = p_0, H_1: p > p_0$ (1-sided alternative)	$H_0: p = p_0, H_1: p < p_0$ (1-sided alternative)
Step 2	$\alpha =?$	$\alpha =?$	$\alpha =?$
Step 3	$z = \dfrac{\hat{p} - p_0}{\sqrt{p_0(1-p_0)/n}}$ Rejection region: $\|z\| \geq z_{\alpha/2}$	$z = \dfrac{\hat{p} - p_0}{\sqrt{p_0(1-p_0)/n}}$ Rejection region: $z \geq z_\alpha$	$z = \dfrac{\hat{p} - p_0}{\sqrt{p_0(1-p_0)/n}}$ Rejection region: $z \leq -z_\alpha$
Step 4	Substitute $\hat{p}$, p_0 and n Calculate z Decision	Substitute $\hat{p}$, p_0 and n Calculate z Decision	Substitute $\hat{p}$, p_0 and n Calculate z Decision
Step 5	$p = 2P(Z \geq \|z\|)$	$p = P(Z \geq z)$	$p = P(Z \leq z)$

EXAMPLE 8.12 A group of concerned citizens wants to show that less than half of the voters support the president's handling of a recent crisis. A random sample of 500 voters gives 228 in support. Does this provide strong evidence for concluding that less than half of all voters support the president's handling of the crisis? Conduct a hypothesis test using $\alpha = 0.01$.

H_0: $p = 0.5$, H_1: $p < 0.5$

$\alpha = 0.01$

The test statistic is

$$Z = \frac{\hat{p} - p_0}{\sqrt{p_0(1 - p_0)/n}}$$

The rejection region is $z \leq -z_\alpha = -z_{0.01} = -2.33$.

Since $\hat{p} = x/n = 228/500 = 0.456$,

$$z = \frac{0.456 - 0.5}{\sqrt{(0.5)(0.5)/500}} = -1.97 > -2.33$$

We do not reject H_0. There is not enough evidence that less than half of all voters support the president's handling of the crisis.

p-value $= P(Z < -1.97) = \Phi(-1.97) = 0.024$

The test in Example 8.12 can be done using R as follows:

```
>prop.test(228, 500, p=0.5, alt="less", correct=F)
```

The output contains the square of the value of z and the p-value. Here "correct=F" is added because no continuity correction is done.

The confidence interval obtained in Example 8.10 can be obtained using R as follows:

```
>prop.test(820, 1500, correct=F)
```

d) Probability of a type II error and size determination

Table 8.8 shows the type II error probability, and Table 8.9 provides required sample size for each of the alternative hypotheses.

TABLE 8.8   Calculation of the probability of a type II error for p

Alternative Hypothesis	$\beta(p')$ for a Level α Test
$H_1: p > p_0$	$\Phi\left(\dfrac{p_0 - p' + z_\alpha\sqrt{p_0(1-p_0)/n}}{\sqrt{p'(1-p')/n}}\right)$
$H_1: p < p_0$	$1 - \Phi\left(\dfrac{p_0 - p' - z_\alpha\sqrt{p_0(1-p_0)/n}}{\sqrt{p'(1-p')/n}}\right)$
$H_1: p \neq p_0$	$\Phi\left(\dfrac{p_0 - p' + z_{\alpha/2}\sqrt{p_0(1-p_0)/n}}{\sqrt{p'(1-p')/n}}\right) - \Phi\left(\dfrac{p_0 - p' - z_{\alpha/2}\sqrt{p_0(1-p_0)/n}}{\sqrt{p'(1-p')/n}}\right)$

TABLE 8.9   Sample size determination for p

Alternative Hypothesis	Required Sample Size
One-sided	$n = \left(\dfrac{z_\alpha\sqrt{p_0(1-p_0)} + z_\beta\sqrt{p'(1-p')}}{p' - p_0}\right)^2$
Two-sided	$n = \left(\dfrac{z_{\alpha/2}\sqrt{p_0(1-p_0)} + z_\beta\sqrt{p'(1-p')}}{p' - p_0}\right)^2$

EXAMPLE 8.13 An airline official claims that at least 85% of the airline's flights arrive on time. Let p denote the true proportion of such flights that arrive on time as claimed and consider the hypotheses $H_0: p = 0.85$ versus $H_1: p < 0.85$.

a. If only 75% of the flights arrive on time, how likely is it that a level 0.01 test based on 100 flights will detect such a departure from H_0?

With $\alpha = 0.01$, $p_0 = 0.85$ and $p' = 0.75$,

$$\beta(0.75) = 1 - \Phi\left(\frac{0.85 - 0.75 - 2.33\sqrt{\frac{(0.85)(0.15)}{100}}}{\sqrt{\frac{(0.75)(0.25)}{100}}}\right) = 1 - \Phi(0.39) = 1 - 0.6517 = 0.3483$$

Thus, the probability that H_0 will be rejected using the test when $p = 0.75$ is $1 - \beta(0.75) = 0.6517$.

b. What should the sample size be in order to ensure that the probability of a type II error at $p = 0.75$ is 0.01?

Using $z_\alpha = z_\beta = 2.33$,

$$n = \left(\frac{2.33\sqrt{(0.85)(0.15)} + 2.33\sqrt{(0.75)(0.25)}}{0.75 - 0.85} \right)^2 = 338.89$$

Therefore, the sample size should be $n = 339$.

3. Inferences Concerning a Population Variance

a) Confidence interval

Let $X_1, X_2, \cdots, X_n$ be a random sample from $N(\mu, \sigma^2)$. Since $(n-1)S^2/\sigma^2$ is distributed as a chi-square with $n-1$ degrees of freedom,

$$P\left(\chi^2_{1-\alpha/2,\, n-1} < \frac{(n-1)S^2}{\sigma^2} < \chi^2_{\alpha/2,\, n-1} \right) = 1 - \alpha$$

where $\chi^2_{\alpha,\, v}$ is defined as the $100(1-\alpha)$-th percentile of the chi-square distribution with v degrees of freedom. Because the chi-square distribution is not symmetric, the lower bound of the confidence interval is not the negative value of the upper bound. The above probability can be rewritten in terms of σ^2 as follows:

$$P\left(\frac{(n-1)S^2}{\chi^2_{\alpha/2,\, n-1}} < \sigma^2 < \frac{(n-1)S^2}{\chi^2_{1-\alpha/2,\, n-1}} \right) = 1 - \alpha$$

Thus, a $100(1-\alpha)\%$ confidence interfal for σ^2 of a normal population is

$$\left(\frac{(n-1)s^2}{\chi^2_{\alpha/2,\, n-1}},\ \frac{(n-1)s^2}{\chi^2_{1-\alpha/2,\, n-1}} \right)$$

A $100(1-\alpha)\%$ confidence interfal for σ^2 of a normal population is

$$\left(\frac{(n-1)s^2}{\chi^2_{\alpha/2,\, n-1}},\ \frac{(n-1)s^2}{\chi^2_{1-\alpha/2,\, n-1}} \right)$$

EXAMPLE 8.14 A random sample of 14 eighth-grade boys in Stony Brook, New York, from a normal population gives the following heights in inches:

60.3, 62.7, 67.3, 62.3, 64.9, 71.2, 59.7, 59.6, 65.5, 68.7, 60.5, 64.2, 65.1, 59.5

The sample mean of these data is equal to 63.7, and the sample variance is equal to 13.562. A 95% confidence interval for the population variance is

$$\left(\frac{(n-1)s^2}{\chi^2_{\frac{\alpha}{2}, n-1}}, \frac{(n-1)s^2}{\chi^2_{1-\frac{\alpha}{2}, n-1}} \right) = \left(\frac{(n-1)s^2}{\chi^2_{0.025, 13}}, \frac{(n-1)s^2}{\chi^2_{0.975, 13}} \right) = \left(\frac{13(13.562)}{24.736}, \frac{13(13.562)}{5.009} \right)$$

$$= (7.13, 35.20)$$

b) Testing hypothesis

For a sample from a normal population, we follow the procedure given in Table 8.10 for testing about σ^2.

TABLE 8.10 Test for a variance

	Case (a)	Case (b)	Case (c)
Step 1	$H_0: \sigma^2 = \sigma_0^2, H_1: \sigma^2 \neq \sigma_0^2$ (2-sided alternative)	$H_0: \sigma^2 = \sigma_0^2, H_1: \sigma^2 > \sigma_0^2$ (1-sided alternative)	$H_0: \sigma^2 = \sigma_0^2, H_1: \sigma^2 < \sigma_0^2$ (1-sided alternative)
Step 2	$\alpha =?$	$\alpha =?$	$\alpha =?$
Step 3	$\chi^2 = \frac{(n-1)s^2}{\sigma_0^2}$ Rejection region: $\chi^2 \geq \chi^2_{\alpha/2, n-1}$ or $\chi^2 \leq \chi^2_{1-\alpha/2, n-1}$	$\chi^2 = \frac{(n-1)s^2}{\sigma_0^2}$ Rejection region: $\chi^2 \geq \chi^2_{\alpha, n-1}$	$\chi^2 = \frac{(n-1)s^2}{\sigma_0^2}$ Rejection region: $\chi^2 \leq \chi^2_{1-\alpha, n-1}$
Step 4	Substitute s^2, σ_0^2 and n Calculate χ^2 Decision	Substitute s^2, σ_0^2 and n Calculate χ^2 Decision	Substitute s^2, σ_0^2 and n Calculate χ^2 Decision
Step 5	$p = 2\min\{P(X \geq \chi^2),$ $P(X \leq \chi^2)\}$ where $X \sim \chi^2_{n-1}$	$p = P(X \geq \chi^2)$ where $X \sim \chi^2_{n-1}$	$p = P(X \leq \chi^2)$ where $X \sim \chi^2_{n-1}$

EXAMPLE 8.15 The length of a certain type of screw is acceptable only if the population standard deviation of the length is at most 0.6 mm. Use the 0.05 level of significance to test the null hypothesis of $\sigma = 0.6$ mm against the alternative hypothesis $\sigma > 0.6$ mm, if the thicknesses of 16 screws of such a type have a standard deviation of 0.75 mm.

$H_0: \sigma = \sigma_0 = 0.6$, $H_1: \sigma > 0.6$

$\alpha = 0.05$

The test statistic is

$$\chi^2 = \frac{(n-1)S^2}{\sigma_0^2}$$

The rejection region is $\chi^2 \geq \chi^2_{\alpha, n-1} = \chi^2_{0.05, 15} = 25.00$.

$$\chi^2 = \frac{15(0.75)^2}{(0.6)^2} = 23.44 < 25.00$$

We do not reject H_0 and conclude that the population standard deviation of the length is not significantly larger than 0.6 mm.

From Table A.5, $\chi^2_{0.10, 15} = 22.31 < 23.44 < 25.00 = \chi^2_{0.05, 15}$.

Thus, $0.05 < p\text{-value} < 0.10$.

Alternatively, the p-value can be obtained using R as follows:

```
>1-pchisq(23.44,15)
```

```
0.07523417
```

This shows that the p-value is 0.075. Note that the p-value is obtained as $1 - P(\chi^2 < 23.44)$, where χ^2 is distributed as χ^2 with 15 degrees of freedom because it is the right-tail probability.

SUMMARY OF CHAPTER 8

1. Direction of the Alternative Hypothesis
    a. One-sided test: Used if the alternative hypothesis is in one direction.
    b. Two-sided test: Used if the alternative hypothesis is in either direction.
2. Relation between CI and Test
    a. CI is the acceptance region of a two-sided test.
    b. In a two-sided test:
        i. do not reject H_0 if the parameter is included in the CI.
        ii. reject H_0 if the parameter is <u>not</u> included in the CI.
3. Frequently used values of $z_{\alpha/2}$

$1 - \alpha$	0.80	0.90	0.95	0.99
$z_{\alpha/2}$	1.28	1.645	1.96	2.58

4. Inferences Concerning a Population Mean
    a. <u>Sample size</u> needed to attain maximum error of estimate E:

$$n = \left(\frac{z_{\alpha/2}\sigma}{E} \right)^2$$

    b. <u>Sample size</u> with β at $\mu = \mu'$
        i. For a one-sided test:

$$n = \left(\frac{\sigma(z_\alpha + z_\beta)}{\mu_0 - \mu'} \right)^2$$

        ii. For a two-sided test:

$$n = \left(\frac{\sigma(z_{\alpha/2} + z_\beta)}{\mu_0 - \mu'} \right)^2$$

    c. <u>Type II error probability</u> $\beta(\mu')$ for a level α test
        i. For $H_1: \mu > \mu_0$:

$$\Phi\left(z_\alpha + \frac{\mu_0 - \mu'}{\sigma/\sqrt{n}} \right)$$

        ii. For $H_1: \mu < \mu_0$:

$$1 - \Phi\left(-z_\alpha + \frac{\mu_0 - \mu'}{\sigma/\sqrt{n}} \right)$$

iii. For $H_1: \mu \neq \mu_0$:

$$\Phi\left(z_{\alpha/2} + \frac{\mu_0 - \mu'}{\sigma / \sqrt{n}}\right) - \Phi\left(-z_{\alpha/2} + \frac{\mu_0 - \mu'}{\sigma / \sqrt{n}}\right)$$

d. Normal population with known σ

    i. A $100(1 - \alpha)\%$ confidence interval:

$$\left(\bar{x} - z_{\alpha/2} \frac{\sigma}{\sqrt{n}} , \bar{x} + z_{\alpha/2} \frac{\sigma}{\sqrt{n}}\right)$$

    ii. Hypothesis testing: See Table 8.2

e. Large sample with unknown σ ($n \geq 30$)

    i. A $100(1 - \alpha)\%$ confidence interval:

$$\left(\bar{x} - z_{\alpha/2} \frac{s}{\sqrt{n}} , \bar{x} + z_{\alpha/2} \frac{s}{\sqrt{n}}\right)$$

    ii. Hypothesis testing: See Table 8.5

f. Small sample with unknown σ ($n < 30$)

    i. Assumption: normal population

    ii. A $100(1 - \alpha)\%$ confidence interval:

$$\left(\bar{x} - t_{\alpha/2, n-1} \frac{s}{\sqrt{n}} , \bar{x} + t_{\alpha/2, n-1} \frac{s}{\sqrt{n}}\right)$$

    iii. Hypothesis testing: See Table 8.6

5. Inferences Concerning a Population Proportion ($n\hat{p} \geq 10$ and $n(1 - \hat{p}) \geq 10$)

a. <u>Sample size</u> needed to attain maximum error of estimate E

    i. When p is known:

$$n = p(1 - p)\left(\frac{z_{\alpha/2}}{E}\right)^2$$

    ii. When p is unknown:

$$n = \frac{1}{4}\left(\frac{z_{\alpha/2}}{E}\right)^2$$

b. <u>Sample size</u> with β at $p = p'$

    i. For a one-sided test:

$$n = \left(\frac{z_\alpha \sqrt{p_0(1 - p_0)} + z_\beta \sqrt{p'(1 - p')}}{p' - p_0}\right)^2$$

ii. For a two sided test:

$$n = \left(\frac{z_{\alpha/2}\sqrt{p_0(1-p_0)} + z_{\beta}\sqrt{p'(1-p')}}{p' - p_0} \right)^2$$

c. <u>Type II error probability</u> $\beta(p')$ for a level α test
   i. For H_1: $p > p_0$:

$$\Phi\left(\frac{p_0 - p' + z_{\alpha}\sqrt{p_0(1-p_0)/n}}{\sqrt{p'(1-p')/n}} \right)$$

   ii. For H_1: $p < p_0$:

$$1 - \Phi\left(\frac{p_0 - p' - z_{\alpha}\sqrt{p_0(1-p_0)/n}}{\sqrt{p'(1-p')/n}} \right)$$

   iii. For H_1: $p \neq p_0$:

$$\Phi\left(\frac{p_0 - p' + z_{\alpha/2}\sqrt{p_0(1-p_0)/n}}{\sqrt{p'(1-p')/n}} \right) - \Phi\left(\frac{p_0 - p' - z_{\alpha/2}\sqrt{p_0(1-p_0)/n}}{\sqrt{p'(1-p')/n}} \right)$$

d. Inference about p
   i. A $100(1-\alpha)\%$ confidence interval:

$$\left(\hat{p} - z_{\alpha/2}\sqrt{\frac{\hat{p}(1-\hat{p})}{n}},\ \hat{p} + z_{\alpha/2}\sqrt{\frac{\hat{p}(1-\hat{p})}{n}} \right), \quad \text{where } \hat{p} = \frac{x}{n}$$

   ii. Hypothesis testing: See Table 8.7
6. <u>Inferences Concerning a Population Variance</u>
   a. Assumption: normal population
   b. A $100(1-\alpha)\%$ confidence interval:

$$\left(\frac{(n-1)s^2}{\chi^2_{\alpha/2,\,n-1}},\ \frac{(n-1)s^2}{\chi^2_{1-\alpha/2,\,n-1}} \right)$$

   c. Hypothesis testing: See Table 8.10

EXERCISES

8.1 For a test concerning a mean, a sample of size $n = 80$ is obtained. Find the p-value for the following tests:
   a. In testing $H_0: \mu \leq \mu_0$ versus $H_1: \mu > \mu_0$, the test statistic is 2.48.
   b. In testing $H_0: \mu \geq \mu_0$ versus $H_1: \mu < \mu_0$, the test statistic is -2.48.
   c. In testing $H_0: \mu = \mu_0$ versus $H_1: \mu \neq \mu_0$, the test statistic is -2.48.

8.2 For a test concerning a mean, a sample of size $n = 60$ is obtained. What is your decision in the following tests?
   a. In testing $H_0: \mu \leq \mu_0$ versus $H_1: \mu > \mu_0$ at $\alpha = 0.05$, the test statistic is 1.75.
   b. In testing $H_0: \mu \geq \mu_0$ versus $H_1: \mu < \mu_0$ at $\alpha = 0.01$, the test statistic is -1.75.
   c. In testing $H_0: \mu = \mu_0$ versus $H_1: \mu \neq \mu_0$ at $\alpha = 0.05$, the test statistic is 1.75.

8.3 For a test concerning a mean, a sample of size $n = 15$ is obtained from a normal population. The population variance is unknown. Find the p-value for the following tests:
   a. In testing $H_0: \mu \leq \mu_0$ versus $H_1: \mu > \mu_0$, the test statistic is 1.345.
   b. In testing $H_0: \mu \geq \mu_0$ versus $H_1: \mu < \mu_0$, the test statistic is -1.345.
   c. In testing $H_0: \mu = \mu_0$ versus $H_1: \mu \neq \mu_0$, the test statistic is -1.345.

8.4 For a test concerning a mean, a sample of size $n = 15$ is obtained from a normal population. The population variance is unknown. What is your decision in the following tests?
   a. In testing $H_0: \mu \leq \mu_0$ versus $H_1: \mu > \mu_0$ at $\alpha = 0.05$, the test statistic is 2.01.
   b. In testing $H_0: \mu \geq \mu_0$ versus $H_1: \mu < \mu_0$ at $\alpha = 0.01$, the test statistic is -2.01.
   c. In testing $H_0: \mu = \mu_0$ versus $H_1: \mu \neq \mu_0$ at $\alpha = 0.05$, the test statistic is 2.01.

8.5 A meteorologist wishes to estimate the carbon dioxide content of air per unit volume in a certain area. It is known from studies that the standard deviation is 15 parts per million (ppm). How many air samples must the meteorologist analyze to be 90% certain that the error of estimate does not exceed 3 ppm?

8.6 A factory manager wants to determine the average time it takes to finish a certain process by workers, and he wants to test it with randomly selected workers. He wants to be able to assert with 95% confidence that the mean of his sample is off by at most 1 minute. If the population standard deviation is 3.2 minutes, how large a sample will he have to take?

8.7   An investigator interested in estimating a population mean wants to be 95% certain that the length of the confidence interval does not exceed 4. Find the required sample size for his study if the population standard deviation is 14.

8.8   The shelf life (duration until the expiration date in months) of certain ointment is known to have a normal distribution. A sample of size 120 tubes of ointment gives $\bar{x} = 36.1$ and $s = 3.7$.
  a.   Construct a 95% confidence interval of the population mean shelf life.
  b.   Suppose a researcher believed before the experiment that $\sigma = 4$. What would be the required sample size to estimate the population mean to be within 0.5 month with 99% confidence?

8.9   The required sample size needs to be changed if the length of the confidence interval for the mean μ of a normal population is changed when the population standard deviation σ is known. Answer the following questions.
  a.   If the length of the confidence interval is doubled, how does the sample size need to be changed?
  b.   If the length of the confidence interval is halved, how does the sample size need to be changed?

8.10   The required sample size needs to be changed depending on the known population standard deviation σ. Suppose the maximum error estimate remains the same. Answer the following questions about how the sample size needs to be changed.
  a.   In case the population standard deviation is tripled
  b.   In case the population standard deviation is halved

8.11   Suppose the average lifetime of a certain type of car battery is known to be 60 months. Consider conducting a two-sided test on it based on a sample of size 25 from a normal distribution with a population standard deviation of 4 months.
  a.   If the true average lifetime is 62 months and $\alpha = 0.01$, what is the probability of a type II error?
  b.   What is the required sample size to satisfy $\alpha = 0.01$ and the type II error probability of $\beta(62) = 0.1$?

8.12   The foot size of each of 16 men was measured, resulting in the sample mean of 27.32 cm. Assume that the distribution of foot sizes is normal with $\sigma = 1.2$ cm.
  a.   Test if the population mean of men's foot sizes is 28.0 cm using $\alpha = 0.01$.
  b.   If $\alpha = 0.01$ is used, what is the probability of a type II error when the population mean is 27.0 cm?
  c.   Find the sample size required to ensure that the type II error probability $\beta(27) = 0.1$ when $\alpha = 0.01$.

8.13   To study the effectiveness of a weight-loss training method, a random sample of 40 females went through the training. After a two-month training period, the weight change is recorded for each participant. The researcher wants to test that the mean weight reduction is larger than 5 pounds at level $\alpha = 0.05$.

    a. The researcher knows that the population standard deviation is approximately 7 pounds. What is the required sample size to detect the power of 80% when the true weight reduction is 8 pounds? Is the sample size of 40 enough?

    b. The average weight loss for the 40 women in the sample was 7.2 pounds with the sample standard deviation of 4.5 pounds. Is there sufficient evidence that the training is effective? Test using $\alpha = 0.05$.

    c. Find the power of the test conducted in part (b) if the true mean weight reduction is 8 pounds.

8.14   The duration of treating a disease by an existing medication has a mean of 14 days. A drug company claims that a new medication can reduce the treatment time. To test this claim, the new medication is tried on 60 patients and their times to recovery are recorded. If the mean recovery time is 13.5 days and the standard deviation is 3 days in this sample, answer the following questions.

    a. Formulate the hypotheses and determine the rejection region of the test with a level of $\alpha = 0.05$.

    b. What is your decision?

    c. Repeat the above test using the p-value.

    d. Using $\sigma = 3$, find the type II error probability of the test for the alternative $\mu' = 13$.

8.15   Use the context of the z test for the mean to a normal population with known variance σ^2 to describe the effect of increasing the size n of a sample on each of the following:

    a. The p-value of a test, when H_0 is false and all facts about the population remain unchanged as n increases.

    b. The probability of a type II error of a level α test, when α, the alternative hypothesis, and all facts about the population remain unchanged.

8.16   In a watermelon farm, the average and standard deviation of the weights of 36 randomly selected sample of watermelons are 20 pounds and 2 pounds, respectively. Construct a 99% confidence interval of the population mean weight.

8.17   To estimate the average starting annual income of engineers with college degrees, an investigator collected income data from a random sample of 60 engineers who graduated from college within a given year. The sample mean is $70,300, and the sample standard deviation is $6,900. Find the following confidence intervals of the mean annual income.
   a.   95% confidence interval
   b.   90% confidence interval

8.18   The punch strengths (in pounds of force) of world-class elite boxers ranging from flyweight to super heavyweight were known to have a mean of 780. A random sample of 32 top boxers was selected from boxing clubs in a certain country, and each one was given a punch strength test. The sample mean was 758, and the sample standard deviation was 38. Are the top boxers in this country below the elite level based on the mean punch strength? Test at the level of $\alpha = 0.05$.

8.19   A sample of 35 speed guns is obtained for checking accuracy in the range between 50 and 60 miles per hour. The sample average of the errors is 1.3 miles per hour, and the sample standard deviation is 1.5 miles per hour.
   a.   Construct a 99% confidence interval for the population mean error.
   b.   Based on the confidence interval obtained in part (a), is there enough evidence that the error is significantly different from zero?

8.20   For the data given below, construct a 95% confidence interval for the population mean.

   53.4 51.6 48.0 49.8 52.8 51.8 48.8 43.4 48.2 51.8 54.6 53.8 54.6 49.6 47.2

8.21   A computer company claims that the batteries in its laptops last 4 hours on average. A consumer report firm gathered a sample of 16 batteries and conducted tests on this claim. The sample mean was 3 hours 50 minutes, and the sample standard deviation was 20 minutes. Assume that the battery time is distributed as normal.
   a.   Test if the average battery time is shorter than 4 hours at $\alpha = 0.05$.
   b.   Construct a 95% confidence interval of the mean battery time.
   c.   If you were to test $H_0: \mu = 240$ minutes versus $H_1: \mu \neq 240$ minutes, what would you conclude from your result in part (b)?
   d.   Suppose that a further study establishes that, in fact, the population mean is 4 hours. Did the test in part (c) make a correct decision? If not, what type of error did it make?

8.22 BMI (Body Mass Index) is obtained as weight (in kg) divided by the square of height (in m^2). Adults with BMI over 25 are considered overweight. A trainer at a health club measured the BMI of the people who registered for his program this week. Assume that the population is normal. The numbers are given below.

29.4 24.2 25.6 23.6 23.0 22.4 27.4 27.8

   a. Construct a 95% confidence interval for the mean BMI.
   b. To find if newly registered people for the program are overweight on average, conduct an appropriate test using $\alpha = 0.05$.
   c. Suppose that a further study establishes that, in fact, the population mean BMI is 25.5. What did the test in part (b) lead to? Was it a correct decision? If not, what type of error did this test make?

8.23 Construct the confidence interval in part (a) and conduct the test in part (b) of Exercise 8.22 using R.

8.24 A manufacturer wishes to set a standard time required by employees to complete a certain process. Times from 16 employees have a mean of 4 hours and a standard deviation of 1.5 hours.
   a. Test if the mean processing time exceeds 3.5 hours at the level of $\alpha = 0.05$. Assume normal population.
   b. Find the p-value.

8.25 System-action temperatures (in degrees F) of sprinkler systems used for fire protection from a random sample of size 13 from a normal population are given below.

130 131 131 132 127 127 128 130 131 131 130 129 128

The system has been designed so that it works when the temperature reaches $130°F$. Do the data fail to meet the manufacturer's goal? Test the relevant hypothesis at $\alpha = 0.05$ using the p-value.

8.26   The heights (in inches) of seventh-grade girl students in a school district on Long Island are known to be normally distributed. The heights of a random sample of 14 seventh-grade girl students in the school district are shown below.

57.8 64.8 61.7 59.8 62.4 68.7 57.2 57.1 63.0 66.2 58.0 62.6 57.0 60.2

a.   Conduct a hypothesis test using $\alpha = 0.01$ to show that the population mean height of the school district is lower than 63.0.
b.   Find the p-value of the test conducted in part (a).
c.   Suppose that a further study establishes that, in fact, the population mean is 61.5. Did the test in part (a) make a correct decision? If not, what type of error did it make?

8.27   A random sample of 18 US adult men had a mean lung capacity of 4.6 liters and a standard deviation of 0.3 liters.
a.   Conduct a 95% confidence interval for the average lung capacity for all US adult men.
b.   A medical researcher claims that the population mean lung capacity of US men is 4.5 liters. Based on this random sample, do you support this researcher's claim? Conduct a two-sided test with $\alpha = 0.1$. What is the p-value of the test?
c.   Test if the population mean lung capacity is greater than 4.5 liters with $\alpha = 0.1$. What is the p-value of the test?

8.28   Answer the questions in Exercise 8.25 and Exercise 8.26 (a) and (b) using R.

8.29   The monthly rents (in dollars) for two-bedroom apartments in a large city are obtained for a sample of 9 apartments.

1,860 1,725 1,950 2,025 2,160 1,650 2,220 2,370 1,830

a.   Conduct a 99% confidence interval for the mean monthly rent for two-bedroom apartments in this city.
b.   Do these data support the claim that the average monthly rent for two-bedroom apartments in this city is lower than $2,200? Test with $\alpha = 0.01$.

8.30   During a recent cold season, 42 people caught a cold out of a random sample of 500 people in a certain city. Find a 95% confidence interval for the rate of people who caught a cold in that city.

8.31 Construct the confidence interval in Exercise 8.30 using R.

8.32 A phone manufacturer wishes to estimate the proportion of people who want to purchase a cell phone which costs more than $800. Find the required sample size to yield a 90% confidence interval whose length is below 0.02.

8.33 A medical study is to be designed to estimate the population proportion p having a certain disease. How many people should be examined if the researcher wishes to be 95% certain that the error of estimation is below 0.1 when
   a. there is no prior knowledge about the value of p?
   b. the proportion is known to be about 0.2?

8.34 A study wishes to estimate the proportion of lung cancer deaths among all cancer deaths. How large a sample of cancer death records must be examined to be 99% certain that the estimate does not differ from the true proportion by more than 0.05?
   a. The American Cancer Society believes that the true proportion is 0.27. Use this information to find the required sample size.
   b. Find the sample size when there is no prior knowledge about the value of p.

8.35 In estimating a population proportion using a large sample, the estimate of p is 0.28, and its 95% error margin is 0.06.
   a. Find a 99% confidence interval for p.
   b. Find the sample size to meet the error margin.

8.36 The National Restaurant Association wishes to estimate the proportion of people who eat out at least twice a week. Find the required sample size to yield 95% confidence interval whose length is below 0.1.

8.37 A toxicologist wishes to investigate the relationship between the probability of a developmental defect and the level of trichlorophenoxyacetic acid in environment. Twenty mg/kg/day of trichlorophenoxyacetic acid is given to A/J strain mice during pregnancy, and the malformation rate of the fetuses is observed.
   a. How large a sample should be chosen to estimate the proportion with a 95% error margin of 0.01? Use $p = 0.2$.
   b. A random sample of 400 mice is taken, and 37 mice were found to have malformed fetuses. Construct a 95% confidence interval for the population proportion.

8.38   In a randomly selected sample of 100 students in a large college, 14 were left-handed.

    a.   Does this provide strong evidence that more than 10% of college students in America are left-handed? Test using $\alpha = 0.05$.

    b.   In part (a), what type of error might you have committed?

    c.   What is the type II error rate of the test conducted in part (a) if the true proportion of left-handers is 0.13 and a sample size of 100 is used?

    d.   How many college students are needed to test that the power $1 - \beta(0.13)$ is 80% for the test of part (a)?

8.39   In a certain industry, about 15 percent of the workers showed some signs of ill effects due to radiation. After management claimed that improvements had been made, 19 of 140 workers tested experienced some ill effects due to radiation.

    a.   Does this support the management's claim? Use $\alpha = 0.05$ to conduct the test.

    b.   Find the p-value of the test done in part (a).

8.40   Conduct the test in Exercise 8.39 using R.

8.41   A political party conducted an election poll for its presidential candidate. In a sample of 500 voters, 272 of them supported this candidate.

    a.   Determine the null and alternative hypotheses.

    b.   What does the test conclude? Use $\alpha = 0.05$.

    c.   Repeat part (b) using the p-value.

8.42   The spokesperson of a political party claims that 70% of New York residents are in favor of same-sex marriage. An opinion poll on same-sex marriage was conducted in New York State. A random selection of 1,180 adult New York State residents are asked if they support same-sex marriage. Suppose 802 people in this sample are in favor of same-sex marriage.

    a.   Do you have enough evidence to reject the party's claim at $\alpha = 0.01$?

    b.   Find the p-value of the test.

8.43   Conduct the test in Exercise 8.42 using R.

8.44   The Dean of the College of Engineering in a state university wishes to test if more than 70% of his students graduate college in four years.

a.   If 42 out of a random selection of 50 students graduated in four years, what does the test conclude? Test at $\alpha = 0.05$.

b.   Find the p-value of the test.

c.   Suppose a further study showed that in fact, 75% of the students in the college graduated in four years. Did the test in part (a) make a correct decision? If not, what type of error did the test make?

8.45   A rental car company decided to conduct a survey on customer satisfaction. Out of 734 customers who participated in the online survey, 342 rated the overall services as excellent.

a.   Test, at level $\alpha = 0.1$, the null hypothesis that the proportion of customers who would rate the overall car rental services as excellent is 0.5 versus a two-sided alternative.

b.   Find the p-value and comment on the strength of evidence.

8.46   In a random sample of 600 American adults, 276 of them want a domestic car for their next purchase. A US auto company wants to find out the percentage of the American adults who want to purchase domestic cars.

a.   Do you have sufficient evidence that less than a half of the Americans want to purchase domestic cars? Test using $\alpha = 0.01$.

b.   Find the p-value of the test.

c.   Suppose 45% of American adults purchased domestic cars the following year. Did your analysis lead to a (i) type I error (ii) type II error, or (iii) correct decision?

8.47   A random sample from a normal population is obtained, and the data are given below. Find a 95% confidence interval for σ.

294 302 338 348 380 400 406 420 438 440 458 476 478 496 500 516 540

8.48   A random sample of size 14 from a normal population gives the standard deviation of 3.27. Find a 90% confidence interval for σ^2.

8.49   A random sample of size 12 from a normal population gives $\bar{x} = 62.5$ and $s^2 = 1597$.

a.   Find a 99% confidence interval for μ.

b.   Find a 95% confidence interval for σ.

8.50   All fifth-grade students are given a test on academic achievement in New York State. Suppose the mean score is 70 for the entire state. A random sample of fifth-grade students is selected from Long Island. Below are the scores in this sample from a normal population.

82 94 66 87 68 85 68 84 70 83 65 70 83 71 82 72 73 81 76 74

    a.   Construct a 95% confidence interval for the population mean score on Long Island.
    b.   Construct a 90% confidence interval for the population standard deviation of the scores on Long Island.
    c.   A teacher at a Long Island high school claims that the mean score on Long Island is higher than the mean for New York State. Conduct a test to see if this claim is reasonable using $\alpha = 0.01$.
    d.   Find the p-value of the test.

8.51   SAT math scores of a random sample of 22 11th-grade students in a certain high school are given below. Assume the population is normally distributed.

460 620 500 770 440 510 510 440 530 550 720
410 550 580 410 580 590 600 400 610 370 660

    a.   The national mean is 511. Test if the mean math SAT score of the students in this school is higher than the national average using $\alpha = 0.05$.
    b.   Construct a 99% confidence interval for the population standard deviation of the scores for this school.

8.52   The weights (in pounds) of 30 randomly selected boy students from a high school are given below.

181 168 150 175 174 154 156 187 161 162 173 163 164 168 147
170 170 163 173 153 174 175 150 175 176 177 178 143 186 157

    a.   Find a 95% confidence interval for σ^2.
    b.   Based on the confidence interval from part (a), what would be your decision for the test $H_0: \sigma^2 = 100$ versus $H_1: \sigma^2 \neq 100$ at the level of $\alpha = 0.05$?
    c.   Find the p-value of the test.

8.53 A random sample of size 13 from a normal population is given below.

69 74 75 76 78 79 80 81 83 83 85 86 87

a. Someone claimed that $\sigma = 10$. Does this data set support the claim? Test at $\alpha = 0.05$.

b. Test if the population mean is 85 using $\alpha = 0.05$.

c. Find the p-value of the test.

8.54 Suppose the 90% confidence interval for σ^2 obtained from a random sample of size 15 from a normal population is (2.9555, 10.6535).

a. Find the sample variance.

b. Find a 95% confidence interval for σ^2.

Inferences Based on Two Samples

1. Inferences Concerning Two Population Means

In this section, we discuss confidence intervals and tests concerning the difference between two means of two different population distributions. Let $X_1, X_2, \cdots, X_m$ be a random sample from a population with mean μ_1 and variance σ_1^2, and $Y_1, Y_2, \cdots, Y_n$ be a random sample from a population with mean μ_2 and variance σ_2^2. The X and Y samples are independent of one another.

> **Setting:**
> 1. $X_1, X_2, \cdots, X_m$ is a random sample from population 1 with mean μ_1 and variance σ_1^2.
> 2. $Y_1, Y_2, \cdots, Y_n$ is a random sample from population 2 with mean μ_2 and variance σ_2^2.
> 3. The above two samples are independent of each other.
>
> Inferences will be made about the difference in means: $\mu_1 - \mu_2 = \Delta$.

A. LARGE, INDEPENDENT SAMPLES (BOTH $m, n \geq 30$)

a. Confidence interval

THEOREM 9.1 If $X_1, X_2, \cdots, X_m$ and $Y_1, Y_2, \cdots, Y_n$ satisfy the above condition, then

1. $E(\bar{X} - \bar{Y}) = \mu_1 - \mu_2$

2. $Var(\bar{X} - \bar{Y}) = \dfrac{\sigma_1^2}{m} + \dfrac{\sigma_2^2}{n}$

PROOF   By the properties of linear combinations of random variables discussed in Chapter 6,

$$E(\bar{X} - \bar{Y}) = E(\bar{X}) - E(\bar{Y}) = \mu_1 - \mu_2$$

and

$$Var(\bar{X} - \bar{Y}) = Var(\bar{X}) + Var(\bar{Y}) = \frac{\sigma_1^2}{n_1} + \frac{\sigma_2^2}{n_2}$$

By the central limit theorem,

$$Z = \frac{(\bar{X} - \bar{Y}) - (\mu_1 - \mu_2)}{\sqrt{\dfrac{\sigma_1^2}{m} + \dfrac{\sigma_2^2}{n}}} \tag{9-1}$$

approaches standard normal as m and n increase. In real problems, however, the population variances are unknown in most of the cases. Therefore, they are replaced with the sample variances for inferences. Throughout this chapter, we assume the variances are unknown.

Statistic (9-1) can be used to approximate to the standard normal when the samples are large enough so that we can apply the central limit theorem and approximate σ_1 and σ_2 with s_1 and s_2, respectively, when m and n are both greater than or equal to 30. A $100(1 - \alpha)\%$ confidence interval of $\mu_1 - \mu_2$ can be constructed by the following approximation.

$$P\left(-z_{\alpha/2} < \frac{(\bar{X} - \bar{Y}) - (\mu_1 - \mu_2)}{\sqrt{\dfrac{S_1^2}{m} + \dfrac{S_2^2}{n}}} < z_{\alpha/2} \right) \approx 1 - \alpha$$

$$P\left(-z_{\alpha/2}\sqrt{\frac{S_1^2}{m} + \frac{S_2^2}{n}} < (\bar{X} - \bar{Y}) - (\mu_1 - \mu_2) < z_{\alpha/2}\sqrt{\frac{S_1^2}{m} + \frac{S_2^2}{n}} \right) \approx 1 - \alpha$$

$$P\left((\bar{X} - \bar{Y}) - z_{\alpha/2}\sqrt{\frac{S_1^2}{m} + \frac{S_2^2}{n}} < \mu_1 - \mu_2 < (\bar{X} - \bar{Y}) + z_{\alpha/2}\sqrt{\frac{S_1^2}{m} + \frac{S_2^2}{n}} \right) \approx 1 - \alpha$$

Therefore, we obtain the following confidence interval.

$$\left((\bar{x} - \bar{y}) - z_{\alpha/2}\sqrt{\frac{s_1^2}{m} + \frac{s_2^2}{n}} < \mu_1 - \mu_2 < (\bar{x} - \bar{y}) + z_{\alpha/2}\sqrt{\frac{s_1^2}{m} + \frac{s_2^2}{n}} \right)$$

For two independent populations, a $100(1 - \alpha)\%$ confidence interval for $\mu_1 - \mu_2$ is

$$\left((\bar{x} - \bar{y}) - z_{\alpha/2}\sqrt{\frac{s_1^2}{m} + \frac{s_2^2}{n}}, (\bar{x} - \bar{y}) + z_{\alpha/2}\sqrt{\frac{s_1^2}{m} + \frac{s_2^2}{n}} \right)$$

if $m \geq 30$ and $n \geq 30$.

EXAMPLE 9.1 In June 2016 chemical analyses were made of 85 water samples (one litter each) taken from various parts of a city lake, and the measurements of chlorine content were recorded. During the next two winters, the use of road salt was substantially reduced in the catchment areas of the lake. In June 2018, 110 water samples were analyzed and their chlorine contents recorded. Calculation of the means and the standard deviations for the two sets of the data gives:

	Chlorine content (in ppm)	
	2016	2018
Mean	18.3	17.8
Standard Deviation	1.2	1.8

Construct a 95% confidence interval for the difference of the population means.

Since both $m = 85$ and $n = 110$ are large, we can use the following formula.

$$\bar{x} - \bar{y} \pm z_{\alpha/2}\sqrt{\frac{s_1^2}{m} + \frac{s_2^2}{n}} = 18.3 - 17.8 \pm z_{0.025}\sqrt{\frac{(1.2)^2}{85} + \frac{(1.8)^2}{110}} = 0.5 \pm 1.96(0.2154)$$

$$= 0.5 \pm 0.422$$

Thus, a 95% confidence interval for $\mu_1 - \mu_2$ is $(0.078, 0.922)$.

As discussed in Chapter 8, the confidence interval is the same as the acceptance region in a two-sided test. If 0 is not included in the confidence interval, we conclude that there is a significant difference between the two means. In the above example, 0 is not included in the confidence interval. Hence, we conclude that the means are significantly different at the level of $\alpha = 0.05$.

b. Testing hypothesis

To test about the difference of two means for large independent samples, we follow the procedure given in Table 9.1.

TABLE 9.1   Test about $\mu_1 - \mu_2$ for large independent samples

	Case (a)	Case (b)	Case (c)
Step 1	$H_0 : \mu_1 - \mu_2 = \Delta_0,$ $H_1 : \mu_1 - \mu_2 \neq \Delta_0$ (2-sided alternative)	$H_0 : \mu_1 - \mu_2 = \Delta_0,$ $H_1 : \mu_1 - \mu_2 > \Delta_0$ (1-sided alternative)	$H_0 : \mu_1 - \mu_2 = \Delta_0,$ $H_1 : \mu_1 - \mu_2 < \Delta_0$ (1-sided alternative)
Step 2	$\alpha = ?$	$\alpha = ?$	$\alpha = ?$
Step 3	$z = \dfrac{(\bar{x} - \bar{y}) - \Delta_0}{\sqrt{\dfrac{s_1^2}{m} + \dfrac{s_2^2}{n}}}$ Rejection region: $\lvert z \rvert \geq z_{\alpha/2}$	$z = \dfrac{(\bar{x} - \bar{y}) - \Delta_0}{\sqrt{\dfrac{s_1^2}{m} + \dfrac{s_2^2}{n}}}$ Rejection region: $z \geq z_\alpha$	$z = \dfrac{(\bar{x} - \bar{y}) - \Delta_0}{\sqrt{\dfrac{s_1^2}{m} + \dfrac{s_2^2}{n}}}$ Rejection region: $z \leq -z_\alpha$
Step 4	Substitute $\bar{x}$, $\bar{y}$, Δ_0, s_1^2, s_2^2, m, and n Calculate z Decision	Substitute $\bar{x}$, $\bar{y}$, Δ_0, s_1^2, s_2^2, m, and n Calculate z Decision	Substitute $\bar{x}$, $\bar{y}$, Δ_0, s_1^2, s_2^2, m and n Calculate z Decision
Step 5	$p = 2P(Z \geq \lvert z \rvert)$	$p = P(Z \geq z)$	$p = P(Z \leq z)$

EXAMPLE 9.2 From Example 9.1, do the data provide strong evidence that there is a reduction of average chlorine level in the lake water in 2018 compared to the level in 2016? Test with $\alpha = 0.05$.

$H_0 : \mu_1 = \mu_2$, $H_1 : \mu_1 > \mu_2$

$\alpha = 0.05$

The test statistic is

$$Z = \frac{\bar{X} - \bar{Y}}{\sqrt{\dfrac{S_1^2}{m} + \dfrac{S_2^2}{n}}}$$

The rejection region is $z \geq z_\alpha = z_{0.05} = 1.645$.

$$z = \frac{18.3 - 17.8}{\sqrt{\dfrac{(1.2)^2}{85} + \dfrac{(1.8)^2}{110}}} = 2.32 > 1.645$$

We reject H_0. The data provide strong evidence that there is a reduction of average chlorine level in the lake water in 2018 compared to 2016.

p-value $= P(Z > 2.32) = 1 - \Phi(2.32) = 0.01$

B. SMALL, INDEPENDENT SAMPLES ($m < 30$ AND/OR $n < 30$)

If the sample sizes are small, the central limit theorem cannot be applied, and thus the normal approximation is not appropriate. For inferences in this case, we are required to assume that both populations are normal and independent of each other. We classify this into two possible cases: (1) The values of σ_1^2 and σ_2^2 are equal, and (2) the values of σ_1^2 and σ_2^2 are unequal.

   a.   When we assume $\sigma_1 = \sigma_2 = \sigma$

Because the two variances are assumed to be equal, it is estimated differently from the large sample case. A weighted average of the two sample variances is used to estimate the common population variance. This is called the pooled variance estimator. If $X_1, X_2, \cdots, X_m$

and $Y_1, Y_2, \cdots, Y_n$ are two independent normal random samples with equal variance from $N(\mu_1, \sigma^2)$ and $N(\mu_2, \sigma^2)$, respectively, we obtain the following results.

1. $E(\bar{X} - \bar{Y}) = E(\bar{X}) - E(\bar{Y}) = \mu_1 - \mu_2$

2. $Var(\bar{X} - \bar{Y}) = Var(\bar{X}) + Var(\bar{Y}) = \sigma^2 \left(\dfrac{1}{m} + \dfrac{1}{n} \right)$

$$Z = \frac{(\bar{X} - \bar{Y}) - (\mu_1 - \mu_2)}{\sigma \sqrt{\left(\dfrac{1}{m} + \dfrac{1}{n} \right)}} \sim N(0,1)$$

The unknown σ^2 is estimated by the following pooled estimator:

$$S_p^2 = \frac{(m-1)S_1^2 + (n-1)S_2^2}{m+n-2}$$

The pooled estimator of σ is $S_p = \sqrt{S_p^2}$. Under the above assumptions, the variable

$$T = \frac{(\bar{X} - \bar{Y}) - (\mu_1 - \mu_2)}{S_p \sqrt{\left(\dfrac{1}{m} + \dfrac{1}{n} \right)}}$$

has a t distribution with $m + n - 2$ degrees of freedom. Based on this, a $100(1 - \alpha)\%$ confidence interval for $\mu_1 - \mu_2$ can be obtained as follows.

For two independent normal populations with equal variance, a $100(1 - \alpha)\%$ confidence interval for $\mu_1 - \mu_2$ is

$$\left(\bar{x} - \bar{y} - t_{\frac{\alpha}{2}, m+n-2} S_p \sqrt{\frac{1}{m} + \frac{1}{n}}, \bar{x} - \bar{y} + t_{\frac{\alpha}{2}, m+n-2} S_p \sqrt{\frac{1}{m} + \frac{1}{n}} \right)$$

EXAMPLE 9.3 A random sample of 15 mothers with low socioeconomic status delivered babies whose average birth weight was 110 ounces and a sample standard deviation was 24 ounces, whereas a random sample of 76 mothers with medium socioeconomic status resulted in a sample average birth weight and sample standard deviation of 120 ounces and 22 ounces, respectively. Assume that the average birth weights are normally distributed in both populations. Find a 95% confidence interval of the difference between the true average birth weights of the two groups.

$$m = 15, \bar{x} = 110, \ s_1 = 24; \ n = 76, \ \bar{y} = 120, \ s_2 = 22$$

Because the sample standard deviations are close to each other, we can assume equal variance. Because the first sample is small, we need to construct a pooled t confidence interval. The pooled sample variance is

$$s_p^2 = \frac{(m-1)s_1^2 + (n-1)s_2^2}{m+n-2} = \frac{14(24)^2 + 75(22)^2}{15+76-2} = 498.5$$

Since $t_{\frac{\alpha}{2},m+n-2} = t_{0.025,89} = 1.99,$

$$\bar{x} - \bar{y} \pm t_{\frac{\alpha}{2},m+n-2} s_p \sqrt{\frac{1}{m} + \frac{1}{n}} = 110 - 120 \pm 1.99\sqrt{498.5}\sqrt{\frac{1}{15} + \frac{1}{76}} = -10 \pm 12.5$$

Thus, a 95% confidence interval for $\mu_1 - \mu_2$ is $(-22.5, 2.5)$.

To test about the difference of two means for small independent samples with equal variance, we follow the procedure given in Table 9.2.

TABLE 9.2   Test about $\mu_1 - \mu_2$ for small independent samples with equal variance

	Case (a)	Case (b)	Case (c)
Step 1	$H_0 : \mu_1 - \mu_2 = \Delta_0,$ $H_1 : \mu_1 - \mu_2 \neq \Delta_0$ (2-sided alternative)	$H_0 : \mu_1 - \mu_2 = \Delta_0,$ $H_1 : \mu_1 - \mu_2 > \Delta_0$ (1-sided alternative)	$H_0 : \mu_1 - \mu_2 = \Delta_0,$ $H_1 : \mu_1 - \mu_2 < \Delta_0$ (1-sided alternative)
Step 2	$\alpha = ?$	$\alpha = ?$	$\alpha = ?$
Step 3	$t = \dfrac{(\bar{x} - \bar{y}) - \Delta_0}{s_p \sqrt{\dfrac{1}{m} + \dfrac{1}{n}}}$ Rejection region: $\|t\| \geq t_{\frac{\alpha}{2}, m+n-2}$	$t = \dfrac{(\bar{x} - \bar{y}) - \Delta_0}{s_p \sqrt{\dfrac{1}{m} + \dfrac{1}{n}}}$ Rejection region: $t \geq t_{\alpha, m+n-2}$	$t = \dfrac{(\bar{x} - \bar{y}) - \Delta_0}{s_p \sqrt{\dfrac{1}{m} + \dfrac{1}{n}}}$ Rejection region: $t \leq -t_{\alpha, m+n-2}$
Step 4	Substitute $\bar{x}$, $\bar{y}$, Δ_0, s_p, m, and n Calculate t Decision	Substitute $\bar{x}$, $\bar{y}$, Δ_0, s_p, m, and n Calculate t Decision	Substitute $\bar{x}$, $\bar{y}$, Δ_0, s_p, m and n Calculate t Decision
Step 5	$p = 2P(T \geq \|t\|)$	$p = P(T \geq t)$	$p = P(T \leq t)$

EXAMPLE 9.4   The intelligence quotients (IQ's) of 16 students from one area of a city showed a mean of 107 and a standard deviation of 10, while the IQ's of 14 students from another area of the city showed a mean of 112 and a standard deviation of 8. The IQ scores are assumed to be normally distributed. Is there a significant difference between the IQs of the two groups at significance level of 0.01?

$$m = 16, \bar{x} = 107, s_1 = 10; \ n = 14, \bar{y} = 112, s_2 = 8$$

Because the two sample standard deviations are close to each other, we assume equal variance.

$$H_0 : \mu_1 = \mu_2, \ H_1 : \mu_1 \neq \mu_2$$

$$\alpha = 0.01$$

The pooled variance is

$$s_p^2 = \frac{(m-1)s_1^2 + (n-1)s_2^2}{m+n-2} = \frac{15(10)^2 + 13(8)^2}{16+14-2} = 83.3$$

The test statistic is

$$T = \frac{\bar{X} - \bar{Y}}{S_p\sqrt{\dfrac{1}{m} + \dfrac{1}{n}}}$$

The rejection region is $|t| \geq t_{\frac{\alpha}{2}, m+n-2} = t_{0.005, 28} = 2.763$.

$$t = \frac{107 - 112}{\sqrt{83.3}\sqrt{\dfrac{1}{16} + \dfrac{1}{14}}} = -1.496$$

Since $|t| = 1.496 < 2.763$, we fail to reject H_0. There is no significant difference between IQ's of the two groups at level 0.01.

$$t_{0.1, 28} = 1.313 < 1.496 < 1.701 = t_{0.05, 28}$$

Therefore, $0.1 = 2 \times 0.05 < p\text{-value} < 2 \times 0.1 = 0.2$.

Alternatively, the p-value can be obtained using R as follows:

>2*(1-pt(1.496, 28))

The output is given below.

0.1458411

This shows that the p-value is 0.146. Note that the p-value is obtained as $2[1 - P(T < 1.496)]$, where T is distributed as t with 28 degrees of freedom, because it is the sum of the left-tail and right-tail probabilities. The probability is equivalent to $2P(T < -1.496)$, which can be obtained using R as follows:

>2*pt(−1.496, 28)

and the answer is the same.

The test in Example 9.4 can be done using R. Suppose x and y are the observations in the two samples. Then the test can be done as follows:

>t.test(x, y, var.equal=T)

The output contains a 95% confidence interval for the mean difference, the value of the t statistic, and the p-value.

b.   When we assume $\sigma_1 \neq \sigma_2$

When the equal variance assumption is not met, we use an alternative test that is more powerful than the test we used for equal variance. Let $X_1, X_2, \cdots, X_m$ and $Y_1, Y_2, \cdots, Y_n$ be two independent normal random samples from $N(\mu_1, \sigma_1^2)$ and $N(\mu_2, \sigma_2^2)$, respectively. If the sample variances of $X_1, X_2, \cdots, X_m$ and $Y_1, Y_2, \cdots, Y_n$ are S_1^2 and S_2^2, respectively, then

$$T = \frac{\bar{X} - \bar{Y} - (\mu_1 - \mu_2)}{\sqrt{\dfrac{S_1^2}{m} + \dfrac{S_2^2}{n}}}$$

is approximately distributed as t with ν degrees of freedom, where

$$\nu = \frac{\left(\dfrac{s_1^2}{m} + \dfrac{s_2^2}{n}\right)^2}{\dfrac{(s_1^2/m)^2}{m-1} + \dfrac{(s_2^2/n)^2}{n-1}}$$

It is rounded down to the nearest integer. Based on this, a $100(1-\alpha)\%$ confidence interval for $\mu_1 - \mu_2$ can be obtained as follows.

For two independent normal populations with different variances, a $100(1-\alpha)\%$ confidence interval for $\mu_1 - \mu_2$ is

$$\left(\bar{x} - \bar{y} - t_{\frac{\alpha}{2}, \nu} \sqrt{\frac{s_1^2}{m} + \frac{s_2^2}{n}}, \; \bar{x} - \bar{y} + t_{\frac{\alpha}{2}, \nu} \sqrt{\frac{s_1^2}{m} + \frac{s_2^2}{n}} \right)$$

where

$$\nu = \frac{\left(\dfrac{s_1^2}{m} + \dfrac{s_2^2}{n} \right)^2}{\dfrac{(s_1^2/m)^2}{m-1} + \dfrac{(s_2^2/n)^2}{n-1}}$$

EXAMPLE 9.5 The breaking load (pound/inch width) for two kinds of fabrics are measured. A summary of the data is given below.

Fabric type	Sample size	Sample mean	Sample sd
Cotton	10	25.9	0.4
Triacetate	10	68.1	1.8

Assuming normality, find a 95% confidence interval for the difference between true average porosity for the two types of fabric.

Since $m = n = 10$, these are two small independent samples. Because the sample standard deviations are very different, we assume different variances.

$$\bar{x} = 25.9, \; \bar{y} = 68.1, \; s_1 = 0.4, \; s_2 = 1.8$$

The degrees of freedom is

$$\nu = \frac{\left(\dfrac{s_1^2}{m} + \dfrac{s_2^2}{n} \right)^2}{\dfrac{(s_1^2/m)^2}{m-1} + \dfrac{(s_2^2/n)^2}{n-1}} = \frac{\left(\dfrac{(0.4)^2}{10} + \dfrac{(1.8)^2}{10} \right)^2}{\dfrac{((0.4)^2/10)^2}{9} + \dfrac{((1.8)^2/10)^2}{9}} = 9.89 \rightarrow \nu = 9$$

$$\bar{x} - \bar{y} \pm t_{0.025,9} \sqrt{\frac{s_1^2}{m} + \frac{s_2^2}{n}} = 25.9 - 68.1 \pm 2.262 \sqrt{\frac{(0.4)^2}{10} + \frac{(1.8)^2}{10}} = -42.2 \pm 1.32$$

Thus, a 95% CI is $(-43.52, -40.88)$.

To test about the difference of two means for small independent samples with unequal variances, we follow the procedure given in Table 9.3.

TABLE 9.3   Test about $\mu_1 - \mu_2$ for small independent samples with different variances

	Case (a)	Case (b)	Case (c)
Step 1	$H_0 : \mu_1 - \mu_2 = \Delta_0,$ $H_1 : \mu_1 - \mu_2 \neq \Delta_0$ (2-sided alternative)	$H_0 : \mu_1 - \mu_2 = \Delta_0,$ $H_1 : \mu_1 - \mu_2 > \Delta_0$ (1-sided alternative)	$H_0 : \mu_1 - \mu_2 = \Delta_0,$ $H_1 : \mu_1 - \mu_2 < \Delta_0$ (1-sided alternative)
Step 2	$\alpha = ?$	$\alpha = ?$	$\alpha = ?$
Step 3	$t = \dfrac{(\bar{x} - \bar{y}) - \Delta_0}{\sqrt{\dfrac{s_1^2}{m} + \dfrac{s_2^2}{n}}}$ Rejection region: $\|t\| \geq t_{\frac{\alpha}{2}, v}$ $v = \dfrac{\left(\dfrac{s_1^2}{m} + \dfrac{s_2^2}{n}\right)^2}{\dfrac{(s_1^2/m)^2}{m-1} + \dfrac{(s_2^2/n)^2}{n-1}}$	$t = \dfrac{(\bar{x} - \bar{y}) - \Delta_0}{\sqrt{\dfrac{s_1^2}{m} + \dfrac{s_2^2}{n}}}$ Rejection region: $t \geq t_{\alpha, v}$ $v = \dfrac{\left(\dfrac{s_1^2}{m} + \dfrac{s_2^2}{n}\right)^2}{\dfrac{(s_1^2/m)^2}{m-1} + \dfrac{(s_2^2/n)^2}{n-1}}$	$t = \dfrac{(\bar{x} - \bar{y}) - \Delta_0}{\sqrt{\dfrac{s_1^2}{m} + \dfrac{s_2^2}{n}}}$ Rejection region: $t \leq -t_{\alpha, v}$ $v = \dfrac{\left(\dfrac{s_1^2}{m} + \dfrac{s_2^2}{n}\right)^2}{\dfrac{(s_1^2/m)^2}{m-1} + \dfrac{(s_2^2/n)^2}{n-1}}$
Step 4	Substitute $\bar{x}, \bar{y}, \Delta_0, s_1^2, s_2^2,$ m, and n Calculate t Decision	Substitute $\bar{x}, \bar{y}, \Delta_0, s_1^2, s_2^2,$ m, and n Calculate t Decision	Substitute $\bar{x}, \bar{y}, \Delta_0, s_1^2, s_2^2,$ m, and n Calculate t Decision
Step 5	$p = 2P(T \geq \|t\|)$	$p = P(T \geq t)$	$p = P(T \leq t)$

EXAMPLE 9.6 The lengths of possession (in months) by first owner were observed for mid-size sedan models from two auto companies. The company making model B cars claims that the owners keep its cars longer than the cars of model A. The data are given below.

Model A	55	54	53	56	50	63	65	62	64	55
Model B	60	67	67	65	62	58	58	58		

Assume unequal variances and carry out a test to see whether the data support this conclusion using $\alpha = 0.05$.

Since $m = 10$ and $n = 8$, these are two small independent samples.

$$\bar{x} = 57.70, \ \bar{y} = 61.88, \ s_1 = 5.29, \ s_2 = 3.98$$

$H_0 : \mu_1 - \mu_2 = 0, \ H_1 : \mu_1 - \mu_2 < 0$

$\alpha = 0.05$

$$\nu = \frac{\left(\dfrac{s_1^2}{m} + \dfrac{s_2^2}{n} \right)^2}{\dfrac{(s_1^2/m)^2}{m-1} + \dfrac{(s_2^2/n)^2}{n-1}} = \frac{\left(\dfrac{(5.29)^2}{10} + \dfrac{(3.98)^2}{8} \right)^2}{\dfrac{((5.29)^2/10)^2}{9} + \dfrac{((3.98)^2/8)^2}{7}} = 15.97 \rightarrow \nu = 15$$

The test statistic is

$$T = \frac{\bar{X} - \bar{Y}}{\sqrt{\dfrac{S_1^2}{m} + \dfrac{S_2^2}{n}}}$$

The rejection region is $t < -t_{\alpha,\nu} = -t_{0.05,15} = -1.753$.

$$t = \frac{57.70 - 61.88}{\sqrt{\dfrac{(5.29)^2}{10} + \dfrac{(3.98)^2}{8}}} = -1.912 < -1.753$$

We reject H_0. The data provide strong evidence that owners keep cars of Model B longer than cars of Model A at level 0.05.

Since $t_{0.05, 15} = 1.753 < 1.912 < 2.131 = t_{0.025, 15}$,

$$0.025 < p\text{-value} < 0.05$$

Alternatively, the p-value can be obtained using R as follows:

>pt(−1.912, 15)

The output is given below.

0.03758497

This shows that the p-value is 0.038.

The test in Example 9.6 can be done using R as follows:

>x=c(55, 54, 53, 56, 50, 63, 65, 62, 64, 55)

>y=c(60, 67, 67, 65, 62, 58, 58, 58)

>t.test(x, y, alt="less",var.equal=F)

The 95% confidence interval found in Example 9.5 can be obtained using R as follows:

>t.test(x, y, var.equal=F)

C. PAIRED DATA

In comparative studies, a treatment is applied to an object. For example, in a study of the effectiveness of physical exercise in weight reduction, a group of people are engaged in a prescribed program of physical exercise for a certain period. The weight of each person is measured before and after the program. In this problem, the two measurements for each person are not independent. In this case, instead of considering all x_i and y_i, we consider the difference $d_i = x_i - y_i$, $i = 1, 2, \cdots, n$. Hence, we have only one sample of size n, sample mean $\bar{d}$, and sample standard deviation s_d. Thus, the problem reduces to an inference based on one sample. We can consider both large sample and small sample inferences, but the sample size is small in most of the real problems because the experiments are expensive and time consuming.

> For paired data, a $100(1-\alpha)\%$ confidence interval for the mean difference δ is
>
> $$\left(\bar{d} - t_{\frac{\alpha}{2},\,n-1}\frac{s_d}{\sqrt{n}}, \; \bar{d} + t_{\alpha/2,\,n-1}\frac{s_d}{\sqrt{n}}\right) \text{ for } n < 30.$$

EXAMPLE 9.7 Two different methods for determining chlorine content were used on samples of Cl_2-demand-free water for various doses and contact times. Observations shown in the following table are in mg/L.

	Subject							
	1	2	3	4	5	6	7	8
Method I	0.39	0.84	1.76	3.35	4.69	7.70	10.52	10.92
Method II	0.36	1.35	2.56	3.92	5.35	8.33	10.70	10.91

a. Construct a 99% confidence interval for the difference in true average residual chlorine readings between the two methods.

	Subject							
	1	2	3	4	5	6	7	8
Method I	0.39	0.84	1.76	3.35	4.69	7.70	10.52	10.92
Method II	0.36	1.35	2.56	3.92	5.35	8.33	10.70	10.91
d	0.03	−0.51	−0.80	−0.57	−0.66	−0.63	−0.18	0.01

$$\bar{d} = -0.414, \; s_d = 0.321$$

$$\bar{d} \pm t_{\frac{\alpha}{2},\,n-1}\frac{s_d}{\sqrt{n}} = \bar{d} \pm t_{0.005,7}\frac{s_d}{\sqrt{n}} = -0.414 \pm 3.499\frac{0.321}{\sqrt{8}} = -0.414 \pm 0.397$$

A 99% confidence interval is $(-0.811, -0.017)$.

b. Based on the confidence interval obtained in part (a), can you say that there is a significant difference (at level $\alpha = 0.01$) in true average chlorine readings between the two methods?

We can answer this question without conducting a hypothesis test. Since the confidence interval obtained in part (a) does not include 0, there is a significant difference.

For paired data, we follow the procedure given in Table 9.4 for testing about the mean difference. The table provides a test for both large sample and small sample, but we will focus on small sample because it is common.

TABLE 9.4   Test about the mean difference $\delta = \mu_1 - \mu_2$ for paired data

	Case (a)	Case (b)	Case (c)
Step 1	$H_0 : \delta = 0,\ H_1 : \delta \neq 0$	$H_0 : \delta = 0,\ H_1 : \delta > 0$	$H_0 : \delta = 0,\ H_1 : \delta < 0$
	(2-sided alternative)	(1-sided alternative)	(1-sided alternative)
Step 2	$\alpha = ?$	$\alpha = ?$	$\alpha = ?$

i. If $n < 30$, conduct the t test with $n - 1$ degrees of freedom.

	Case (a)	Case (b)	Case (c)
Step 3	$t = \dfrac{\bar{d}}{s_d/\sqrt{n}}$	$t = \dfrac{\bar{d}}{s_d/\sqrt{n}}$	$t = \dfrac{\bar{d}}{s_d/\sqrt{n}}$
	Rejection region:	Rejection region:	Rejection region:
	$|t| \geq t_{\alpha/2}$	$t \geq t_\alpha$	$t \leq -t_\alpha$
Step 4	Substitute $\bar{d}$, s_d, and n	Substitute $\bar{d}$, s_d, and n	Substitute $\bar{d}$, s_d, and n
	Calculate t	Calculate t	Calculate t
	Decision	Decision	Decision
Step 5	$p = 2P(T \geq |t|)$	$p = P(T \geq t)$	$p = P(T \leq t)$

ii.   If $n \geq 30$, conduct the z test.

Step 3	$z = \dfrac{\bar{d}}{s_d/\sqrt{n}}$	$z = \dfrac{\bar{d}}{s_d/\sqrt{n}}$	$z = \dfrac{\bar{d}}{s_d/\sqrt{n}}$
	Rejection region:	Rejection region:	Rejection region:
	$\lvert z \rvert \geq z_{\alpha/2}$	$z \geq z_{\alpha}$	$z \leq -z_{\alpha}$
Step 4	Substitute $\bar{d}$, s_d, and n	Substitute $\bar{d}$, s_d, and n	Substitute $\bar{d}$, s_d, and n
	Calculate z	Calculate z	Calculate z
	Decision	Decision	Decision
Step 5	$p = 2P(Z \geq \lvert z \rvert)$	$p = P(Z \geq z)$	$p = P(Z \leq z)$

EXAMPLE 9.8   A test is designed to compare the wearing qualities of two brands of motorcycle tires. One tire of each brand is placed on each of six motorcycles that are driven for a specified mileage. The two tires are then exchanged (front and rear) and driven for an equal number of miles. As a conclusion of this test, wear is measured (in thousandths of an inch). The data are given here:

	Motorcycle					
	1	2	3	4	5	6
Brand A	98	61	38	117	88	109
Brand B	102	60	46	125	94	111
d	−4	1	−8	−8	−6	−2

Is there a significant difference of wearing between the two brands? Use $\alpha = 0.05$.

$H_0 : \delta = 0$, $H_1 : \delta \neq 0$

$\alpha = 0.05$

The test statistic is

$$T = \frac{\bar{D}}{S_d/\sqrt{n}}$$

$$n = 6, \bar{d} = -4.5, s_d = 3.56$$

The rejection region is $|t| \geq t_{\alpha/2,\,n-1} = t_{0.025,5} = 2.571$.

$$t = \frac{-4.5}{3.56/\sqrt{6}} = -3.096$$

Since $|t| = 3.096 > 2.571$, we reject H_0 and conclude that the difference of wearing between the two brands is significant.

From Table A.4, $t_{0.025,5} = 2.571 < 3.096 < 3.365 = t_{0.01,5}$.

Thus, $0.01 \times 2 = 0.02 < p\text{-value} < 0.05 = 0.025 \times 2$.

Alternatively, the p-value can be obtained using R as follows:

>2*pt(−3.096, 5)

The output is given below.

0.02697522

This shows that the p-value is 0.027. Note that the p-value is obtained as $2P(T < -3.096)$, where T is distributed as t with 5 degrees of freedom, because it is a two-tailed probability.

The test in Example 9.8 can be done using R in two ways. Suppose x contains the first observations and y contains the second observations. The first method is that one sample t test is done to $x-y$ as given below.

>x−y

>t.test(x)

Alternatively, a paired t test can be done by

>t.test(x, y, paired=T)

To help identify an appropriate method for inferences concerning two population means, a flowchart is given in Figure 9.1.

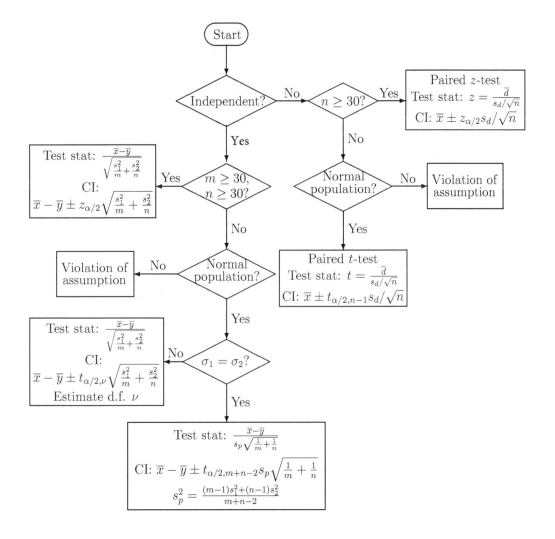

FIGURE 9.1   Flowchart for searching an appropriate method for inferences concerning two population means.

2. *Inferences Concerning Two Population Proportions*

When we compare the cure rates of two medications for the same disease or when we compare the consumer response to two different products (or in other similar situations), we are interested in testing whether two binomial populations have the same proportion. Let $X \sim \text{Bin}(m, p_1)$ and $Y \sim \text{Bin}(n, p_2)$ such that X and Y are independent random variables. Then as we observed in Chapter 6, $\hat{p}_1 = x/m$ and $\hat{p}_2 = y/n$ are unbiased estimators of p_1 and p_2, respectively. The estimator $\hat{p}_1 - \hat{p}_2$ can be used for estimating $p_1 - p_2$. The variance of $\hat{p}_1 - \hat{p}_2$ is

$$Var\left(\frac{X}{m} - \frac{Y}{n}\right) = Var\left(\frac{X}{m}\right) + Var\left(\frac{Y}{n}\right) = \frac{1}{m^2}Var(X) + \frac{1}{n^2}Var(Y) = \frac{p_1(1-p_1)}{m} + \frac{p_2(1-p_2)}{n}$$

By the central limit theorem,

$$Z = \frac{\hat{p}_1 - \hat{p}_2 - (p_1 - p_2)}{\sqrt{\dfrac{\hat{p}_1(1-\hat{p}_1)}{m} + \dfrac{\hat{p}_2(1-\hat{p}_2)}{n}}}$$

approaches standard normal.

A $100(1-\alpha)\%$ confidence interval for $p_1 - p_2$ can be constructed by the following approximation:

$$P\left(-z_{\alpha/2} < \frac{\hat{p}_1 - \hat{p}_2 - (p_1 - p_2)}{\sqrt{\dfrac{p_1(1-p_1)}{m} + \dfrac{p_2(1-p_2)}{n}}} < z_{\alpha/2}\right) \approx 1 - \alpha$$

This can be written as

$$P\left(-z_{\alpha/2}\sqrt{\frac{p_1(1-p_1)}{m} + \frac{p_2(1-p_2)}{n}} < \hat{p}_1 - \hat{p}_2 - (p_1 - p_2) < z_{\alpha/2}\sqrt{\frac{p_1(1-p_1)}{m} + \frac{p_2(1-p_2)}{n}}\right)$$

$$\approx 1 - \alpha$$

and this can be written as

$$P\left(\hat{p}_1 - \hat{p}_2 - z_{\frac{\alpha}{2}} \sqrt{\frac{p_1(1-p_1)}{m} + \frac{p_2(1-p_2)}{n}} < p_1 - p_2 < \hat{p}_1 - \hat{p}_2 + z_{\alpha/2} \sqrt{\frac{p_1(1-p_1)}{m} + \frac{p_2(1-p_2)}{n}} \right)$$

$$\approx 1 - \alpha$$

Therefore, we can construct the following confidence interval.

A $100(1-\alpha)\%$ confidence interval for $p_1 - p_2$ is

$$\left(\hat{p}_1 - \hat{p}_2 - z_{\frac{\alpha}{2}} \sqrt{\frac{\hat{p}_1(1-\hat{p}_1)}{m} + \frac{\hat{p}_2(1-\hat{p}_2)}{n}}, \hat{p}_1 - \hat{p}_2 + z_{\alpha/2} \sqrt{\frac{\hat{p}_1(1-\hat{p}_1)}{m} + \frac{\hat{p}_2(1-\hat{p}_2)}{n}} \right)$$

where $\hat{p}_1 = x/m$ and $\hat{p}_2 = y/n$.

EXAMPLE 9.9 The summary data reporting incidence of lung cancers among smokers and nonsmokers in a certain area are given below.

	Smoker	Nonsmoker
Sample size	1548	3176
Number of lung cancers	42	24

Find a 99% confidence interval for $p_1 - p_2$.

From the above table, $m = 1548$, $x = 42$, $n = 3176$, $y = 24$.

Thus, $\hat{p}_1 = \dfrac{x}{m} = \dfrac{42}{1548} = 0.0271$ and $\hat{p}_2 = \dfrac{y}{n} = \dfrac{24}{3176} = 0.0076$.

$$\hat{p}_1 - \hat{p}_2 \pm z_{\frac{\alpha}{2}} \sqrt{\frac{\hat{p}_1(1-\hat{p}_1)}{m} + \frac{\hat{p}_2(1-\hat{p}_2)}{n}}$$

$$= 0.0271 - 0.0076 \pm z_{0.005} \sqrt{\frac{(0.0271)(0.9729)}{1548} + \frac{(0.0076)(0.9924)}{3176}}$$

$$= 0.0195 \pm 2.58 \cdot 0.0044 = 0.0195 \pm 0.0114$$

Therefore, a 99% confidence interval for $p_1 - p_2$ is $(0.0081, 0.0309)$.

In hypothesis testing, we assume $p_1 = p_2 = p$ under H_0 and the estimator

$$\hat{p} = \frac{x + y}{m + n} = \frac{m}{m + n}\hat{p}_1 + \frac{n}{m + n}\hat{p}_2$$

is used for the common proportion. We follow the procedure given in Table 9.5 for testing about the difference between two proportions.

TABLE 9.5   Test about $p_1 - p_2$ for two independent samples

	Case (a)	Case (b)	Case (c)
Step 1	$H_0 : p_1 - p_2 = 0,$ $H_1 : p_1 - p_2 \neq 0$ (2-sided alternative)	$H_0 : p_1 - p_2 = 0,$ $H_1 : p_1 - p_2 > 0$ (1-sided alternative)	$H_0 : p_1 - p_2 = 0,$ $H_1 : p_1 - p_2 < 0$ (1-sided alternative)
Step 2	$\alpha = ?$	$\alpha = ?$	$\alpha = ?$
Step 3	$z = \dfrac{\hat{p}_1 - \hat{p}_2}{\sqrt{\hat{p}(1-\hat{p})\left(\dfrac{1}{m}+\dfrac{1}{n}\right)}}$ Rejection region: $\lvert z \rvert \geq z_{\frac{\alpha}{2}}$	$z = \dfrac{\hat{p}_1 - \hat{p}_2}{\sqrt{\hat{p}(1-\hat{p})\left(\dfrac{1}{m}+\dfrac{1}{n}\right)}}$ Rejection region: $z \geq z_\alpha$	$z = \dfrac{\hat{p}_1 - \hat{p}_2}{\sqrt{\hat{p}(1-\hat{p})\left(\dfrac{1}{m}+\dfrac{1}{n}\right)}}$ Rejection region: $z \leq -z_\alpha$
Step 4	Substitute $\hat{p}_1, \hat{p}_2, \hat{p}, m,$ and n Calculate z Decision	Substitute $\hat{p}_1, \hat{p}_2, \hat{p}, m,$ and n Calculate z Decision	Substitute $\hat{p}_1, \hat{p}_2, \hat{p}, m,$ and n Calculate z Decision
Step 5	$p = 2P(Z \geq \lvert z \rvert)$	$p = P(Z \geq z)$	$p = P(Z \leq z)$

EXAMPLE 9.10 A certain washing machine manufacturer claims that the fraction p_1 of his washing machines that need repairs in the first five years of operation is less than the fraction p_2 of another brand. To test this claim, we observe 200 machines of each brand and find that 21 machines of his brand and

37 machines of the other brand need repairs. Do these data support the manufacturer's claim? Use $\alpha = 0.05$.

$H_0 : p_1 = p_2$, $H_1 : p_1 < p_2$

$\alpha = 0.05$

The test statistic is

$$Z = \frac{\hat{p}_1 - \hat{p}_2}{\sqrt{\hat{p}(1-\hat{p})\left(\dfrac{1}{m} + \dfrac{1}{n}\right)}}$$

Since $m = n = 200$, $x = 21$, $y = 37$,

$$\hat{p}_1 = \frac{x}{m} = \frac{21}{200} = 0.105, \hat{p}_2 = \frac{y}{n} = \frac{37}{200} = 0.185$$

and

$$\hat{p} = \frac{x+y}{m+n} = \frac{21+37}{200+200} = 0.145$$

The rejection region is $z \le -z_\alpha = -z_{0.05} = -1.645$.

$$z = \frac{0.105 - 0.185}{\sqrt{(0.145)(0.855)\left(\dfrac{1}{200} + \dfrac{1}{200}\right)}} = -2.27 < -1.645$$

We reject H_0. These data support the manufacturer's claim.

p-value $= P(Z < -2.27) = 0.012$

The above test can be conducted using R as

>prop.test(x=c(21, 37), n=c(200, 200), alt="less", correct=F)

The output is given below.

2-sample test for equality of proportions without continuity correction

data: c(21, 37) out of c(200, 200)

X-squared = 5.1623, df = 1, p-value = 0.01154

alternative hypothesis: less

95 percent confidence interval:

 -1.0000000 -0.0224595

sample estimates:

prop 1 prop 2

This shows that the p-value is 0.01154. The outcome includes a 95% confidence interval for the difference between the two proportions as (-1.000, -0.022).

3. *Inferences Concerning Two Population Variances*

If independent random samples of size m and n are taken from normal populations having the same variance, it follows from Theorem 6.6 that $F = S_1^2/S_2^2$, the ratio of the two sample variances, has the F distribution with $m - 1$ and $n - 1$ degrees of freedom. If we take the reciprocal of F, then

$$\frac{1}{F} = \frac{S_2^2}{S_1^2}$$

has the F distribution with $n - 1$ and $m - 1$ degrees of freedom. Define F_{α, v_1, v_2} as the $100(1 - \alpha)$-th percentile of F_{v_1, v_2}. Then as discussed in Chapter 6, we can find the following relationship:

$$F_{1-\alpha, v_1, v_2} = \frac{1}{F_{\alpha, v_2, v_1}}$$

Since the variance is a measure of dispersion, the ratio of two variances is more meaningful than the difference of the two in comparison. Let $X_1, X_2, \cdots, X_m$ be a random sample from a population with mean μ_1 and variance σ_1^2, and $Y_1, Y_2, \cdots, Y_n$ be a random sample from a population with mean μ_2 and variance σ_2^2. The X and Y samples are independent of one another. Then

$$\frac{S_1^2/\sigma_1^2}{S_2^2/\sigma_2^2} \sim F_{m-1,\, n-1}$$

Thus, a $100(1-\alpha)\%$ confidence interval for σ_1^2/σ_2^2 can be constructed as follows.

$$P\left(F_{1-\frac{\alpha}{2},\, m-1,\, n-1} < \frac{S_1^2/\sigma_1^2}{S_2^2/\sigma_2^2} < F_{\frac{\alpha}{2},\, m-1,\, n-1} \right) = 1 - \alpha$$

and this can be written as

$$P\left(\frac{S_1^2/S_2^2}{F_{\frac{\alpha}{2},\, m-1,\, n-1}} < \frac{\sigma_1^2}{\sigma_2^2} < \frac{S_1^2/S_2^2}{F_{1-\frac{\alpha}{2},\, m-1,\, n-1}} \right) = 1 - \alpha$$

As discussed above, this probability is equivalent to the following probability:

$$P\left(\frac{S_1^2/S_2^2}{F_{\frac{\alpha}{2},\, m-1,\, n-1}} < \frac{\sigma_1^2}{\sigma_2^2} < \frac{S_1^2}{S_2^2} F_{\frac{\alpha}{2},\, n-1,\, m-1} \right) = 1 - \alpha$$

Accordingly, we have the following result.

A $100(1-\alpha)\%$ confidence interval for σ_1^2/σ_2^2 is

$$\left(\frac{s_1^2/s_2^2}{F_{\frac{\alpha}{2},\, m-1,\, n-1}} ,\; \frac{s_1^2}{s_2^2} F_{\frac{\alpha}{2},\, n-1,\, m-1} \right)$$

For two variances, we test for the ratio of the two. Most of the time, we test for equal variance. Table 9.6 shows the procedure of the test.

TABLE 9.6   Test for comparing two variances for independent samples

	Case (a)	Case (b)	Case (c)
Step 1	$H_0 : \sigma_1^2 = \sigma_2^2,$ $H_1 : \sigma_1^2 \neq \sigma_2^2$ (2-sided alternative)	$H_0 : \sigma_1^2 = \sigma_2^2,$ $H_1 : \sigma_1^2 > \sigma_2^2$ (1-sided alternative)	$H_0 : \sigma_1^2 = \sigma_2^2,$ $H_1 : \sigma_1^2 < \sigma_2^2$ (1-sided alternative)
Step 2	$\alpha = ?$	$\alpha = ?$	$\alpha = ?$
Step 3	$f = \dfrac{s_1^2}{s_2^2}$ Rejection region: $f \geq F_{\frac{\alpha}{2}, m-1, n-1}$ or $f \leq 1 / F_{\frac{\alpha}{2}, n-1, m-1}$	$f = \dfrac{s_1^2}{s_2^2}$ Rejection region: $f \geq F_{\alpha, m-1, n-1}$	$f = \dfrac{s_1^2}{s_2^2}$ Rejection region: $f \leq 1 / F_{\alpha, n-1, m-1}$
Step 4	Substitute s_1^2 and s_2^2 Calculate f Decision	Substitute s_1^2 and s_2^2 Calculate f Decision	Substitute s_1^2 and s_2^2 Calculate f Decision
Step 5	$p = 2\min\{P(F \geq f), \ P(F \leq f)\}$ where $F \sim F_{m-1, n-1}$	$p = P(F \geq f)$ where $F \sim F_{m-1, n-1}$	$p = P(F \leq f)$ where $F \sim F_{m-1, n-1}$

EXAMPLE 9.11 A survey was done on the delivery time for pizza orders. Consumers were randomly divided into group A of 25 people and group B of 25 people. The sample in group A ordered pizza from pizza chain A, and the sample in group B ordered pizza from pizza chain B. The sample standard deviations were 11 minutes for chain A and 7 minutes for chain B. Do these data support the claim that one chain is more consistent in delivery time than the other chain? Test using $\alpha = 0.01$.

$H_0 : \sigma_1^2 = \sigma_2^2, H_1 : \sigma_1^2 \neq \sigma_2^2$

$\alpha = 0.01$

The test statistic is

$$F = \frac{S_1^2}{S_2^2}$$

The rejection region is $f \geq F_{\frac{\alpha}{2}, m-1, n-1} = F_{0.005, 24, 24} = 2.967$

or $f \leq \dfrac{1}{F_{\frac{\alpha}{2}, n-1, m-1}} = \dfrac{1}{F_{0.005, 24, 24}} = 0.337.$

$$f = \frac{11^2}{9^2} = \frac{121}{81} = 1.494$$

Because $0.337 < 1.494 < 2.967$, we do not reject H_0. There is not enough evidence that the variances are different.

Using R, we obtain
>pf(1.494, 24, 24)
0.8339702

Since $P(F < 1.49) = 0.834$ and thus $P(F > 1.49) = 0.166$, the p-value is $2(0.166) = 0.332$.

EXAMPLE 9.12 For the data in Example 9.11, find a 90% confidence interval for the ratio between the two standard deviations.

A 90% confidence interval for σ_1^2 / σ_2^2 is

$$\left(\frac{s_1^2/s_2^2}{F_{\frac{\alpha}{2}, m-1, n-1}}, \frac{s_1^2}{s_2^2} F_{\frac{\alpha}{2}, n-1, m-1} \right) = \left(\frac{s_1^2/s_2^2}{F_{0.05, 24, 24}}, \frac{s_1^2}{s_2^2} F_{0.05, 24, 24} \right) = \left(\frac{121/81}{1.984}, \frac{121}{81}(1.984) \right)$$

$$= \left(0.753, 2.964 \right)$$

Hence, a 90% confidence interval for σ_1 / σ_2 is

$$\left(\sqrt{0.753}, \sqrt{2.964} \right) = (0.868, 1.722)$$

The test in Example 9.11 can be done using R. Suppose x and y are the observations in the two samples. Then the test can be done as follows:

>var.test(x, y, conf.level=0.99)

EXAMPLE 9.13 The effectiveness of two training methods is compared. A class of 24 students is randomly divided into two groups, and each group is taught according to a different method. Their test scores at the end of the semester show the following characteristics:

$$m = 13, \bar{x} = 74.5, s_1^2 = 82.6$$

and

$$n = 11, \bar{y} = 71.8, s_2^2 = 112.6$$

a. Assuming underlying normal distributions with $\sigma_1^2 = \sigma_2^2$, find a 95% confidence interval for $\mu_1 - \mu_2$.

$$s_p^2 = \frac{(m-1)s_1^2 + (n-1)s_2^2}{m+n-2} = \frac{12(82.6)+10(112.6)}{22} = 96.24$$

$$t_{\frac{\alpha}{2}, m+n-2} = t_{0.025,22} = 2.074$$

$$\bar{x} - \bar{y} \pm t_{\frac{\alpha}{2}, m+n-2} s_p \sqrt{\frac{1}{m} + \frac{1}{n}} = 74.5 - 71.8 \pm (2.074)\sqrt{96.24}\sqrt{\frac{1}{13} + \frac{1}{11}} = 2.7 \pm 8.335$$

A 95% confidence interval is $(-5.635, 11.035)$.

b. Find a 98% confidence interval for σ_1/σ_2.

A 98% confidence interval for σ_1^2/σ_2^2 is

$$\left(\frac{s_1^2/s_2^2}{F_{\frac{\alpha}{2}, m-1, n-1}}, \frac{s_1^2}{s_2^2} F_{\frac{\alpha}{2}, n-1, m-1} \right) = \left(\frac{s_1^2/s_2^2}{F_{0.01,12, 10}}, \frac{s_1^2}{s_2^2} F_{0.01,10, 12} \right) = \left(\frac{82.6}{(112.6)(4.71)}, \frac{82.6}{112.6}(4.30) \right)$$

$$= (0.156, 3.154)$$

Hence, a 98% confidence interval for σ_1/σ_2 is

$$(\sqrt{0.156}, \sqrt{3.154}) = (0.395, 1.776)$$

c. Does the assumption of $\sigma_1^2 = \sigma_2^2$ in part (a) seem justified?

Because 1 is included in the confidence interval of $(0.395, 1.776)$ obtained in part (b), the assumption seems to be justified.

SUMMARY OF CHAPTER 9

1. Inferences Concerning Two Population Means
    a. Setting
        i. $X_1, X_2, \cdots, X_m$ is a random sample with mean μ_1 and variance σ_1^2.
        ii. $Y_1, Y_2, \cdots, Y_n$ is a random sample with mean μ_2 and variance σ_2^2.
        iii. The above two samples are independent of each other.
        iv. Inferences are made about the difference in means: $\mu_1 - \mu_2$.
    b. Large, independent samples (both $m, n \geq 30$)
        i.   A $100(1 - \alpha)\%$ confidence interval:

$$\left((\bar{x} - \bar{y}) - z_{\frac{\alpha}{2}}\sqrt{\frac{s_1^2}{m} + \frac{s_2^2}{n}}, (\bar{x} - \bar{y}) + z_{\alpha/2}\sqrt{\frac{s_1^2}{m} + \frac{s_2^2}{n}} \right)$$

        ii. Hypothesis testing: See Table 9.1
    c. Small, independent samples ($m < 30$ and/or $n < 30$)
        i.   Assumption: normal populations
        ii. When we assume $\sigma_1 = \sigma_2 = \sigma$
            1. A $100(1 - \alpha)\%$ confidence interval:

$$\left(\bar{x} - \bar{y} - t_{\frac{\alpha}{2}, m+n-2} \, s_p\sqrt{\frac{1}{m} + \frac{1}{n}}, \bar{x} - \bar{y} + t_{\frac{\alpha}{2}, m+n-2} \, s_p\sqrt{\frac{1}{m} + \frac{1}{n}} \right)$$

            where $s_p^2 = \dfrac{(m-1)s_1^2 + (n-1)s_2^2}{m+n-2}$

  2. Hypothesis testing: See Table 9.2
iii. When we assume $\sigma_1 \neq \sigma_2$
   1. A $100(1-\alpha)\%$ confidence interval:

$$\left(\bar{x} - \bar{y} - t_{\frac{\alpha}{2}, v} \sqrt{\frac{s_1^2}{m} + \frac{s_2^2}{n}}, \ \bar{x} - \bar{y} + t_{\frac{\alpha}{2}, v} \sqrt{\frac{s_1^2}{m} + \frac{s_2^2}{n}} \right)$$

   where $v = \dfrac{\left(\dfrac{s_1^2}{m} + \dfrac{s_2^2}{n} \right)^2}{\dfrac{\left(s_1^2/n_1 \right)^2}{m-1} + \dfrac{\left(s_2^2/n_2 \right)^2}{n-1}}$ (round down to the nearest integer)

  2. Hypothesis testing: See Table 9.3
d. Paired data ($n < 30$)
  i. Use the difference d of two observations for each subject
  ii. A $100(1-\alpha)\%$ confidence interval:

$$\left(\bar{d} - t_{\frac{\alpha}{2}, n-1} \frac{s_d}{\sqrt{n}}, \ \bar{d} + t_{\alpha/2, n-1} \frac{s_d}{\sqrt{n}} \right) \text{ for } n < 30$$

  iii. Hypothesis testing: See Table 9.4
2. Inferences Concerning Two Population Proportions
  a. Setting
    i. $X \sim \text{Bin}(m, p_1)$ and $Y \sim \text{Bin}(n, p_2)$ are independent random variables.
    ii. $\hat{p}_1 = x / m$ and $\hat{p}_2 = y/n$
    iii. Inferences are made about the difference in means: $p_1 - p_2$.
  b. A $100(1-\alpha)\%$ confidence interval:

$$\left(\hat{p}_1 - \hat{p}_2 - z_{\frac{\alpha}{2}} \sqrt{\frac{\hat{p}_1(1-\hat{p}_1)}{m} + \frac{\hat{p}_2(1-\hat{p}_2)}{n}}, \hat{p}_1 - \hat{p}_2 + z_{\alpha/2} \sqrt{\frac{\hat{p}_1(1-\hat{p}_1)}{m} + \frac{\hat{p}_2(1-\hat{p}_2)}{n}} \right)$$

  c. Hypothesis testing: See Table 9.5
3. Inferences Concerning Two Population Variances
  a. Setting
    i. $X_1, X_2, \cdots, X_m$ is a random sample with mean μ_1 and variance σ_1^2.
    ii. $Y_1, Y_2, \cdots, Y_n$ is a random sample with mean μ_2 and variance σ_2^2.
    iii. The above two samples are independent of each other.
    iv. Inferences are made about the ratio of the variances: σ_1^2/σ_2^2

b. A $100(1 - \alpha)\%$ confidence interval:

$$\left(\frac{s_1^2 / s_2^2}{F_{\frac{\alpha}{2}, \ m-1, \ n-1}}, \ \frac{s_1^2}{s_2^2} F_{\frac{\alpha}{2}, \ n-1, \ m-1} \right)$$

c. Hypothesis testing: See Table 9.6

EXERCISES

9.1   Let $\bar{x}$ and $\bar{y}$ be the means of random samples of sizes $m = 12$ and $n = 16$ from the respective normal distributions $N(\mu_1, \sigma_1^2)$ and $N(\mu_2, \sigma_2^2)$, where it is known that $\sigma_1^2 = 24$ and $\sigma_2^2 = 19$.

a. Find $Var(\bar{X} - \bar{Y})$.
b. Give a test statistic suitable for testing $H_0 : \mu_1 = \mu_2$ versus $H_1 : \mu_1 < \mu_2$ and say how it may be used.
c. Give the rejection region of your test at level $\alpha = 0.05$.

9.2   A test is conducted to compare breaking strength of cell phones manufactured by two companies. Summary data are given below.

Company A: $m = 65, \bar{x} = 107$ pounds, $s_1 = 10$ pounds

Company B: $n = 60, \bar{y} = 113$ pounds, $s_2 = 13$ pounds

Construct a 95% confidence interval for the difference of means.

9.3   A manufacturer of paints (company A) claimed that the drying time of its paint is shorter than that of another company (company B). A market research firm conducted a test to find out if the manufacturer's claim is true. Paints produced by the two companies were randomly selected, and the drying times were measured. Summary data on drying time are given below.

Company A: $m = 45$, $\bar{x} = 63.5$ minutes, $s_1 = 5.4$ minutes
Company B: $n = 60$, $\bar{y} = 66.2$ minutes, $s_2 = 5.8$ minutes

Conduct a test to find out if the drying time of the paints made by company A is shorter. Use $\alpha = 0.01$.

9.4   A test is conducted to compare the tread wear of certain type of tires on highways paved with asphalt and highways paved with concrete. Road tests were conducted with the same type of tires on two types of highways. Summary data on the mileage of the tires up to a certain level of wear are given below.

	Sample size	Sample mean	Sample sd
Asphalt	35	29,700	9,700
Concrete	35	25,500	7,800

Does this information suggest that tires wear faster on concrete-paved highways than asphalt-paved highways? Test using $\alpha = 0.1$.

9.5   Total dissolved solids (TDS) level is a measure of water purity. Researchers measured the TDS level of water samples from random locations in each of two lakes. The data are given below.

	Sample size	Sample mean	Sample sd
Lake A	10	108	25
Lake B	9	97	21

Assume equal variance and construct a 99% confidence interval for the difference in mean TDS levels.

9.6 The fuel efficiency (in miles per gallon) of midsize sedans made by two auto companies is compared. Thirty test drivers are randomly divided into two groups of 15 people. One group of drivers took turn driving a car made by company A on the same route of a highway, and the other group of drivers did the same with a car made by company B. The company A car yielded an average gas mileage of 34.6 and a standard deviation of 2.5, whereas the company B car yielded an average gas mileage of 37.2 and a standard deviation of 2.2. Construct a 95% confidence interval for the difference in gas mileage between the two models. Assume equal variance.

9.7 During a study, researchers conducted an experiment to determine whether noise can result in weight loss. Twenty-seven adult mice of a certain strain were randomly allocated to two groups. Mice in one group were in cages with normal noise level, while the mice in the other group were in cages with excessive noise. After two weeks, the weight of each mouse was measured, with the results given in the table below.

	Sample size	Sample mean	Sample sd
Control	14	61.18	4.26
Noise	13	57.56	3.46

a. Do the data support the researcher's expectation? Assume equal variance and perform a test using $\alpha = 0.05$.
b. How would this conclusion change if $\alpha = 0.01$?
c. What type of error could possibly have been made in part (a)?

9.8 The effectiveness of two teaching methods is compared. A class of 26 students is randomly divided into two groups and each group is taught according to a different method. A summary of their test scores at the end of the session are given below.

Method 1: $m = 14$, $\bar{x} = 75.5$, $s_1^2 = 83.4$
Method 2: $n = 12$, $\bar{y} = 83.8$, $s_2^2 = 111.8$

Assuming underlying normal distributions with equal variance, answer the following questions:
a. Find a 99% confidence interval for the difference of the two means.
b. Find a 95% confidence interval for the difference of the two means.
c. At what significance level can you say that the difference is significant? Answer without conducting a hypothesis testing.

9.9   In a study on the healing process of bone fractures, researchers observed the time required to completely heal a bone fracture. Ten rats were randomly divided into two groups of five. No treatment was done to one group and a medical treatment was given to the other group. The time spent (in days) to completely heal the fracture for each rat is given below.

Control: 30, 22, 28, 35, 45
Treatment: 33, 40, 24, 25, 24

Test if the mean duration of healing can be reduced by the medical treatment at $\alpha = 0.1$. Assume equal variance.

9.10   Conduct the test in Exercise 9.9 using R.

9.11   A study has been done to test the effect of classical music on the brain. Twenty sixth-grade students were randomly divided into two groups of 10 students, and the same math test was given to these students. The first group of students listened to classical music for 20 minutes right before taking the test, and the other group of students took the test without listening to the music. Here are their scores.
Group with music: 91 77 58 89 83 78 74 81 91 88
Group without music: 81 65 69 69 67 61 67 87 64 81
Assume equal variance for performing the following calculations:
a.   Construct a 95% confidence interval for the mean difference.
b.   Test if the students who listened to classical music before the test scored higher than the other group of students using $\alpha = 0.05$.
c.   Conduct the same test as in part (b) using $\alpha = 0.01$.

9.12   Find the confidence interval and conduct the tests in Exercise 9.11 using R.

9.13   The following table shows summary data on mercury concentration in salmons (in ppm) from two different areas of the Pacific Ocean.

Area 1: $m = 15$, $\bar{x} = 0.0860$, $s_1 = 0.0032$
Area 2: $n = 15$, $\bar{y} = 0.0884$, $s_2 = 0.0028$

Do the data suggest that the mercury concentration is higher in salmons from area 2? Assume equal variance and use $\alpha = 0.05$.

9.14   The statewide math test scores of 16 students from one school district showed a mean of 68 and a standard deviation of 10, while the scores of 14 students from another school district showed a mean of 73 and a standard deviation of 8. Is there a significant difference between the mean scores of the two school districts at significance level of 0.05? Assume equal variance.

9.15   The following are the number of sales that a sample of 8 life insurance salespeople of an insurance company in New York and a sample of 7 salespeople in New Jersey made over a certain year.

New York: 49, 58, 34, 61, 53, 60, 59, 44
New Jersey: 40, 26, 52, 42, 36, 31, 38

Assuming that the populations sampled can be approximated closely with normal distributions having the same variance, is there a difference in the number of sales between the New York salespeople and the New Jersey salespeople? Conduct a hypothesis test at the significance level of 0.01.

9.16   Independent random samples are selected from two populations. The summary statistics are given below. Assume unequal variances for the following questions.

$$m = 5, \quad \bar{x} = 12.7, \quad s_1 = 3.2$$
$$n = 7, \quad \bar{y} = 9.9, \quad s_2 = 2.1$$

a.   Construct a 95% confidence interval for the difference of the means.
b.   Test if the means of the two populations are different based on the result from part (a).

9.17   Starting salaries (in thousand dollars) of recent college graduates in computer technology and automotive industries were compared. Summary statistics are given below.

Computer technology: $m = 22, \bar{x} = 55.7, \ s_1 = 12.2$
Automotive: $n = 20, \bar{y} = 52.8, s_2 = 3.7$

Assume unequal variances and perform the following calculations.
a.   Construct a 90% confidence interval for difference of the means.
b.   Test if the computer technology industry pays more than the automotive industry to recent college graduates. Use $\alpha = 0.1$.

9.18 A manufacturer of furniture claims that the company's new model of bookshelves is easier to assemble than the old model. To test the claim, 7 people were assigned to assemble the bookshelves from each of the two models. Here are the times (in minutes) they needed to assemble the bookshelves.

	Person						
	1	2	3	4	5	6	7
Old model	17	22	20	14	15	21	24
New model	15	19	18	13	15	19	23

a. Construct a 95% confidence interval for the difference in population means.
b. Test if the mean assembly time for the new model is shorter than that of the old model using $\alpha = 0.05$.
c. Consider the two samples are independent and conduct the test in part (b). Assume equal variance. Is the decision reversed? Why or why not?

9.19 Answer the questions in Exercise 9.18 using R.

9.20 A random sample of 14 men who participated in a weight-loss program measured their weight (in pounds) on the first day and at the end of the 10-week session. The data are given below.

	Subject													
	1	2	3	4	5	6	7	8	9	10	11	12	13	14
Day 1	180	222	240	192	216	210	252	228	222	204	216	228	204	204
Last day	132	162	186	174	198	156	174	168	192	150	210	180	138	210

a. Find a 99% confidence interval for the mean weight change.
b. Can you conclude that the program is effective? Use $\alpha = 0.01$.

9.21   Some people believe from their observations that the first child is more fluent than the second in their parents' native language among second-generation Chinese Americans. To test this claim, a Chinese comprehension test was given to 6 randomly selected second-generation siblings in Chinese American families. The scores are given below.

	Sibling					
	1	2	3	4	5	6
First child	43	27	57	53	71	72
Second child	42	24	39	41	61	61

Do the data support the above claim? Test using $\alpha = 0.01$.

9.22   A college prep program compared the practice SAT scores (math and reading combined) given before and after an eight-week instruction for each student. The scores are given below.

	Student								
	1	2	3	4	5	6	7	8	9
Before	1280	1200	1050	1190	1250	1290	1220	1270	1260
After	1380	1310	1090	1240	1290	1360	1270	1330	1310

a.   Test if the average score has been raised by 50 points using a paired t test with $\alpha = 0.05$.
b.   If you assume independence and conduct the test in the wrong way, do you obtain the same result as in part (a)?

9.23   A random sample of size 12 is selected from women with hypertension. For each person, systolic blood pressure was measured right before and one hour after taking the medicine. The mean difference of the blood pressure was −16.5, and the standard deviation of the difference was 10.7. Is there sufficient evidence to conclude that the hypertension medicine lowered blood pressure? Test using $\alpha = 0.05$.

9.24 To investigate if caffeine increases heart rate, a study was conducted with 8 male adults who volunteered. Heart rate per minute was measured on each person before and a few minutes after drinking 3 cups of coffee containing 300 mg of caffeine in total. The heart rates were measured while resting. The numbers are given below.

	Subject							
	1	2	3	4	5	6	7	8
Before	62	62	57	51	76	59	56	58
After	69	79	71	85	102	70	61	69

Do the data suggest that caffeine leads to a higher heart rate? Test using $\alpha = 0.01$.

9.25 An auto company claims that the fuel efficiency of its SUV has been substantially improved. A consumer advocate organization wishes to compare the fuel efficiencies of its SUV between the 2016 and 2019 models. Each of 12 drivers drove both the 2016 and 2019 models on a highway and the fuel efficiency was measured (in miles per gallon). The same driver drove the same path for both the models. The test results are given in the table below.

	Driver											
Model	1	2	3	4	5	6	7	8	9	10	11	12
2019	31.8	30.9	31.5	32.4	31.4	32.6	31.7	31.3	32.0	32.8	31.5	33.9
2016	28.9	32.0	30.6	30.7	31.8	32.3	31.3	32.5	29.5	30.9	31.0	31.9

Use the 0.05 level of significance to test the claim that the average fuel efficiency of the SVU has improved.

9.26 A certain refrigerator manufacturer claims that the fraction p_1 of his refrigerators that need repairs in the first 10 years of operation is less than the fraction p_2 of another brand. To test this claim, researchers observed 150 refrigerators from each brand and found that $x = 20$ and $y = 36$ refrigerators needed repairs. Do these data support the manufacturer's claim? Conduct a test using $\alpha = 0.05$.

9.27   Public surveys were conducted on environmental issues in 2017 and 2018. One of the questions on the survey was, "How serious do you think the atmospheric contamination by exhaust gas is?" In 2017, 420 out of 1,090 people surveyed said it is very serious, and in 2018, 1,063 out of 2,600 people surveyed said it is very serious. Find a 90% confidence interval for the difference between the two proportions.

9.28   A study is undertaken to compare the cure rates of certain lethal disease by drug A and drug B. Among 190 patients who took drug A, 100 were cured, and among 65 patients who took drug B, 55 were cured. Do the data provide strong evidence that the cure rate is different between the two drugs? Test using $\alpha = 0.01$.

9.29   In animal carcinogenicity research, the carcinogenic potential of drugs and other chemical substances used by humans are studied. Such bioassays are conducted in animals at doses that are generally well above human exposure levels, in order to detect carcinogenicity with a relatively small number of animals. Four hundred ppm of benzidine dihydrochloride is given to each of the male and female mice from a certain strain. In one of these experiments, tumors were found in 54 out of 484 male mice and 127 out of 429 female mice.

    a.   Find a 99% confidence interval for the difference between the tumor rates of male mice and female mice.

    b.   Test if the tumor rate of female mice is higher than the tumor rate of male mice using $\alpha = 0.01$.

9.30   Incidence rates of asthma from samples obtained from smokers and nonsmokers are given below.

	Occurrence of asthma	
Smoker	$m = 993$	$x = 209$
Nonsmoker	$n = 1012$	$y = 172$

    a.   Find a 95% confidence interval for the difference in asthma incidence rates between smokers and nonsmokers.

    b.   Does the confidence interval obtained in part (a) suggest that the asthma incidence rate is different between smokers and nonsmokers at the level of $\alpha = 0.05$?

    c.   Find a 99% confidence interval for the difference in asthma incidence rates between smokers and nonsmokers.

    d.   Does the confidence interval obtained in part (c) suggest that the asthma incidence rate is different between smokers and nonsmokers at the level of $\alpha = 0.01$?

    e.   Test if smokers have a higher incidence rate of asthma than nonsmokers using $\alpha = 0.05$.

9.31 A survey on computers requiring repairs within two years was conducted. Twenty-one out of 200 computers from company A and 37 out of 200 computers from company B needed repairs. Do these data show that computers from company A are more reliable than computers from company B? Test using $\alpha = 0.05$.

9.32 In a sociology course, one group of students had access to a set of lectures online in addition to the classroom lectures, and the other group of students did not have access to the online lectures. In each case an exam was given after the lectures. Here are the scores.

Classroom and online: 61 71 61 100 94
Classroom only: 64 61 56 53 54

a. Test if the variances of the scores are different between the two groups using $\alpha = 0.05$.
b. Do the data suggest that the online access improved the average test scores? Conduct a test using $\alpha = 0.05$ based on the result on equality of variances obtained in part (a).

9.33 In a US senatorial election, 9 voters were randomly selected from those who voted for a candidate from a conservative party, and 9 voters were randomly selected from those who voted for a candidate from a liberal party. Their ages are given below.

Voted for the candidate from a conservative party: 51, 76, 62, 55, 39, 43, 46, 49, 56
Voted for the candidate from a liberal party: 44, 62, 60, 51, 35, 41, 39, 39, 36

a. Test for equal variance between the two groups using $\alpha = 0.05$.
b. Based on the result obtained in part (a), construct a 90% confidence interval for the difference in mean age between the two groups of voters.
c. Do the data suggest that the candidate from the liberal party received votes from younger voters on average? Conduct a test using $\alpha = 0.1$ based on the result on equality of variances obtained in part (a).
d. Conduct the same test as in part (c) using $\alpha = 0.05$.

9.34 Answer Exercise 9.33 using R.

9.35 A feeding test is conducted on a random sample of 23 Rhode Island Red roosters. The chickens are randomly divided into two feeding groups. The first group contains 12 chickens and is fed type 1 diet, and the second group contains 11 chickens and is fed type 2 diet. The weights of the chickens are measured when they turn one year old. A summary of the weights (in pounds) is given below.

	Sample size	Sample mean	Sample sd
Type 1 diet	12	9.0	1.6
Type 2 diet	11	8.4	1.7

a. Assume that the population distributions are normal with equal variance. Do these data support that type 2 diet yields lower weight than type 1 diet? Test with $\alpha = 0.01$.
b. Find a 95% confidence interval for $\frac{\sigma_1}{\sigma_2}$.
c. Does the assumption of equal variance in part (a) seem justified?

9.36 The annual salaries (in US dollars) of supermarket clerks in two cities are given below.

	Sample size	Sample mean	Sample sd
City 1	10	47,000	1,200
City 2	10	49,000	2,600

a. Assuming underlying normal distributions with unequal variances, find a 95% confidence interval for the difference in population mean salaries between the two cities.
b. Based on the result from part (a), is there a significant difference in the mean salaries between the two cities?
c. Test if the mean salary of supermarket clerks in city 1 is lower than that in city 2 using $\alpha = 0.05$.
d. Does the assumption of unequal variance in part (a) seem justified? Test using $\alpha = 0.05$.

9.37 A market research firm studied the waiting time to order meals during lunchtime at two fast-food restaurant chains. For a sample of 28 locations of chain A, the sample standard deviation of waiting time was 263 seconds, and for a sample of 26 locations of chain B, the sample standard deviation was 421 seconds. Do these data suggest that the waiting time is more consistent in chain A? Test at $\alpha = 0.01$. What is the p-value?

9.38   Random samples are obtained from two normal populations with variances $\sigma_1^2 = 10$ and $\sigma_2^2 = 15$, respectively. If the sample sizes are 20 and 18, respectively, find the 90th percentile of the ratio of the sample variances $\frac{S_1^2}{S_2^2}$ as instructed below.

a.   Find the percentile using the distribution table.

b.   Find the percentile using R.

Appendix

Table A.1: Source: http://www.stat.ufl.edu/~athienit/Tables/tables.pdf.
Table A.2: Source: http://math.usask.ca/~laverty/S245/Tables/BinTables.pdf.
Table A.3: Source: https://mat.iitm.ac.in/home/vetri/public_html/statistics/poisson.pdf.
Table A.4: Source: http://www.stat.ufl.edu/~athienit/Tables/tables.pdf.
Table A.5: Source: http://www.stat.ufl.edu/~athienit/Tables/tables.pdf.
Table A.6: Source: http://www.stat.ufl.edu/~athienit/Tables/tables.pdf.

Table A.1: Standard Normal Distribution Table

Shaded area = $Pr(Z \leq z)$

z	0.00	0.01	0.02	0.03	0.04	0.05	0.06	0.07	0.08	0.09
− 3.4	0.0003	0.0003	0.0003	0.0003	0.0003	0.0003	0.0003	0.0003	0.0003	0.0002
− 3.3	0.0005	0.0005	0.0005	0.0004	0.0004	0.0004	0.0004	0.0004	0.0004	0.0003
− 3.2	0.0007	0.0007	0.0006	0.0006	0.0006	0.0006	0.0006	0.0005	0.0005	0.0005
− 3.1	0.0010	0.0009	0.0009	0.0009	0.0008	0.0008	0.0008	0.0008	0.0007	0.0007
− 3.0	0.0013	0.0013	0.0013	0.0012	0.0012	0.0011	0.0011	0.0011	0.0010	0.0010
− 2.9	0.0019	0.0018	0.0018	0.0017	0.0016	0.0016	0.0015	0.0015	0.0014	0.0014
− 2.8	0.0026	0.0025	0.0024	0.0023	0.0023	0.0022	0.0021	0.0021	0.0020	0.0019
− 2.7	0.0035	0.0034	0.0033	0.0032	0.0031	0.0030	0.0029	0.0028	0.0027	0.0026
− 2.6	0.0047	0.0045	0.0044	0.0043	0.0041	0.0040	0.0039	0.0038	0.0037	0.0036
− 2.5	0.0062	0.0060	0.0059	0.0057	0.0055	0.0054	0.0052	0.0051	0.0049	0.0048
− 2.4	0.0082	0.0080	0.0078	0.0075	0.0073	0.0071	0.0069	0.0068	0.0066	0.0064
− 2.3	0.0107	0.0104	0.0102	0.0099	0.0096	0.0094	0.0091	0.0089	0.0087	0.0084
− 2.2	0.0139	0.0136	0.0132	0.0129	0.0125	0.0122	0.0119	0.0116	0.0113	0.0110
− 2.1	0.0179	0.0174	0.0170	0.0166	0.0162	0.0158	0.0154	0.0150	0.0146	0.0143
− 2.0	0.0228	0.0222	0.0217	0.0212	0.0207	0.0202	0.0197	0.0192	0.0188	0.0183
− 1.9	0.0287	0.0281	0.0274	0.0268	0.0262	0.0256	0.0250	0.0244	0.0239	0.0233
− 1.8	0.0359	0.0351	0.0344	0.0336	0.0329	0.0322	0.0314	0.0307	0.0301	0.0294
− 1.7	0.0446	0.0436	0.0427	0.0418	0.0409	0.0401	0.0392	0.0384	0.0375	0.0367
− 1.6	0.0548	0.0537	0.0526	0.0516	0.0505	0.0495	0.0485	0.0475	0.0465	0.0455
− 1.5	0.0668	0.0655	0.0643	0.0630	0.0618	0.0606	0.0594	0.0582	0.0571	0.0559
− 1.4	0.0808	0.0793	0.0778	0.0764	0.0749	0.0735	0.0721	0.0708	0.0694	0.0681
− 1.3	0.0968	0.0951	0.0934	0.0918	0.0901	0.0885	0.0869	0.0853	0.0838	0.0823
− 1.2	0.1151	0.1131	0.1112	0.1093	0.1075	0.1056	0.1038	0.1020	0.1003	0.0985
− 1.1	0.1357	0.1335	0.1314	0.1292	0.1271	0.1251	0.1230	0.1210	0.1190	0.1170
− 1.0	0.1587	0.1562	0.1539	0.1515	0.1492	0.1469	0.1446	0.1423	0.1401	0.1379
− 0.9	0.1841	0.1814	0.1788	0.1762	0.1736	0.1711	0.1685	0.1660	0.1635	0.1611
− 0.8	0.2119	0.2090	0.2061	0.2033	0.2005	0.1977	0.1949	0.1922	0.1894	0.1867
− 0.7	0.2420	0.2389	0.2358	0.2327	0.2296	0.2266	0.2236	0.2206	0.2177	0.2148
− 0.6	0.2743	0.2709	0.2676	0.2643	0.2611	0.2578	0.2546	0.2514	0.2483	0.2451
− 0.5	0.3085	0.3050	0.3015	0.2981	0.2946	0.2912	0.2877	0.2843	0.2810	0.2776
− 0.4	0.3446	0.3409	0.3372	0.3336	0.3300	0.3264	0.3228	0.3192	0.3156	0.3121
− 0.3	0.3821	0.3783	0.3745	0.3707	0.3669	0.3632	0.3594	0.3557	0.3520	0.3483
− 0.2	0.4207	0.4168	0.4129	0.4090	0.4052	0.4013	0.3974	0.3936	0.3897	0.3859
− 0.1	0.4602	0.4562	0.4522	0.4483	0.4443	0.4404	0.4364	0.4325	0.4286	0.4247
− 0.0	0.5000	0.4960	0.4920	0.4880	0.4840	0.4801	0.4761	0.4721	0.4681	0.4641

z	Area
− 3.50	0.00023263
− 4.00	0.00003167
− 4.50	0.00000340
− 5.00	0.00000029

Source: Computed by M. Longnecker using Splus.

z	0.00	0.01	0.02	0.03	0.04	0.05	0.06	0.07	0.08	0.09
0.0	0.5000	0.5040	0.5080	0.5120	0.5160	0.5199	0.5239	0.5279	0.5319	0.5359
0.1	0.5398	0.5438	0.5478	0.5517	0.5557	0.5596	0.5636	0.5675	0.5714	0.5753
0.2	0.5793	0.5832	0.5871	0.5910	0.5948	0.5987	0.6026	0.6064	0.6103	0.6141
0.3	0.6179	0.6217	0.6255	0.6293	0.6331	0.6368	0.6406	0.6443	0.6480	0.6517
0.4	0.6554	0.6591	0.6628	0.6664	0.6700	0.6736	0.6772	0.6808	0.6844	0.6879
0.5	0.6915	0.6950	0.6985	0.7019	0.7054	0.7088	0.7123	0.7157	0.7190	0.7224
0.6	0.7257	0.7291	0.7324	0.7357	0.7389	0.7422	0.7454	0.7486	0.7517	0.7549
0.7	0.7580	0.7611	0.7642	0.7673	0.7704	0.7734	0.7764	0.7794	0.7823	0.7852
0.8	0.7881	0.7910	0.7939	0.7967	0.7995	0.8023	0.8051	0.8078	0.8106	0.8133
0.9	0.8159	0.8186	0.8212	0.8238	0.8264	0.8289	0.8315	0.8340	0.8365	0.8389
1.0	0.8413	0.8438	0.8461	0.8485	0.8508	0.8531	0.8554	0.8577	0.8599	0.8621
1.1	0.8643	0.8665	0.8686	0.8708	0.8729	0.8749	0.8770	0.8790	0.8810	0.8830
1.2	0.8849	0.8869	0.8888	0.8907	0.8925	0.8944	0.8962	0.8980	0.8997	0.9015
1.3	0.9032	0.9049	0.9066	0.9082	0.9099	0.9115	0.9131	0.9147	0.9162	0.9177
1.4	0.9192	0.9207	0.9222	0.9236	0.9251	0.9265	0.9279	0.9292	0.9306	0.9319
1.5	0.9332	0.9345	0.9357	0.9370	0.9382	0.9394	0.9406	0.9418	0.9429	0.9441
1.6	0.9452	0.9463	0.9474	0.9484	0.9495	0.9505	0.9515	0.9525	0.9535	0.9545
1.7	0.9554	0.9564	0.9573	0.9582	0.9591	0.9599	0.9608	0.9616	0.9625	0.9633
1.8	0.9641	0.9649	0.9656	0.9664	0.9671	0.9678	0.9686	0.9693	0.9699	0.9706
1.9	0.9713	0.9719	0.9726	0.9732	0.9738	0.9744	0.9750	0.9756	0.9761	0.9767
2.0	0.9772	0.9778	0.9783	0.9788	0.9793	0.9798	0.9803	0.9808	0.9812	0.9817
2.1	0.9821	0.9826	0.9830	0.9834	0.9838	0.9842	0.9846	0.9850	0.9854	0.9857
2.2	0.9861	0.9864	0.9868	0.9871	0.9875	0.9878	0.9881	0.9884	0.9887	0.9890
2.3	0.9893	0.9896	0.9898	0.9901	0.9904	0.9906	0.9909	0.9911	0.9913	0.9916
2.4	0.9918	0.9920	0.9922	0.9925	0.9927	0.9929	0.9931	0.9932	0.9934	0.9936
2.5	0.9938	0.9940	0.9941	0.9943	0.9945	0.9946	0.9948	0.9949	0.9951	0.9952
2.6	0.9953	0.9955	0.9956	0.9957	0.9959	0.9960	0.9961	0.9962	0.9963	0.9964
2.7	0.9965	0.9966	0.9967	0.9968	0.9969	0.9970	0.9971	0.9972	0.9973	0.9974
2.8	0.9974	0.9975	0.9976	0.9977	0.9977	0.9978	0.9979	0.9979	0.9980	0.9981
2.9	0.9981	0.9982	0.9982	0.9983	0.9984	0.9984	0.9985	0.9985	0.9986	0.9986
3.0	0.9987	0.9987	0.9987	0.9988	0.9988	0.9989	0.9989	0.9989	0.9990	0.9990
3.1	0.9990	0.9991	0.9991	0.9991	0.9992	0.9992	0.9992	0.9992	0.9993	0.9993
3.2	0.9993	0.9993	0.9994	0.9994	0.9994	0.9994	0.9994	0.9995	0.9995	0.9995
3.3	0.9995	0.9995	0.9995	0.9996	0.9996	0.9996	0.9996	0.9996	0.9996	0.9997
3.4	0.9997	0.9997	0.9997	0.9997	0.9997	0.9997	0.9997	0.9997	0.9997	0.9998

z	Area
3.50	0.99976737
4.00	0.99996833
4.50	0.99999660
5.00	0.99999971

Table A.2: Binomial Distribution Table

$$P[X \le c] = \sum_{x=0}^{c} \binom{n}{x} p^x (1-p)^{n-x}$$

	c	0.05	0.10	0.20	0.30	0.40	0.50	0.60	0.70	0.80	0.90	0.95
n = 1	0	0.950	0.900	0.800	0.700	0.600	0.500	0.400	0.300	0.200	0.100	0.050
	1	1.000	1.000	1.000	1.000	1.000	1.000	1.000	1.000	1.000	1.000	1.000
n = 2	0	0.903	0.810	0.640	0.490	0.360	0.250	0.160	0.090	0.040	0.010	0.003
	1	0.998	0.990	0.960	0.910	0.840	0.750	0.640	0.510	0.360	0.190	0.098
	2	1.000	1.000	1.000	1.000	1.000	1.000	1.000	1.000	1.000	1.000	1.000
n = 3	0	0.857	0.729	0.512	0.343	0.216	0.125	0.064	0.027	0.008	0.001	0.000
	1	0.993	0.972	0.896	0.784	0.648	0.500	0.352	0.216	0.104	0.028	0.007
	2	1.000	0.999	0.992	0.973	0.936	0.875	0.784	0.657	0.488	0.271	0.143
	3	1.000	1.000	1.000	1.000	1.000	1.000	1.000	1.000	1.000	1.000	1.000
n = 4	0	0.815	0.656	0.410	0.240	0.130	0.063	0.026	0.008	0.002	0.000	0.000
	1	0.986	0.948	0.819	0.652	0.475	0.313	0.179	0.084	0.027	0.004	0.000
	2	1.000	0.996	0.973	0.916	0.821	0.688	0.525	0.348	0.181	0.052	0.014
	3	1.000	1.000	0.998	0.992	0.974	0.938	0.870	0.760	0.590	0.344	0.185
	4	1.000	1.000	1.000	1.000	1.000	1.000	1.000	1.000	1.000	1.000	1.000
n = 5	0	0.774	0.590	0.328	0.168	0.078	0.031	0.010	0.002	0.000	0.000	0.000
	1	0.977	0.919	0.737	0.528	0.337	0.188	0.087	0.031	0.007	0.000	0.000
	2	0.999	0.991	0.942	0.837	0.683	0.500	0.317	0.163	0.058	0.009	0.001
	3	1.000	1.000	0.993	0.969	0.913	0.813	0.663	0.472	0.263	0.081	0.023
	4	1.000	1.000	1.000	0.998	0.990	0.969	0.922	0.832	0.672	0.410	0.226
	5	1.000	1.000	1.000	1.000	1.000	1.000	1.000	1.000	1.000	1.000	1.000
n = 6	0	0.735	0.531	0.262	0.118	0.047	0.016	0.004	0.001	0.000	0.000	0.000
	1	0.967	0.886	0.655	0.420	0.233	0.109	0.041	0.011	0.002	0.000	0.000
	2	0.998	0.984	0.901	0.744	0.544	0.344	0.179	0.070	0.017	0.001	0.000
	3	1.000	0.999	0.983	0.930	0.821	0.656	0.456	0.256	0.099	0.016	0.002
	4	1.000	1.000	0.998	0.989	0.959	0.891	0.767	0.580	0.345	0.114	0.033
	5	1.000	1.000	1.000	0.999	0.996	0.984	0.953	0.882	0.738	0.469	0.265
	6	1.000	1.000	1.000	1.000	1.000	1.000	1.000	1.000	1.000	1.000	1.000
n = 7	0	0.698	0.478	0.210	0.082	0.028	0.008	0.002	0.000	0.000	0.000	0.000
	1	0.956	0.850	0.577	0.329	0.159	0.063	0.019	0.004	0.000	0.000	0.000
	2	0.996	0.974	0.852	0.647	0.420	0.227	0.096	0.029	0.005	0.000	0.000
	3	1.000	0.997	0.967	0.874	0.710	0.500	0.290	0.126	0.033	0.003	0.000
	4	1.000	1.000	0.995	0.971	0.904	0.773	0.580	0.353	0.148	0.026	0.004
	5	1.000	1.000	1.000	0.996	0.981	0.938	0.841	0.671	0.423	0.150	0.044
	6	1.000	1.000	1.000	1.000	0.998	0.992	0.972	0.918	0.790	0.522	0.302
	7	1.000	1.000	1.000	1.000	1.000	1.000	1.000	1.000	1.000	1.000	1.000

(continued)

						p						
	c	0.05	0.10	0.20	0.30	0.40	0.50	0.60	0.70	0.80	0.90	0.95
n = 8	0	0.663	0.430	0.168	0.058	0.017	0.004	0.001	0.000	0.000	0.000	0.000
	1	0.943	0.813	0.503	0.255	0.106	0.035	0.009	0.001	0.000	0.000	0.000
	2	0.994	0.962	0.797	0.552	0.315	0.145	0.050	0.011	0.001	0.000	0.000
	3	1.000	0.995	0.944	0.806	0.594	0.363	0.174	0.058	0.010	0.000	0.000
	4	1.000	1.000	0.990	0.942	0.826	0.637	0.406	0.194	0.056	0.005	0.000
	5	1.000	1.000	0.999	0.989	0.950	0.855	0.685	0.448	0.203	0.038	0.006
	6	1.000	1.000	1.000	0.999	0.991	0.965	0.894	0.745	0.497	0.187	0.057
	7	1.000	1.000	1.000	1.000	0.999	0.996	0.983	0.942	0.832	0.570	0.337
	8	1.000	1.000	1.000	1.000	1.000	1.000	1.000	1.000	1.000	1.000	1.000
n = 9	0	0.630	0.387	0.134	0.040	0.010	0.002	0.000	0.000	0.000	0.000	0.000
	1	0.929	0.775	0.436	0.196	0.071	0.020	0.004	0.000	0.000	0.000	0.000
	2	0.992	0.947	0.738	0.463	0.232	0.090	0.025	0.004	0.000	0.000	0.000
	3	0.999	0.992	0.914	0.730	0.483	0.254	0.099	0.025	0.003	0.000	0.000
	4	1.000	0.999	0.980	0.901	0.733	0.500	0.267	0.099	0.020	0.001	0.000
	5	1.000	1.000	0.997	0.975	0.901	0.746	0.517	0.270	0.086	0.008	0.001
	6	1.000	1.000	1.000	0.996	0.975	0.910	0.768	0.537	0.262	0.053	0.008
	7	1.000	1.000	1.000	1.000	0.996	0.980	0.929	0.804	0.564	0.225	0.071
	8	1.000	1.000	1.000	1.000	1.000	0.998	0.990	0.960	0.866	0.613	0.370
	9	1.000	1.000	1.000	1.000	1.000	1.000	1.000	1.000	1.000	1.000	1.000
n = 10	0	0.599	0.349	0.107	0.028	0.006	0.001	0.000	0.000	0.000	0.000	0.000
	1	0.914	0.736	0.376	0.149	0.046	0.011	0.002	0.000	0.000	0.000	0.000
	2	0.988	0.930	0.678	0.383	0.167	0.055	0.012	0.002	0.000	0.000	0.000
	3	0.999	0.987	0.879	0.650	0.382	0.172	0.055	0.011	0.001	0.000	0.000
	4	1.000	0.998	0.967	0.850	0.633	0.377	0.166	0.047	0.006	0.000	0.000
	5	1.000	1.000	0.994	0.953	0.834	0.623	0.367	0.150	0.033	0.002	0.000
	6	1.000	1.000	0.999	0.989	0.945	0.828	0.618	0.350	0.121	0.013	0.001
	7	1.000	1.000	1.000	0.998	0.988	0.945	0.833	0.617	0.322	0.070	0.012
	8	1.000	1.000	1.000	1.000	0.998	0.989	0.954	0.851	0.624	0.264	0.086
	9	1.000	1.000	1.000	1.000	1.000	0.999	0.994	0.972	0.893	0.651	0.401
	10	1.000	1.000	1.000	1.000	1.000	1.000	1.000	1.000	1.000	1.000	1.000
n = 11	0	0.569	0.314	0.086	0.020	0.004	0.000	0.000	0.000	0.000	0.000	0.000
	1	0.898	0.697	0.322	0.113	0.030	0.006	0.001	0.000	0.000	0.000	0.000
	2	0.985	0.910	0.617	0.313	0.119	0.033	0.006	0.001	0.000	0.000	0.000
	3	0.998	0.981	0.839	0.570	0.296	0.113	0.029	0.004	0.000	0.000	0.000
	4	1.000	0.997	0.950	0.790	0.533	0.274	0.099	0.022	0.002	0.000	0.000
	5	1.000	1.000	0.988	0.922	0.753	0.500	0.247	0.078	0.012	0.000	0.000
	6	1.000	1.000	0.998	0.978	0.901	0.726	0.467	0.210	0.050	0.003	0.000
	7	1.000	1.000	1.000	0.996	0.971	0.887	0.704	0.430	0.161	0.019	0.002
	8	1.000	1.000	1.000	0.999	0.994	0.967	0.881	0.687	0.383	0.090	0.015
	9	1.000	1.000	1.000	1.000	0.999	0.994	0.970	0.887	0.678	0.303	0.102
	10	1.000	1.000	1.000	1.000	1.000	1.000	0.996	0.980	0.914	0.686	0.431
	11	1.000	1.000	1.000	1.000	1.000	1.000	1.000	1.000	1.000	1.000	1.000

(continued)

	c	0.05	0.10	0.20	0.30	0.40	0.50	0.60	0.70	0.80	0.90	0.95
n = 12	0	0.540	0.282	0.069	0.014	0.002	0.000	0.000	0.000	0.000	0.000	0.000
	1	0.882	0.659	0.275	0.085	0.020	0.003	0.000	0.000	0.000	0.000	0.000
	2	0.980	0.889	0.558	0.253	0.083	0.019	0.003	0.000	0.000	0.000	0.000
	3	0.998	0.974	0.795	0.493	0.225	0.073	0.015	0.002	0.000	0.000	0.000
	4	1.000	0.996	0.927	0.724	0.438	0.194	0.057	0.009	0.001	0.000	0.000
	5	1.000	0.999	0.981	0.882	0.665	0.387	0.158	0.039	0.004	0.000	0.000
	6	1.000	1.000	0.996	0.961	0.842	0.613	0.335	0.118	0.019	0.001	0.000
	7	1.000	1.000	0.999	0.991	0.943	0.806	0.562	0.276	0.073	0.004	0.000
	8	1.000	1.000	1.000	0.998	0.985	0.927	0.775	0.507	0.205	0.026	0.002
	9	1.000	1.000	1.000	1.000	0.997	0.981	0.917	0.747	0.442	0.111	0.020
	10	1.000	1.000	1.000	1.000	1.000	0.997	0.980	0.915	0.725	0.341	0.118
	11	1.000	1.000	1.000	1.000	1.000	1.000	0.998	0.986	0.931	0.718	0.460
	12	1.000	1.000	1.000	1.000	1.000	1.000	1.000	1.000	1.000	1.000	1.000
n = 13	0	0.513	0.254	0.055	0.010	0.001	0.000	0.000	0.000	0.000	0.000	0.000
	1	0.865	0.621	0.234	0.064	0.013	0.002	0.000	0.000	0.000	0.000	0.000
	2	0.975	0.866	0.502	0.202	0.058	0.011	0.001	0.000	0.000	0.000	0.000
	3	0.997	0.966	0.747	0.421	0.169	0.046	0.008	0.001	0.000	0.000	0.000
	4	1.000	0.994	0.901	0.654	0.353	0.133	0.032	0.004	0.000	0.000	0.000
	5	1.000	0.999	0.970	0.835	0.574	0.291	0.098	0.018	0.001	0.000	0.000
	6	1.000	1.000	0.993	0.938	0.771	0.500	0.229	0.062	0.007	0.000	0.000
	7	1.000	1.000	0.999	0.982	0.902	0.709	0.426	0.165	0.030	0.001	0.000
	8	1.000	1.000	1.000	0.996	0.968	0.867	0.647	0.346	0.099	0.006	0.000
	9	1.000	1.000	1.000	0.999	0.992	0.954	0.831	0.579	0.253	0.034	0.003
	10	1.000	1.000	1.000	1.000	0.999	0.989	0.942	0.798	0.498	0.134	0.025
	11	1.000	1.000	1.000	1.000	1.000	0.998	0.987	0.936	0.766	0.379	0.135
	12	1.000	1.000	1.000	1.000	1.000	1.000	0.999	0.990	0.945	0.746	0.487
	13	1.000	1.000	1.000	1.000	1.000	1.000	1.000	1.000	1.000	1.000	1.000
n = 14	0	0.488	0.229	0.044	0.007	0.001	0.000	0.000	0.000	0.000	0.000	0.000
	1	0.847	0.585	0.198	0.047	0.008	0.001	0.000	0.000	0.000	0.000	0.000
	2	0.970	0.842	0.448	0.161	0.040	0.006	0.001	0.000	0.000	0.000	0.000
	3	0.996	0.956	0.698	0.355	0.124	0.029	0.004	0.000	0.000	0.000	0.000
	4	1.000	0.991	0.870	0.584	0.279	0.090	0.018	0.002	0.000	0.000	0.000
	5	1.000	0.999	0.956	0.781	0.486	0.212	0.058	0.008	0.000	0.000	0.000
	6	1.000	1.000	0.988	0.907	0.692	0.395	0.150	0.031	0.002	0.000	0.000
	7	1.000	1.000	0.998	0.969	0.850	0.605	0.308	0.093	0.012	0.000	0.000
	8	1.000	1.000	1.000	0.992	0.942	0.788	0.514	0.219	0.044	0.001	0.000
	9	1.000	1.000	1.000	0.998	0.982	0.910	0.721	0.416	0.130	0.009	0.000
	10	1.000	1.000	1.000	1.000	0.996	0.971	0.876	0.645	0.302	0.044	0.004
	11	1.000	1.000	1.000	1.000	0.999	0.994	0.960	0.839	0.552	0.158	0.030
	12	1.000	1.000	1.000	1.000	1.000	0.999	0.992	0.953	0.802	0.415	0.153
	13	1.000	1.000	1.000	1.000	1.000	1.000	0.999	0.993	0.956	0.771	0.512
	14	1.000	1.000	1.000	1.000	1.000	1.000	1.000	1.000	1.000	1.000	1.000

(*continued*)

	c	0.05	0.10	0.20	0.30	0.40	0.50	0.60	0.70	0.80	0.90	0.95
n = 15	0	0.463	0.206	0.035	0.005	0.000	0.000	0.000	0.000	0.000	0.000	0.000
	1	0.829	0.549	0.167	0.035	0.005	0.000	0.000	0.000	0.000	0.000	0.000
	2	0.964	0.816	0.398	0.127	0.027	0.004	0.000	0.000	0.000	0.000	0.000
	3	0.995	0.944	0.648	0.297	0.091	0.018	0.002	0.000	0.000	0.000	0.000
	4	0.999	0.987	0.836	0.515	0.217	0.059	0.009	0.001	0.000	0.000	0.000
	5	1.000	0.998	0.939	0.722	0.403	0.151	0.034	0.004	0.000	0.000	0.000
	6	1.000	1.000	0.982	0.869	0.610	0.304	0.095	0.015	0.001	0.000	0.000
	7	1.000	1.000	0.996	0.950	0.787	0.500	0.213	0.050	0.004	0.000	0.000
	8	1.000	1.000	0.999	0.985	0.905	0.696	0.390	0.131	0.018	0.000	0.000
	9	1.000	1.000	1.000	0.996	0.966	0.849	0.597	0.278	0.061	0.002	0.000
	10	1.000	1.000	1.000	0.999	0.991	0.941	0.783	0.485	0.164	0.013	0.001
	11	1.000	1.000	1.000	1.000	0.998	0.982	0.909	0.703	0.352	0.056	0.005
	12	1.000	1.000	1.000	1.000	1.000	0.996	0.973	0.873	0.602	0.184	0.036
	13	1.000	1.000	1.000	1.000	1.000	1.000	0.995	0.965	0.833	0.451	0.171
	14	1.000	1.000	1.000	1.000	1.000	1.000	1.000	0.995	0.965	0.794	0.537
	15	1.000	1.000	1.000	1.000	1.000	1.000	1.000	1.000	1.000	1.000	1.000
n = 16	0	0.440	0.185	0.028	0.003	0.000	0.000	0.000	0.000	0.000	0.000	0.000
	1	0.811	0.515	0.141	0.026	0.003	0.000	0.000	0.000	0.000	0.000	0.000
	2	0.957	0.789	0.352	0.099	0.018	0.002	0.000	0.000	0.000	0.000	0.000
	3	0.993	0.932	0.598	0.246	0.065	0.011	0.001	0.000	0.000	0.000	0.000
	4	0.999	0.983	0.798	0.450	0.167	0.038	0.005	0.000	0.000	0.000	0.000
	5	1.000	0.997	0.918	0.660	0.329	0.105	0.019	0.002	0.000	0.000	0.000
	6	1.000	0.999	0.973	0.825	0.527	0.227	0.058	0.007	0.000	0.000	0.000
	7	1.000	1.000	0.993	0.926	0.716	0.402	0.142	0.026	0.001	0.000	0.000
	8	1.000	1.000	0.999	0.974	0.858	0.598	0.284	0.074	0.007	0.000	0.000
	9	1.000	1.000	1.000	0.993	0.942	0.773	0.473	0.175	0.027	0.001	0.000
	10	1.000	1.000	1.000	0.998	0.981	0.895	0.671	0.340	0.082	0.003	0.000
	11	1.000	1.000	1.000	1.000	0.995	0.962	0.833	0.550	0.202	0.017	0.001
	12	1.000	1.000	1.000	1.000	0.999	0.989	0.935	0.754	0.402	0.068	0.007
	13	1.000	1.000	1.000	1.000	1.000	0.998	0.982	0.901	0.648	0.211	0.043
	14	1.000	1.000	1.000	1.000	1.000	1.000	0.997	0.974	0.859	0.485	0.189
	15	1.000	1.000	1.000	1.000	1.000	1.000	1.000	0.997	0.972	0.815	0.560
	16	1.000	1.000	1.000	1.000	1.000	1.000	1.000	1.000	1.000	1.000	1.000

(continued)

	c	0.05	0.10	0.20	0.30	0.40	0.50	0.60	0.70	0.80	0.90	0.95
							p					
n = 17	0	0.418	0.167	0.023	0.002	0.000	0.000	0.000	0.000	0.000	0.000	0.000
	1	0.792	0.482	0.118	0.019	0.002	0.000	0.000	0.000	0.000	0.000	0.000
	2	0.950	0.762	0.310	0.077	0.012	0.001	0.000	0.000	0.000	0.000	0.000
	3	0.991	0.917	0.549	0.202	0.046	0.006	0.000	0.000	0.000	0.000	0.000
	4	0.999	0.978	0.758	0.389	0.126	0.025	0.003	0.000	0.000	0.000	0.000
	5	1.000	0.995	0.894	0.597	0.264	0.072	0.011	0.001	0.000	0.000	0.000
	6	1.000	0.999	0.962	0.775	0.448	0.166	0.035	0.003	0.000	0.000	0.000
	7	1.000	1.000	0.989	0.895	0.641	0.315	0.092	0.013	0.000	0.000	0.000
	8	1.000	1.000	0.997	0.960	0.801	0.500	0.199	0.040	0.003	0.000	0.000
	9	1.000	1.000	1.000	0.987	0.908	0.685	0.359	0.105	0.011	0.000	0.000
	10	1.000	1.000	1.000	0.997	0.965	0.834	0.552	0.225	0.038	0.001	0.000
	11	1.000	1.000	1.000	0.999	0.989	0.928	0.736	0.403	0.106	0.005	0.000
	12	1.000	1.000	1.000	1.000	0.997	0.975	0.874	0.611	0.242	0.022	0.001
	13	1.000	1.000	1.000	1.000	1.000	0.994	0.954	0.798	0.451	0.083	0.009
	14	1.000	1.000	1.000	1.000	1.000	0.999	0.988	0.923	0.690	0.238	0.050
	15	1.000	1.000	1.000	1.000	1.000	1.000	0.998	0.981	0.882	0.518	0.208
	16	1.000	1.000	1.000	1.000	1.000	1.000	1.000	0.998	0.977	0.833	0.582
	17	1.000	1.000	1.000	1.000	1.000	1.000	1.000	1.000	1.000	1.000	1.000
n = 18	0	0.397	0.150	0.018	0.002	0.000	0.000	0.000	0.000	0.000	0.000	0.000
	1	0.774	0.450	0.099	0.014	0.001	0.000	0.000	0.000	0.000	0.000	0.000
	2	0.942	0.734	0.271	0.060	0.008	0.001	0.000	0.000	0.000	0.000	0.000
	3	0.989	0.902	0.501	0.165	0.033	0.004	0.000	0.000	0.000	0.000	0.000
	4	0.998	0.972	0.716	0.333	0.094	0.015	0.001	0.000	0.000	0.000	0.000
	5	1.000	0.994	0.867	0.534	0.209	0.048	0.006	0.000	0.000	0.000	0.000
	6	1.000	0.999	0.949	0.722	0.374	0.119	0.020	0.001	0.000	0.000	0.000
	7	1.000	1.000	0.984	0.859	0.563	0.240	0.058	0.006	0.000	0.000	0.000
	8	1.000	1.000	0.996	0.940	0.737	0.407	0.135	0.021	0.001	0.000	0.000
	9	1.000	1.000	0.999	0.979	0.865	0.593	0.263	0.060	0.004	0.000	0.000
	10	1.000	1.000	1.000	0.994	0.942	0.760	0.437	0.141	0.016	0.000	0.000
	11	1.000	1.000	1.000	0.999	0.980	0.881	0.626	0.278	0.051	0.001	0.000
	12	1.000	1.000	1.000	1.000	0.994	0.952	0.791	0.466	0.133	0.006	0.000
	13	1.000	1.000	1.000	1.000	0.999	0.985	0.906	0.667	0.284	0.028	0.002
	14	1.000	1.000	1.000	1.000	1.000	0.996	0.967	0.835	0.499	0.098	0.011
	15	1.000	1.000	1.000	1.000	1.000	0.999	0.992	0.940	0.729	0.266	0.058
	16	1.000	1.000	1.000	1.000	1.000	1.000	0.999	0.986	0.901	0.550	0.226
	17	1.000	1.000	1.000	1.000	1.000	1.000	1.000	0.998	0.982	0.850	0.603
	18	1.000	1.000	1.000	1.000	1.000	1.000	1.000	1.000	1.000	1.000	1.000

(*continued*)

	c	0.05	0.10	0.20	0.30	0.40	0.50	0.60	0.70	0.80	0.90	0.95
							p					
n = 19	0	0.377	0.135	0.014	0.001	0.000	0.000	0.000	0.000	0.000	0.000	0.000
	1	0.755	0.420	0.083	0.010	0.001	0.000	0.000	0.000	0.000	0.000	0.000
	2	0.933	0.705	0.237	0.046	0.005	0.000	0.000	0.000	0.000	0.000	0.000
	3	0.987	0.885	0.455	0.133	0.023	0.002	0.000	0.000	0.000	0.000	0.000
	4	0.998	0.965	0.673	0.282	0.070	0.010	0.001	0.000	0.000	0.000	0.000
	5	1.000	0.991	0.837	0.474	0.163	0.032	0.003	0.000	0.000	0.000	0.000
	6	1.000	0.998	0.932	0.666	0.308	0.084	0.012	0.001	0.000	0.000	0.000
	7	1.000	1.000	0.977	0.818	0.488	0.180	0.035	0.003	0.000	0.000	0.000
	8	1.000	1.000	0.993	0.916	0.667	0.324	0.088	0.011	0.000	0.000	0.000
	9	1.000	1.000	0.998	0.967	0.814	0.500	0.186	0.033	0.002	0.000	0.000
	10	1.000	1.000	1.000	0.989	0.912	0.676	0.333	0.084	0.007	0.000	0.000
	11	1.000	1.000	1.000	0.997	0.965	0.820	0.512	0.182	0.023	0.000	0.000
	12	1.000	1.000	1.000	0.999	0.988	0.916	0.692	0.334	0.068	0.002	0.000
	13	1.000	1.000	1.000	1.000	0.997	0.968	0.837	0.526	0.163	0.009	0.000
	14	1.000	1.000	1.000	1.000	0.999	0.990	0.930	0.718	0.327	0.035	0.002
	15	1.000	1.000	1.000	1.000	1.000	0.998	0.977	0.867	0.545	0.115	0.013
	16	1.000	1.000	1.000	1.000	1.000	1.000	0.995	0.954	0.763	0.295	0.067
	17	1.000	1.000	1.000	1.000	1.000	1.000	0.999	0.990	0.917	0.580	0.245
	18	1.000	1.000	1.000	1.000	1.000	1.000	1.000	0.999	0.986	0.865	0.623
	19	1.000	1.000	1.000	1.000	1.000	1.000	1.000	1.000	1.000	1.000	1.000
n = 20	0	0.358	0.122	0.012	0.001	0.000	0.000	0.000	0.000	0.000	0.000	0.000
	1	0.736	0.392	0.069	0.008	0.001	0.000	0.000	0.000	0.000	0.000	0.000
	2	0.925	0.677	0.206	0.035	0.004	0.000	0.000	0.000	0.000	0.000	0.000
	3	0.984	0.867	0.411	0.107	0.016	0.001	0.000	0.000	0.000	0.000	0.000
	4	0.997	0.957	0.630	0.238	0.051	0.006	0.000	0.000	0.000	0.000	0.000
	5	1.000	0.989	0.804	0.416	0.126	0.021	0.002	0.000	0.000	0.000	0.000
	6	1.000	0.998	0.913	0.608	0.250	0.058	0.006	0.000	0.000	0.000	0.000
	7	1.000	1.000	0.968	0.772	0.416	0.132	0.021	0.001	0.000	0.000	0.000
	8	1.000	1.000	0.990	0.887	0.596	0.252	0.057	0.005	0.000	0.000	0.000
	9	1.000	1.000	0.997	0.952	0.755	0.412	0.128	0.017	0.001	0.000	0.000
	10	1.000	1.000	0.999	0.983	0.872	0.588	0.245	0.048	0.003	0.000	0.000
	11	1.000	1.000	1.000	0.995	0.943	0.748	0.404	0.113	0.010	0.000	0.000
	12	1.000	1.000	1.000	0.999	0.979	0.868	0.584	0.228	0.032	0.000	0.000
	13	1.000	1.000	1.000	1.000	0.994	0.942	0.750	0.392	0.087	0.002	0.000
	14	1.000	1.000	1.000	1.000	0.998	0.979	0.874	0.584	0.196	0.011	0.000
	15	1.000	1.000	1.000	1.000	1.000	0.994	0.949	0.762	0.370	0.043	0.003
	16	1.000	1.000	1.000	1.000	1.000	0.999	0.984	0.893	0.589	0.133	0.016
	17	1.000	1.000	1.000	1.000	1.000	1.000	0.996	0.965	0.794	0.323	0.075
	18	1.000	1.000	1.000	1.000	1.000	1.000	0.999	0.992	0.931	0.608	0.264
	19	1.000	1.000	1.000	1.000	1.000	1.000	1.000	0.999	0.988	0.878	0.642
	20	1.000	1.000	1.000	1.000	1.000	1.000	1.000	1.000	1.000	1.000	1.000

(continued)

	c	0.05	0.10	0.20	0.30	0.40	0.50	0.60	0.70	0.80	0.90	0.95
n = 25	0	0.277	0.072	0.004	0.000	0.000	0.000	0.000	0.000	0.000	0.000	0.000
	1	0.642	0.271	0.027	0.002	0.000	0.000	0.000	0.000	0.000	0.000	0.000
	2	0.873	0.537	0.098	0.009	0.000	0.000	0.000	0.000	0.000	0.000	0.000
	3	0.966	0.764	0.234	0.033	0.002	0.000	0.000	0.000	0.000	0.000	0.000
	4	0.993	0.902	0.421	0.090	0.009	0.000	0.000	0.000	0.000	0.000	0.000
	5	0.999	0.967	0.617	0.193	0.029	0.002	0.000	0.000	0.000	0.000	0.000
	6	1.000	0.991	0.780	0.341	0.074	0.007	0.000	0.000	0.000	0.000	0.000
	7	1.000	0.998	0.891	0.512	0.154	0.022	0.001	0.000	0.000	0.000	0.000
	8	1.000	1.000	0.953	0.677	0.274	0.054	0.004	0.000	0.000	0.000	0.000
	9	1.000	1.000	0.983	0.811	0.425	0.115	0.013	0.000	0.000	0.000	0.000
	10	1.000	1.000	0.994	0.902	0.586	0.212	0.034	0.002	0.000	0.000	0.000
	11	1.000	1.000	0.998	0.956	0.732	0.345	0.078	0.006	0.000	0.000	0.000
	12	1.000	1.000	1.000	0.983	0.846	0.500	0.154	0.017	0.000	0.000	0.000
	13	1.000	1.000	1.000	0.994	0.922	0.655	0.268	0.044	0.002	0.000	0.000
	14	1.000	1.000	1.000	0.998	0.966	0.788	0.414	0.098	0.006	0.000	0.000
	15	1.000	1.000	1.000	1.000	0.987	0.885	0.575	0.189	0.017	0.000	0.000
	16	1.000	1.000	1.000	1.000	0.996	0.946	0.726	0.323	0.047	0.000	0.000
	17	1.000	1.000	1.000	1.000	0.999	0.978	0.846	0.488	0.109	0.002	0.000
	18	1.000	1.000	1.000	1.000	1.000	0.993	0.926	0.659	0.220	0.009	0.000
	19	1.000	1.000	1.000	1.000	1.000	0.998	0.971	0.807	0.383	0.033	0.001
	20	1.000	1.000	1.000	1.000	1.000	1.000	0.991	0.910	0.579	0.098	0.007
	21	1.000	1.000	1.000	1.000	1.000	1.000	0.998	0.967	0.766	0.236	0.034
	22	1.000	1.000	1.000	1.000	1.000	1.000	1.000	0.991	0.902	0.463	0.127
	23	1.000	1.000	1.000	1.000	1.000	1.000	1.000	0.998	0.973	0.729	0.358
	24	1.000	1.000	1.000	1.000	1.000	1.000	1.000	1.000	0.996	0.928	0.723
	25	1.000	1.000	1.000	1.000	1.000	1.000	1.000	1.000	1.000	1.000	1.000

Table A.3: Poisson Distribution Table

The table below gives the probability of that a Poisson random variable X with mean $= \lambda$ is less than or equal to x. That is, the table gives

$$P\left(X \le x\right) = \sum_{r=0}^{x} \lambda^r \frac{e^{-\lambda}}{r!}$$

$\lambda =$		0.1	0.2	0.3	0.4	0.5	0.6	0.7	0.8	0.9	1.0	1.2	1.4	1.6	1.8
$x=$	0	0.9048	0.8187	0.7408	0.6703	0.6065	0.5488	0.4966	0.4493	0.4066	0.3679	0.3012	0.2466	0.2019	0.1653
	1	0.9953	0.9825	0.9631	0.9384	0.9098	0.8781	0.8442	0.8088	0.7725	0.7358	0.6626	0.5918	0.5249	0.4628
	2	0.9998	0.9989	0.9964	0.9921	0.9856	0.9769	0.9659	0.9526	0.9371	0.9197	0.8795	0.8335	0.7834	0.7306
	3	1.0000	0.9999	0.9997	0.9992	0.9982	0.9966	0.9942	0.9909	0.9865	0.9810	0.9662	0.9463	0.9212	0.8913
	4	1.0000	1.0000	1.0000	0.9999	0.9998	0.9996	0.9992	0.9986	0.9977	0.9963	0.9923	0.9857	0.9763	0.9636
	5	1.0000	1.0000	1.0000	1.0000	1.0000	1.0000	0.9999	0.9998	0.9997	0.9994	0.9985	0.9968	0.9940	0.9896
	6	1.0000	1.0000	1.0000	1.0000	1.0000	1.0000	1.0000	1.0000	1.0000	0.9999	0.9997	0.9994	0.9987	0.9974
	7	1.0000	1.0000	1.0000	1.0000	1.0000	1.0000	1.0000	1.0000	1.0000	1.0000	1.0000	0.9999	0.9997	0.9994
	8	1.0000	1.0000	1.0000	1.0000	1.0000	1.0000	1.0000	1.0000	1.0000	1.0000	1.0000	1.0000	1.0000	0.9999
	9	1.0000	1.0000	1.0000	1.0000	1.0000	1.0000	1.0000	1.0000	1.0000	1.0000	1.0000	1.0000	1.0000	1.0000

$\lambda =$		2.0	2.2	2.4	2.6	2.8	3.0	3.2	3.4	3.6	3.8	4.0	4.5	5.0	5.5
$x=$	0	0.1353	0.1108	0.0907	0.0743	0.0608	0.0498	0.0408	0.0334	0.0273	0.0224	0.0183	0.0111	0.0067	0.0041
	1	0.4060	0.3546	0.3084	0.2674	0.2311	0.1991	0.1712	0.1468	0.1257	0.1074	0.0916	0.0611	0.0404	0.0266
	2	0.6767	0.6227	0.5697	0.5184	0.4695	0.4232	0.3799	0.3397	0.3027	0.2689	0.2381	0.1736	0.1247	0.0884
	3	0.8571	0.8194	0.7787	0.7360	0.6919	0.6472	0.6025	0.5584	0.5152	0.4735	0.4335	0.3423	0.2650	0.2017
	4	0.9473	0.9275	0.9041	0.8774	0.8477	0.8153	0.7806	0.7442	0.7064	0.6678	0.6288	0.5321	0.4405	0.3575
	5	0.9834	0.9751	0.9643	0.9510	0.9349	0.9161	0.8946	0.8705	0.8441	0.8156	0.7851	0.7029	0.6160	0.5289
	6	0.9955	0.9925	0.9884	0.9828	0.9756	0.9665	0.9554	0.9421	0.9267	0.9091	0.8893	0.8311	0.7622	0.6860
	7	0.9989	0.9980	0.9967	0.9947	0.9919	0.9881	0.9832	0.9769	0.9692	0.9599	0.9489	0.9134	0.8666	0.8095
	8	0.9998	0.9995	0.9991	0.9985	0.9976	0.9962	0.9943	0.9917	0.9883	0.9840	0.9786	0.9597	0.9319	0.8944
	9	1.0000	0.9999	0.9998	0.9996	0.9993	0.9989	0.9982	0.9973	0.9960	0.9942	0.9919	0.9829	0.9682	0.9462
	10	1.0000	1.0000	1.0000	0.9999	0.9998	0.9997	0.9995	0.9992	0.9987	0.9981	0.9972	0.9933	0.9863	0.9747
	11	1.0000	1.0000	1.0000	1.0000	1.0000	0.9999	0.9999	0.9998	0.9996	0.9994	0.9991	0.9976	0.9945	0.9890
	12	1.0000	1.0000	1.0000	1.0000	1.0000	1.0000	1.0000	0.9999	0.9999	0.9998	0.9997	0.9992	0.9980	0.9955
	13	1.0000	1.0000	1.0000	1.0000	1.0000	1.0000	1.0000	1.0000	1.0000	1.0000	0.9999	0.9997	0.9993	0.9983
	14	1.0000	1.0000	1.0000	1.0000	1.0000	1.0000	1.0000	1.0000	1.0000	1.0000	1.0000	0.9999	0.9998	0.9994
	15	1.0000	1.0000	1.0000	1.0000	1.0000	1.0000	1.0000	1.0000	1.0000	1.0000	1.0000	1.0000	0.9999	0.9998
	16	1.0000	1.0000	1.0000	1.0000	1.0000	1.0000	1.0000	1.0000	1.0000	1.0000	1.0000	1.0000	1.0000	0.9999
	17	1.0000	1.0000	1.0000	1.0000	1.0000	1.0000	1.0000	1.0000	1.0000	1.0000	1.0000	1.0000	1.0000	1.0000

$\lambda =$		6.0	6.5	7.0	7.5	8.0	8.5	9.0	9.5	10.0	11.0	10.0	12.0	14.0	15.0
$x=$	0	0.0025	0.0015	0.0009	0.0006	0.0003	0.0002	0.0001	0.0001	0.0000	0.0000	0.0000	0.0000	0.0000	0.0000
	1	0.0174	0.0113	0.0073	0.0047	0.0030	0.0019	0.0012	0.0008	0.0005	0.0002	0.0005	0.0001	0.0000	0.0000
	2	0.0620	0.0430	0.0296	0.0203	0.0138	0.0093	0.0062	0.0042	0.0028	0.0012	0.0028	0.0005	0.0001	0.0000
	3	0.1512	0.1118	0.0818	0.0591	0.0424	0.0301	0.0212	0.0149	0.0103	0.0049	0.0103	0.0023	0.0005	0.0002
	4	0.2851	0.2237	0.1730	0.1321	0.0996	0.0744	0.0550	0.0403	0.0293	0.0151	0.0293	0.0076	0.0018	0.0009
	5	0.4457	0.3690	0.3007	0.2414	0.1912	0.1496	0.1157	0.0885	0.0671	0.0375	0.0671	0.0203	0.0055	0.0028
	6	0.6063	0.5265	0.4497	0.3782	0.3134	0.2562	0.2068	0.1649	0.1301	0.0786	0.1301	0.0458	0.0142	0.0076
	7	0.7440	0.6728	0.5987	0.5246	0.4530	0.3856	0.3239	0.2687	0.2202	0.1432	0.2202	0.0895	0.0316	0.0180
	8	0.8472	0.7916	0.7291	0.6620	0.5925	0.5231	0.4557	0.3918	0.3328	0.2320	0.3328	0.1550	0.0621	0.0374
	9	0.9161	0.8774	0.8305	0.7764	0.7166	0.6530	0.5874	0.5218	0.4579	0.3405	0.4579	0.2424	0.1094	0.0699
	10	0.9574	0.9332	0.9015	0.8622	0.8159	0.7634	0.7060	0.6453	0.5830	0.4599	0.5830	0.3472	0.1757	0.1185
	11	0.9799	0.9661	0.9467	0.9208	0.8881	0.8487	0.8030	0.7520	0.6968	0.5793	0.6968	0.4616	0.2600	0.1848
	12	0.9912	0.9840	0.9730	0.9573	0.9362	0.9091	0.8758	0.8364	0.7916	0.6887	0.7916	0.5760	0.3585	0.2676
	13	0.9964	0.9929	0.9872	0.9784	0.9658	0.9486	0.9261	0.8981	0.8645	0.7813	0.8645	0.6815	0.4644	0.3632
	14	0.9986	0.9970	0.9943	0.9897	0.9827	0.9726	0.9585	0.9400	0.9165	0.8540	0.9165	0.7720	0.5704	0.4657
	15	0.9995	0.9988	0.9976	0.9954	0.9918	0.9862	0.9780	0.9665	0.9513	0.9074	0.9513	0.8444	0.6694	0.5681
	16	0.9998	0.9996	0.9990	0.9980	0.9963	0.9934	0.9889	0.9823	0.9730	0.9441	0.9730	0.8987	0.7559	0.6641
	17	0.9999	0.9998	0.9996	0.9992	0.9984	0.9970	0.9947	0.9911	0.9857	0.9678	0.9857	0.9370	0.8272	0.7489
	18	1.0000	0.9999	0.9999	0.9997	0.9993	0.9987	0.9976	0.9957	0.9928	0.9823	0.9928	0.9626	0.8826	0.8195
	19	1.0000	1.0000	1.0000	0.9999	0.9997	0.9995	0.9989	0.9980	0.9965	0.9907	0.9965	0.9787	0.9235	0.8752
	20	1.0000	1.0000	1.0000	1.0000	0.9999	0.9998	0.9996	0.9991	0.9984	0.9953	0.9984	0.9884	0.9521	0.9170
	21	1.0000	1.0000	1.0000	1.0000	1.0000	0.9999	0.9998	0.9996	0.9993	0.9977	0.9993	0.9939	0.9712	0.9469
	22	1.0000	1.0000	1.0000	1.0000	1.0000	1.0000	0.9999	0.9999	0.9997	0.9990	0.9997	0.9970	0.9833	0.9673
	23	1.0000	1.0000	1.0000	1.0000	1.0000	1.0000	1.0000	0.9999	0.9999	0.9995	0.9999	0.9985	0.9907	0.9805
	24	1.0000	1.0000	1.0000	1.0000	1.0000	1.0000	1.0000	1.0000	1.0000	0.9998	1.0000	0.9993	0.9950	0.9888
	25	1.0000	1.0000	1.0000	1.0000	1.0000	1.0000	1.0000	1.0000	1.0000	0.9999	1.0000	0.9997	0.9974	0.9938
	26	1.0000	1.0000	1.0000	1.0000	1.0000	1.0000	1.0000	1.0000	1.0000	1.0000	1.0000	0.9999	0.9987	0.9967
	27	1.0000	1.0000	1.0000	1.0000	1.0000	1.0000	1.0000	1.0000	1.0000	1.0000	1.0000	0.9999	0.9994	0.9983
	28	1.0000	1.0000	1.0000	1.0000	1.0000	1.0000	1.0000	1.0000	1.0000	1.0000	1.0000	1.0000	0.9997	0.9991
	29	1.0000	1.0000	1.0000	1.0000	1.0000	1.0000	1.0000	1.0000	1.0000	1.0000	1.0000	1.0000	0.9999	0.9996
	30	1.0000	1.0000	1.0000	1.0000	1.0000	1.0000	1.0000	1.0000	1.0000	1.0000	1.0000	1.0000	0.9999	0.9998
	31	1.0000	1.0000	1.0000	1.0000	1.0000	1.0000	1.0000	1.0000	1.0000	1.0000	1.0000	1.0000	1.0000	0.9999
	32	1.0000	1.0000	1.0000	1.0000	1.0000	1.0000	1.0000	1.0000	1.0000	1.0000	1.0000	1.0000	1.0000	1.0000

Table A.4: T Distribution Table

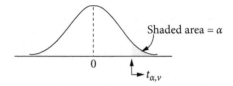

Shaded area = α

$t_{\alpha,\nu}$

df/α =	.40	.25	.10	.05	.025	.01	.005	.001	.0005
1	0.325	1.000	3.078	6.314	12.706	31.821	63.657	318.309	636.619
2	0.289	0.816	1.886	2.920	4.303	6.965	9.925	22.327	31.599
3	0.277	0.765	1.638	2.353	3.182	4.541	5.841	10.215	12.924
4	0.271	0.741	1.533	2.132	2.776	3.747	4.604	7.173	8.610
5	0.267	0.727	1.476	2.015	2.571	3.365	4.032	5.893	6.869
6	0.265	0.718	1.440	1.943	2.447	3.143	3.707	5.208	5.959
7	0.263	0.711	1.415	1.895	2.365	2.998	3.499	4.785	5.408
8	0.262	0.706	1.397	1.860	2.306	2.896	3.355	4.501	5.041
9	0.261	0.703	1.383	1.833	2.262	2.821	3.250	4.297	4.781
10	0.260	0.700	1.372	1.812	2.228	2.764	3.169	4.144	4.587
11	0.260	0.697	1.363	1.796	2.201	2.718	3.106	4.025	4.437
12	0.259	0.695	1.356	1.782	2.179	2.681	3.055	3.930	4.318
13	0.259	0.694	1.350	1.771	2.160	2.650	3.012	3.852	4.221
14	0.258	0.692	1.345	1.761	2.145	2.624	2.977	3.787	4.140
15	0.258	0.691	1.341	1.753	2.131	2.602	2.947	3.733	4.073
16	0.258	0.690	1.337	1.746	2.120	2.583	2.921	3.686	4.015
17	0.257	0.689	1.333	1.740	2.110	2.567	2.898	3.646	3.965
18	0.257	0.688	1.330	1.734	2.101	2.552	2.878	3.610	3.922
19	0.257	0.688	1.328	1.729	2.093	2.539	2.861	3.579	3.883
20	0.257	0.687	1.325	1.725	2.086	2.528	2.845	3.552	3.850
21	0.257	0.686	1.323	1.721	2.080	2.518	2.831	3.527	3.819
22	0.256	0.686	1.321	1.717	2.074	2.508	2.819	3.505	3.792
23	0.256	0.685	1.319	1.714	2.069	2.500	2.807	3.485	3.768
24	0.256	0.685	1.318	1.711	2.064	2.492	2.797	3.467	3.745
25	0.256	0.684	1.316	1.708	2.060	2.485	2.787	3.450	3.725
26	0.256	0.684	1.315	1.706	2.056	2.479	2.779	3.435	3.707
27	0.256	0.684	1.314	1.703	2.052	2.473	2.771	3.421	3.690
28	0.256	0.683	1.313	1.701	2.048	2.467	2.763	3.408	3.674
29	0.256	0.683	1.311	1.699	2.045	2.462	2.756	3.396	3.659
30	0.256	0.683	1.310	1.697	2.042	2.457	2.750	3.385	3.646
35	0.255	0.682	1.306	1.690	2.030	2.438	2.724	3.340	3.591
40	0.255	0.681	1.303	1.684	2.021	2.423	2.704	3.307	3.551
50	0.255	0.679	1.299	1.676	2.009	2.403	2.678	3.261	3.496
60	0.254	0.679	1.296	1.671	2.000	2.390	2.660	3.232	3.460
120	0.254	0.677	1.289	1.658	1.980	2.358	2.617	3.160	3.373
inf.	0.253	0.674	1.282	1.645	1.960	2.326	2.576	3.090	3.291

Source: Computed by M. Longnecker using Splus.

Table A.5: Chi-square Distribution Table

df	$\alpha =$ .999	.995	.99	.975	.95	.90
1	.000002	.000039	.000157	.000982	.003932	.01579
2	.002001	.01003	.02010	.05064	.1026	.2107
3	.02430	.07172	.1148	.2158	.3518	.5844
4	.09080	.2070	.2971	.4844	.7107	1.064
5	.2102	.4117	.5543	.8312	1.145	1.610
6	.3811	.6757	.8721	1.237	1.635	2.204
7	.5985	.9893	1.239	1.690	2.167	2.833
8	.8571	1.344	1.646	2.180	2.733	3.490
9	1.152	1.735	2.088	2.700	3.325	4.168
10	1.479	2.156	2.558	3.247	3.940	4.865
11	1.834	2.603	3.053	3.816	4.575	5.578
12	2.214	3.074	3.571	4.404	5.226	6.304
13	2.617	3.565	4.107	5.009	5.892	7.042
14	3.041	4.075	4.660	5.629	6.571	7.790
15	3.483	4.601	5.229	6.262	7.261	8.547
16	3.942	5.142	5.812	6.908	7.962	9.312
17	4.416	5.697	6.408	7.564	8.672	10.09
18	4.905	6.265	7.015	8.231	9.390	10.86
19	5.407	6.844	7.633	8.907	10.12	11.65
20	5.921	7.434	8.260	9.591	10.85	12.44
21	6.447	8.034	8.897	10.28	11.59	13.24
22	6.983	8.643	9.542	10.98	12.34	14.04
23	7.529	9.260	10.20	11.69	13.09	14.85
24	8.085	9.886	10.86	12.40	13.85	15.66
25	8.649	10.52	11.52	13.12	14.61	16.47
26	9.222	11.16	12.20	13.84	15.38	17.29
27	9.803	11.81	12.88	14.57	16.15	18.11
28	10.39	12.46	13.56	15.31	16.93	18.94
29	10.99	13.12	14.26	16.06	17.71	19.77
30	11.59	13.79	14.95	16.79	18.49	20.60
40	17.92	20.71	22.16	24.43	26.51	29.05
50	24.67	27.99	29.71	32.36	34.76	37.69
60	31.74	35.53	37.48	40.48	43.19	46.46
70	39.04	43.28	45.44	48.76	51.74	55.33
80	46.52	51.17	53.54	57.15	60.39	64.28
90	54.16	59.20	61.75	65.65	69.13	73.29
100	61.92	67.33	70.06	74.22	77.93	82.36
120	77.76	83.85	86.92	91.57	95.70	100.62
240	177.95	187.32	191.99	198.98	205.14	212.39

(*continued*)

$\alpha =$ .10	.05	.025	.01	.005	.001	df
2.706	3.841	5.024	6.635	7.879	10.83	1
4.605	5.991	7.378	9.210	10.60	13.82	2
6.251	7.815	9.348	11.34	12.84	16.27	3
7.779	9.488	11.14	13.28	14.86	18.47	4
9.236	11.07	12.83	15.09	16.75	20.52	5
10.64	12.59	14.45	16.81	18.55	22.46	6
12.02	14.07	16.01	18.48	20.28	24.32	7
13.36	15.51	17.53	20.09	21.95	26.12	8
14.68	16.92	19.02	21.67	23.59	27.88	9
15.99	18.31	20.48	23.21	25.19	29.59	10
17.28	19.68	21.92	24.72	26.76	31.27	11
18.55	21.03	23.34	26.22	28.30	32.91	12
19.81	22.36	24.74	27.69	29.82	34.53	13
21.06	23.68	26.12	29.14	31.32	36.12	14
22.31	25.00	27.49	30.58	32.80	37.70	15
23.54	26.30	28.85	32.00	34.27	39.25	16
24.77	27.59	30.19	33.41	35.72	40.79	17
25.99	28.87	31.53	34.81	37.16	42.31	18
27.20	30.14	32.85	36.19	38.58	43.82	19
28.41	31.41	34.17	37.57	40.00	45.31	20
29.62	32.67	35.48	38.93	41.40	46.80	21
30.81	33.92	36.78	40.29	42.80	48.27	22
32.01	35.17	38.08	41.64	44.18	49.73	23
33.20	36.42	39.36	42.98	45.56	51.18	24
34.38	37.65	40.65	44.31	46.93	52.62	25
35.56	38.89	41.92	45.64	48.29	54.05	26
36.74	40.11	43.19	46.96	49.65	55.48	27
37.92	41.34	44.46	48.28	50.99	56.89	28
39.09	42.56	45.72	49.59	52.34	58.30	29
40.26	43.77	46.98	50.89	53.67	59.70	30
51.81	55.76	59.34	63.69	66.77	73.40	40
63.17	67.50	71.42	76.15	79.49	86.66	50
74.40	79.08	83.30	88.38	91.95	99.61	60
85.53	90.53	95.02	100.43	104.21	112.32	70
96.58	101.88	106.63	112.33	116.32	124.84	80
107.57	113.15	118.14	124.12	128.30	137.21	90
118.50	124.34	129.56	135.81	140.17	149.45	100
140.23	146.57	152.21	158.95	163.65	173.62	120
268.47	277.14	284.80	293.89	300.18	313.44	240

Source: Computed by P. J. Hildebrand.

Table A.6: F Distribution Table

Percentage points of the F distribution (df_2 between 1 and 6)

df$_2$	α	1	2	3	4	5	6	7	8	9	10
											df$_1$
1	.25	5.83	7.50	8.20	8.58	8.82	8.98	9.10	9.19	9.26	9.32
	.10	39.86	49.50	53.59	55.83	57.24	58.20	58.91	59.44	59.86	60.19
	.05	161.4	199.5	215.7	224.6	230.2	234.0	236.8	238.9	240.5	241.9
	.025	647.8	799.5	864.2	899.6	921.8	937.1	948.2	956.7	963.3	968.6
	.01	4052	5000	5403	5625	5764	5859	5928	5981	6022	6056
2	.25	2.57	3.00	3.15	3.23	3.28	3.31	3.34	3.35	3.37	3.38
	.10	8.53	9.00	9.16	9.24	9.29	9.33	9.35	9.37	9.38	9.39
	.05	18.51	19.00	19.16	19.25	19.30	19.33	19.35	19.37	19.38	19.40
	.025	38.51	39.00	39.17	39.25	39.30	39.33	39.36	39.37	39.39	39.40
	.01	98.50	99.00	99.17	99.25	99.30	99.33	99.36	99.37	99.39	99.40
	.005	198.5	199.0	199.2	199.2	199.3	199.3	199.4	199.4	199.4	199.4
	.001	998.5	999.0	999.2	999.2	999.3	999.3	999.4	999.4	999.4	999.4
3	.25	2.02	2.28	2.36	2.39	2.41	2.42	2.43	2.44	2.44	2.44
	.10	5.54	5.46	5.39	5.34	5.31	5.28	5.27	5.25	5.24	5.23
	.05	10.13	9.55	9.28	9.12	9.01	8.94	8.89	8.85	8.81	8.79
	.025	17.44	16.04	15.44	15.10	14.88	14.73	14.62	14.54	14.47	14.42
	.01	34.12	30.82	29.46	28.71	28.24	27.91	27.67	27.49	27.35	27.23
	.005	55.55	49.80	47.47	46.19	45.39	44.84	44.43	44.13	43.88	43.69
	.001	167.0	148.5	141.1	137.1	134.6	132.8	131.6	130.6	129.9	129.2
4	.25	1.81	2.00	2.05	2.06	2.07	2.08	2.08	2.08	2.08	2.08
	.10	4.54	4.32	4.19	4.11	4.05	4.01	3.98	3.95	3.94	3.92
	.05	7.71	6.94	6.59	6.39	6.26	6.16	6.09	6.04	6.00	5.96
	.025	12.22	10.65	9.98	9.60	9.36	9.20	9.07	8.98	8.90	8.84
	.01	21.20	18.00	16.69	15.98	15.52	15.21	14.98	14.80	14.66	14.55
	.005	31.33	26.28	24.26	23.15	22.46	21.97	21.62	21.35	21.14	20.97
	.001	74.14	61.25	56.18	53.44	51.71	50.53	49.66	49.00	48.47	48.05
5	.25	1.69	1.85	1.88	1.89	1.89	1.89	1.89	1.89	1.89	1.89
	.10	4.06	3.78	3.62	3.52	3.45	3.40	3.37	3.34	3.32	3.30
	.05	6.61	5.79	5.41	5.19	5.05	4.95	4.88	4.82	4.77	4.74
	.025	10.01	8.43	7.76	7.39	7.15	6.98	6.85	6.76	6.68	6.62
	.01	16.26	13.27	12.06	11.39	10.97	10.67	10.46	10.29	10.16	10.05
	.005	22.78	18.31	16.53	15.56	14.94	14.51	14.20	13.96	13.77	13.62
	.001	47.18	37.12	33.20	31.09	29.75	28.83	28.16	27.65	27.24	26.92
6	.25	1.62	1.76	1.78	1.79	1.79	1.78	1.78	1.78	1.77	1.77
	.10	3.78	3.46	3.29	3.18	3.11	3.05	3.01	2.98	2.96	2.94
	.05	5.99	5.14	4.76	4.53	4.39	4.28	4.21	4.15	4.10	4.06
	.025	8.81	7.26	6.60	6.23	5.99	5.82	5.70	5.60	5.52	5.46
	.01	13.75	10.92	9.78	9.15	8.75	8.47	8.26	8.10	7.98	7.87
	.005	18.63	14.54	12.92	12.03	11.46	11.07	10.79	10.57	10.39	10.25
	.001	35.51	27.00	23.70	21.92	20.80	20.03	19.46	19.03	18.69	18.41

TABLE 8
(*continued*)

12	15	20	24	30	40	60	120	240	inf.	α	df_2
										df_1	
9.41	9.49	9.58	9.63	9.67	9.71	9.76	9.80	9.83	9.85	.25	1
60.71	61.22	61.74	62.00	62.26	62.53	62.79	63.06	63.19	63.33	.10	
243.9	245.9	248.0	249.1	250.1	251.1	252.2	253.3	253.8	254.3	.05	
976.7	984.9	993.1	997.2	1001	1006	1010	1014	1016	1018	.025	
6106	6157	6209	6235	6261	6287	6313	6339	6353	6366	.01	
3.39	3.41	3.43	3.43	3.44	3.45	3.46	3.47	3.47	3.48	.25	2
9.41	9.42	9.44	9.45	9.46	9.47	9.47	9.48	9.49	9.49	.10	
19.41	19.43	19.45	19.45	19.46	19.47	19.48	19.49	19.49	19.50	.05	
39.41	39.43	39.45	39.46	39.46	39.47	39.48	39.49	39.49	39.50	.025	
99.42	99.43	99.45	99.46	99.47	99.47	99.48	99.49	99.50	99.50	.01	
199.4	199.4	199.4	199.5	199.5	199.5	199.5	199.5	199.5	199.5	.005	
999.4	999.4	999.4	999.5	999.5	999.5	999.5	999.5	999.5	999.5	.001	
2.45	2.46	2.46	2.46	2.47	2.47	2.47	2.47	2.47	2.47	.25	3
5.22	5.20	5.18	5.18	5.17	5.16	5.15	5.14	5.14	5.13	.10	
8.74	8.70	8.66	8.64	8.62	8.59	8.57	8.55	8.54	8.53	.05	
14.34	14.25	14.17	14.12	14.08	14.04	13.99	13.95	13.92	13.90	.025	
27.05	26.87	26.69	26.60	26.50	26.41	26.32	26.22	26.17	26.13	.01	
43.39	43.08	42.78	42.62	42.47	42.31	42.15	41.99	41.91	41.83	.005	
128.3	127.4	126.4	125.9	125.4	125.0	124.5	124.0	123.7	123.5	.001	
2.08	2.08	2.08	2.08	2.08	2.08	2.08	2.08	2.08	2.08	.25	4
3.90	3.87	3.84	3.83	3.82	3.80	3.79	3.78	3.77	3.76	.10	
5.91	5.86	5.80	5.77	5.75	5.72	5.69	5.66	5.64	5.63	.05	
8.75	8.66	8.56	8.51	8.46	8.41	8.36	8.31	8.28	8.26	.025	
14.37	14.20	14.02	13.93	13.84	13.75	13.65	13.56	13.51	13.46	.01	
20.70	20.44	20.17	20.03	19.89	19.75	19.61	19.47	19.40	19.32	.005	
47.41	46.76	46.10	45.77	45.43	45.09	44.75	44.40	44.23	44.05	.001	
1.89	1.89	1.88	1.88	1.88	1.88	1.87	1.87	1.87	1.87	.25	5
3.27	3.24	3.21	3.19	3.17	3.16	3.14	3.12	3.11	3.10	.10	
4.68	4.62	4.56	4.53	4.50	4.46	4.43	4.40	4.38	4.36	.05	
6.52	6.43	6.33	6.28	6.23	6.18	6.12	6.07	6.04	6.02	.025	
9.89	9.72	9.55	9.47	9.38	9.29	9.20	9.11	9.07	9.02	.01	
13.38	13.15	12.90	12.78	12.66	12.53	12.40	12.27	12.21	12.14	.005	
26.42	25.91	25.39	25.13	24.87	24.60	24.33	24.06	23.92	23.79	.001	
1.77	1.76	1.76	1.75	1.75	1.75	1.74	1.74	1.74	1.74	.25	6
2.90	2.87	2.84	2.82	2.80	2.78	2.76	2.74	2.73	2.72	.10	
4.00	3.94	3.87	3.84	3.81	3.77	3.74	3.70	3.69	3.67	.05	
5.37	5.27	5.17	5.12	5.07	5.01	4.96	4.90	4.88	4.85	.025	
7.72	7.56	7.40	7.31	7.23	7.14	7.06	6.97	6.92	6.88	.01	
10.03	9.81	9.59	9.47	9.36	9.24	9.12	9.00	8.94	8.88	.005	
17.99	17.56	17.12	16.90	16.67	16.44	16.21	15.98	15.86	15.75	.001	

TABLE 8
Percentage points of the F distribution (df$_2$ between 7 and 12)

df$_2$	α	1	2	3	4	5	6	7	8	9	10
							df$_1$				
7	.25	1.57	1.70	1.72	1.72	1.71	1.71	1.70	1.70	1.69	1.69
	.10	3.59	3.26	3.07	2.96	2.88	2.83	2.78	2.75	2.72	2.70
	.05	5.59	4.74	4.35	4.12	3.97	3.87	3.79	3.73	3.68	3.64
	.025	8.07	6.54	5.89	5.52	5.29	5.12	4.99	4.90	4.82	4.76
	.01	12.25	9.55	8.45	7.85	7.46	7.19	6.99	6.84	6.72	6.62
	.005	16.24	12.40	10.88	10.05	9.52	9.16	8.89	8.68	8.51	8.38
	.001	29.25	21.69	18.77	17.20	16.21	15.52	15.02	14.63	14.33	14.08
8	.25	1.54	1.66	1.67	1.66	1.66	1.65	1.64	1.64	1.63	1.63
	.10	3.46	3.11	2.92	2.81	2.73	2.67	2.62	2.59	2.56	2.54
	.05	5.32	4.46	4.07	3.84	3.69	3.58	3.50	3.44	3.39	3.35
	.025	7.57	6.06	5.42	5.05	4.82	4.65	4.53	4.43	4.36	4.30
	.01	11.26	8.65	7.59	7.01	6.63	6.37	6.18	6.03	5.91	5.81
	.005	14.69	11.04	9.60	8.81	8.30	7.95	7.69	7.50	7.34	7.21
	.001	25.41	18.49	15.83	14.39	13.48	12.86	12.40	12.05	11.77	11.54
9	.25	1.51	1.62	1.63	1.63	1.62	1.61	1.60	1.60	1.59	1.59
	.10	3.36	3.01	2.81	2.69	2.61	2.55	2.51	2.47	2.44	2.42
	.05	5.12	4.26	3.86	3.63	3.48	3.37	3.29	3.23	3.18	3.14
	.025	7.21	5.71	5.08	4.72	4.48	4.32	4.20	4.10	4.03	3.96
	.01	10.56	8.02	6.99	6.42	6.06	5.80	5.61	5.47	5.35	5.26
	.005	13.61	10.11	8.72	7.96	7.47	7.13	6.88	6.69	6.54	6.42
	.001	22.86	16.39	13.90	12.56	11.71	11.13	10.70	10.37	10.11	9.89
10	.25	1.49	1.60	1.60	1.59	1.59	1.58	1.57	1.56	1.56	1.55
	.10	3.29	2.92	2.73	2.61	2.52	2.46	2.41	2.38	2.35	2.32
	.05	4.96	4.10	3.71	3.48	3.33	3.22	3.14	3.07	3.02	2.98
	.025	6.94	5.46	4.83	4.47	4.24	4.07	3.95	3.85	3.78	3.72
	.01	10.04	7.56	6.55	5.99	5.64	5.39	5.20	5.06	4.94	4.85
	.005	12.83	9.43	8.08	7.34	6.87	6.54	6.30	6.12	5.97	5.85
	.001	21.04	14.91	12.55	11.28	10.48	9.93	9.52	9.20	8.96	8.75
11	.25	1.47	1.58	1.58	1.57	1.56	1.55	1.54	1.53	1.53	1.52
	.10	3.23	2.86	2.66	2.54	2.45	2.39	2.34	2.30	2.27	2.25
	.05	4.84	3.98	3.59	3.36	3.20	3.09	3.01	2.95	2.90	2.85
	.025	6.72	5.26	4.63	4.28	4.04	3.88	3.76	3.66	3.59	3.53
	.01	9.65	7.21	6.22	5.67	5.32	5.07	4.89	4.74	4.63	4.54
	.005	12.23	8.91	7.60	6.88	6.42	6.10	5.86	5.68	5.54	5.42
	.001	19.69	13.81	11.56	10.35	9.58	9.05	8.66	8.35	8.12	7.92
12	.25	1.46	1.56	1.56	1.55	1.54	1.53	1.52	1.51	1.51	1.50
	.10	3.18	2.81	2.61	2.48	2.39	2.33	2.28	2.24	2.21	2.19
	.05	4.75	3.89	3.49	3.26	3.11	3.00	2.91	2.85	2.80	2.75
	.025	6.55	5.10	4.47	4.12	3.89	3.73	3.61	3.51	3.44	3.37
	.01	9.33	6.93	5.95	5.41	5.06	4.82	4.64	4.50	4.39	4.30
	.005	11.75	8.51	7.23	6.52	6.07	5.76	5.52	5.35	5.20	5.09
	.001	18.64	12.97	10.80	9.63	8.89	8.38	8.00	7.71	7.48	7.29

(*continued*)

12	15	20	24	30	40	60	120	240	inf.	α	df₂
											df₁
1.68	1.68	1.67	1.67	1.66	1.66	1.65	1.65	1.65	1.65	.25	7
2.67	2.63	2.59	2.58	2.56	2.54	2.51	2.49	2.48	2.47	.10	
3.57	3.51	3.44	3.41	3.38	3.34	3.30	3.27	3.25	3.23	.05	
4.67	4.57	4.47	4.41	4.36	4.31	4.25	4.20	4.17	4.14	.025	
6.47	6.31	6.16	6.07	5.99	5.91	5.82	5.74	5.69	5.65	.01	
8.18	7.97	7.75	7.64	7.53	7.42	7.31	7.19	7.13	7.08	.005	
13.71	13.32	12.93	12.73	12.53	12.33	12.12	11.91	11.80	11.70	.001	
1.62	1.62	1.61	1.60	1.60	1.59	1.59	1.58	1.58	1.58	.25	8
2.50	2.46	2.42	2.40	2.38	2.36	2.34	2.32	2.30	2.29	.10	
3.28	3.22	3.15	3.12	3.08	3.04	3.01	2.97	2.95	2.93	.05	
4.20	4.10	4.00	3.95	3.89	3.84	3.78	3.73	3.70	3.67	.025	
5.67	5.52	5.36	5.28	5.20	5.12	5.03	4.95	4.90	4.86	.01	
7.01	6.81	6.61	6.50	6.40	6.29	6.18	6.06	6.01	5.95	.005	
11.19	10.84	10.48	10.30	10.11	9.92	9.73	9.53	9.43	9.33	.001	
1.58	1.57	1.56	1.56	1.55	1.54	1.64	1.53	1.53	1.53	.25	9
2.38	2.34	2.30	2.28	2.25	2.23	2.21	2.18	2.17	2.16	.10	
3.07	3.01	2.94	2.90	2.86	2.83	2.79	2.75	2.73	2.71	.05	
3.87	3.77	3.67	3.61	3.56	3.51	3.45	3.39	3.36	3.33	.025	
5.11	4.96	4.81	4.73	4.65	4.57	4.48	4.40	4.35	4.31	.01	
6.23	6.03	5.83	5.73	5.62	5.52	5.41	5.30	5.24	5.19	.005	
9.57	9.24	8.90	8.72	8.55	8.37	8.19	8.00	7.91	7.81	.001	
1.54	1.53	1.52	1.52	1.51	1.51	1.50	1.49	1.49	1.48	.25	10
2.28	2.24	2.20	2.18	2.16	2.13	2.11	2.08	2.07	2.06	.10	
2.91	2.85	2.77	2.74	2.70	2.66	2.62	2.58	2.56	2.54	.05	
3.62	3.52	3.42	3.37	3.31	3.26	3.20	3.14	3.11	3.08	.025	
4.71	4.56	4.41	4.33	4.25	4.17	4.08	4.00	3.95	3.91	.01	
5.66	5.47	5.27	5.17	5.07	4.97	4.86	4.75	4.69	4.64	.005	
8.45	8.13	7.80	7.64	7.47	7.30	7.12	6.94	6.85	6.76	.001	
1.51	1.50	1.49	1.49	1.48	1.47	1.47	1.46	1.45	1.45	.25	11
2.21	2.17	2.12	2.10	2.08	2.05	2.03	2.00	1.99	1.97	.10	
2.79	2.72	2.65	2.61	2.57	2.53	2.49	2.45	2.43	2.40	.05	
3.43	3.33	3.23	3.17	3.12	3.06	3.00	2.94	2.91	2.88	.025	
4.40	4.25	4.10	4.02	3.94	3.86	3.78	3.69	3.65	3.60	.01	
5.24	5.05	4.86	4.76	4.65	4.55	4.45	4.34	4.28	4.23	.005	
7.63	7.32	7.01	6.85	6.68	6.52	6.35	6.18	6.09	6.00	.001	
1.49	1.48	1.47	1.46	1.45	1.45	1.44	1.43	1.43	1.42	.25	12
2.15	2.10	2.06	2.04	2.01	1.99	1.96	1.93	1.92	1.90	.10	
2.69	2.62	2.54	2.51	2.47	2.43	2.38	2.34	2.32	2.30	.05	
3.28	3.18	3.07	3.02	2.96	2.91	2.85	2.79	2.76	2.72	.025	
4.16	4.01	3.86	3.78	3.70	3.62	3.54	3.45	3.41	3.36	.01	
4.91	4.72	4.53	4.43	4.33	4.23	4.12	4.01	3.96	3.90	.005	
7.00	6.71	6.40	6.25	6.09	5.93	5.76	5.59	5.51	5.42	.001	

Percentage points of the F distribution (df$_2$ between 13 and 18)

df$_2$	α	1	2	3	4	5	6	7	8	9	10
						df$_1$					
13	.25	1.45	1.55	1.55	1.53	1.52	1.51	1.50	1.49	1.49	1.48
	.10	3.14	2.76	2.56	2.43	2.35	2.28	2.23	2.20	2.16	2.14
	.05	4.67	3.81	3.41	3.18	3.03	2.92	2.83	2.77	2.71	2.67
	.025	6.41	4.97	4.35	4.00	3.77	3.60	3.48	3.39	3.31	3.25
	.01	9.07	6.70	5.74	5.21	4.86	4.62	4.44	4.30	4.19	4.10
	.005	11.37	8.19	6.93	6.23	5.79	5.48	5.25	5.08	4.94	4.82
	.001	17.82	12.31	10.21	9.07	8.35	7.86	7.49	7.21	6.98	6.80
14	.25	1.44	1.53	1.53	1.52	1.51	1.50	1.49	1.48	1.47	1.46
	.10	3.10	2.73	2.52	2.39	2.31	2.24	2.19	2.15	2.12	2.10
	.05	4.60	3.74	3.34	3.11	2.96	2.85	2.76	2.70	2.65	2.60
	.025	6.30	4.86	4.24	3.89	3.66	3.50	3.38	3.29	3.21	3.15
	.01	8.86	6.51	5.56	5.04	4.69	4.46	4.28	4.14	4.03	3.94
	.005	11.06	7.92	6.68	6.00	5.56	5.26	5.03	4.86	4.72	4.60
	.001	17.14	11.78	9.73	8.62	7.92	7.44	7.08	6.80	6.58	6.40
15	.25	1.43	1.52	1.52	1.51	1.49	1.48	1.47	1.46	1.46	1.45
	.10	3.07	2.70	2.49	2.36	2.27	2.21	2.16	2.12	2.09	2.06
	.05	4.54	3.68	3.29	3.06	2.90	2.79	2.71	2.64	2.59	2.54
	.025	6.20	4.77	4.15	3.80	3.58	3.41	3.29	3.20	3.12	3.06
	.01	8.68	6.36	5.42	4.89	4.56	4.32	4.14	4.00	3.89	3.80
	.005	10.80	7.70	6.48	5.80	5.37	5.07	4.85	4.67	4.54	4.42
	.001	16.59	11.34	9.34	8.25	7.57	7.09	6.74	6.47	6.26	6.08
16	.25	1.42	1.51	1.51	1.50	1.48	1.47	1.46	1.45	1.44	1.44
	.10	3.05	2.67	2.46	2.33	2.24	2.18	2.13	2.09	2.06	2.03
	.05	4.49	3.63	3.24	3.01	2.85	2.74	2.66	2.59	2.54	2.49
	.025	6.12	4.69	4.08	3.73	3.50	3.34	3.22	3.12	3.05	2.99
	.01	8.53	6.23	5.29	4.77	4.44	4.20	4.03	3.89	3.78	3.69
	.005	10.58	7.51	6.30	5.64	5.21	4.91	4.69	4.52	4.38	4.27
	.001	16.12	10.97	9.01	7.94	7.27	6.80	6.46	6.19	5.98	5.81
17	.25	1.42	1.51	1.50	1.49	1.47	1.46	1.45	1.44	1.43	1.43
	.10	3.03	2.64	2.44	2.31	2.22	2.15	2.10	2.06	2.03	2.00
	.05	4.45	3.59	3.20	2.96	2.81	2.70	2.61	2.55	2.49	2.45
	.025	6.04	4.62	4.01	3.66	3.44	3.28	3.16	3.06	2.98	2.92
	.01	8.40	6.11	5.18	4.67	4.34	4.10	3.93	3.79	3.68	3.59
	.005	10.38	7.35	6.16	5.50	5.07	4.78	4.56	4.39	4.25	4.14
	.001	15.72	10.66	8.73	7.68	7.02	6.56	6.22	5.96	5.75	5.58
18	.25	1.41	1.50	1.49	1.48	1.46	1.45	1.44	1.43	1.42	1.42
	.10	3.01	2.62	2.42	2.29	2.20	2.13	2.08	2.04	2.00	1.98
	.05	4.41	3.55	3.16	2.93	2.77	2.66	2.58	2.51	2.46	2.41
	.025	5.98	4.56	3.95	3.61	3.38	3.22	3.10	3.01	2.93	2.87
	.01	8.29	6.01	5.09	4.58	4.25	4.01	3.84	3.71	3.60	3.51
	.005	10.22	7.21	6.03	5.37	4.96	4.66	4.44	4.28	4.14	4.03
	.001	15.38	10.39	8.49	7.46	6.81	6.35	6.02	5.76	5.56	5.39

(continued)

					df_1						
12	**15**	**20**	**24**	**30**	**40**	**60**	**120**	**240**	**inf.**	α	df_2
1.47	1.46	1.45	1.44	1.43	1.42	1.42	1.41	1.40	1.40	.25	**13**
2.10	2.05	2.01	1.98	1.96	1.93	1.90	1.88	1.86	1.85	.10	
2.60	2.53	2.46	2.42	2.38	2.34	2.30	2.25	2.23	2.21	.05	
3.15	3.05	2.95	2.89	2.84	2.78	2.72	2.66	2.63	2.60	.025	
3.96	3.82	3.66	3.59	3.51	3.43	3.34	3.25	3.21	3.17	.01	
4.64	4.46	4.27	4.17	4.07	3.97	3.87	3.76	3.70	3.65	.005	
6.52	6.23	5.93	5.78	5.63	5.47	5.30	5.14	5.05	4.97	.001	
1.45	1.44	1.43	1.42	1.41	1.41	1.40	1.39	1.38	1.38	.25	**14**
2.05	2.01	1.96	1.94	1.91	1.89	1.86	1.83	1.81	1.80	.10	
2.53	2.46	2.39	2.35	2.31	2.27	2.22	2.18	2.15	2.13	.05	
3.05	2.95	2.84	2.79	2.73	2.67	2.61	2.55	2.52	2.49	.025	
3.80	3.66	3.51	3.43	3.35	3.27	3.18	3.09	3.05	3.00	.01	
4.43	4.25	4.06	3.96	3.86	3.76	3.66	3.55	3.49	3.44	.005	
6.13	5.85	5.56	5.41	5.25	5.10	4.94	4.77	4.69	4.60	.001	
1.44	1.43	1.41	1.41	1.40	1.39	1.38	1.37	1.36	1.36	.25	**15**
2.02	1.97	1.92	1.90	1.87	1.85	1.82	1.79	1.77	1.76	.10	
2.48	2.40	2.33	2.29	2.25	2.20	2.16	2.11	2.09	2.07	.05	
2.96	2.86	2.76	2.70	2.64	2.59	2.52	2.46	2.43	2.40	.025	
3.67	3.52	3.37	3.29	3.21	3.13	3.05	2.96	2.91	2.87	.01	
4.25	4.07	3.88	3.79	3.69	3.58	3.48	3.37	3.32	3.26	.005	
5.81	5.54	5.25	5.10	4.95	4.80	4.64	4.47	4.39	4.31	.001	
1.43	1.41	1.40	1.39	1.38	1.37	1.36	1.35	1.35	1.34	.25	**16**
1.99	1.94	1.89	1.87	1.84	1.81	1.78	1.75	1.73	1.72	.10	
2.42	2.35	2.28	2.24	2.19	2.15	2.11	2.06	2.03	2.01	.05	
2.89	2.79	2.68	2.63	2.57	2.51	2.45	2.38	2.35	2.32	.025	
3.55	3.41	3.26	3.18	3.10	3.02	2.93	2.84	2.80	2.75	.01	
4.10	3.92	3.73	3.64	3.54	3.44	3.33	3.22	3.17	3.11	.005	
5.55	5.27	4.99	4.85	4.70	4.54	4.39	4.23	4.14	4.06	.001	
1.41	1.40	1.39	1.38	1.37	1.36	1.35	1.34	1.33	1.33	.25	**17**
1.96	1.91	1.86	1.84	1.81	1.78	1.75	1.72	1.70	1.69	.10	
2.38	2.31	2.23	2.19	2.15	2.10	2.06	2.01	1.99	1.96	.05	
2.82	2.72	2.62	2.56	2.50	2.44	2.38	2.32	2.28	2.25	.025	
3.46	3.31	3.16	3.08	3.00	2.92	2.83	2.75	2.70	2.65	.01	
3.97	3.79	3.61	3.51	3.41	3.31	3.21	3.10	3.04	2.98	.005	
5.32	5.05	4.78	4.63	4.48	4.33	4.18	4.02	3.93	3.85	.001	
1.40	1.39	1.38	1.37	1.36	1.35	1.34	1.33	1.32	1.32	.25	**18**
1.93	1.89	1.84	1.81	1.78	1.75	1.72	1.69	1.67	1.66	.10	
2.34	2.27	2.19	2.15	2.11	2.06	2.02	1.97	1.94	1.92	.05	
2.77	2.67	2.56	2.50	2.44	2.38	2.32	2.26	2.22	2.19	.025	
3.37	3.23	3.08	3.00	2.92	2.84	2.75	2.66	2.61	2.57	.01	
3.86	3.68	3.50	3.40	3.30	3.20	3.10	2.99	2.93	2.87	.005	
5.13	4.87	4.59	4.45	4.30	4.15	4.00	3.84	3.75	3.67	.001	

Percentage points of the F distribution (df_2 between 19 and 24)

df_2	α	**df$_1$**									
		1	**2**	**3**	**4**	**5**	**6**	**7**	**8**	**9**	**10**
19	.25	1.41	1.49	1.49	1.47	1.46	1.44	1.43	1.42	1.41	1.41
	.10	2.99	2.61	2.40	2.27	2.18	2.11	2.06	2.02	1.98	1.96
	.05	4.38	3.52	3.13	2.90	2.74	2.63	2.54	2.48	2.42	2.38
	.025	5.92	4.51	3.90	3.56	3.33	3.17	3.05	2.96	2.88	2.82
	.01	8.18	5.93	5.01	4.50	4.17	3.94	3.77	3.63	3.52	3.43
	.005	10.07	7.09	5.92	5.27	4.85	4.56	4.34	4.18	4.04	3.93
	.001	15.08	10.16	8.28	7.27	6.62	6.18	5.85	5.59	5.39	5.22
20	.25	1.40	1.49	1.48	1.47	1.45	1.44	1.43	1.42	1.41	1.40
	.10	2.97	2.59	2.38	2.25	2.16	2.09	2.04	2.00	1.96	1.94
	.05	4.35	3.49	3.10	2.87	2.71	2.60	2.51	2.45	2.39	2.35
	.025	5.87	4.46	3.86	3.51	3.29	3.13	3.01	2.91	2.84	2.77
	.01	8.10	5.85	4.94	4.43	4.10	3.87	3.70	3.56	3.46	3.37
	.005	9.94	6.99	5.82	5.17	4.76	4.47	4.26	4.09	3.96	3.85
	.001	14.82	9.95	8.10	7.10	6.46	6.02	5.69	5.44	5.24	5.08
21	.25	1.40	1.48	1.48	1.46	1.44	1.43	1.42	1.41	1.40	1.39
	.10	2.96	2.57	2.36	2.23	2.14	2.08	2.02	1.98	1.95	1.92
	.05	4.32	3.47	3.07	2.84	2.68	2.57	2.49	2.42	2.37	2.32
	.025	5.83	4.42	3.82	3.48	3.25	3.09	2.97	2.87	2.80	2.73
	.01	8.02	5.78	4.87	4.37	4.04	3.81	3.64	3.51	3.40	3.31
	.005	9.83	6.89	5.73	5.09	4.68	4.39	4.18	4.01	3.88	3.77
	.001	14.59	9.77	7.94	6.95	6.32	5.88	5.56	5.31	5.11	4.95
22	.25	1.40	1.48	1.47	1.45	1.44	1.42	1.41	1.40	1.39	1.39
	.10	2.95	2.56	2.35	2.22	2.13	2.06	2.01	1.97	1.93	1.90
	.05	4.30	3.44	3.05	2.82	2.66	2.55	2.46	2.40	2.34	2.30
	.025	5.79	4.38	3.78	3.44	3.22	3.05	2.93	2.84	2.76	2.70
	.01	7.95	5.72	4.82	4.31	3.99	3.76	3.59	3.45	3.35	3.26
	.005	9.73	6.81	5.65	5.02	4.61	4.32	4.11	3.94	3.81	3.70
	.001	14.38	9.61	7.80	6.81	6.19	5.76	5.44	5.19	4.99	4.83
23	.25	1.39	1.47	1.47	1.45	1.43	1.42	1.41	1.40	1.39	1.38
	.10	2.94	2.55	2.34	2.21	2.11	2.05	1.99	1.95	1.92	1.89
	.05	4.28	3.42	3.03	2.80	2.64	2.53	2.44	2.37	2.32	2.27
	.025	5.75	4.35	3.75	3.41	3.18	3.02	2.90	2.81	2.73	2.67
	.01	7.88	5.66	4.76	4.26	3.94	3.71	3.54	3.41	3.30	3.21
	.005	9.63	6.73	5.58	4.95	4.54	4.26	4.05	3.88	3.75	3.64
	.001	14.20	9.47	7.67	6.70	6.08	5.65	5.33	5.09	4.89	4.73
24	.25	1.39	1.47	1.46	1.44	1.43	1.41	1.40	1.39	1.38	1.38
	.10	2.93	2.54	2.33	2.19	2.10	2.04	1.98	1.94	1.91	1.88
	.05	4.26	3.40	3.01	2.78	2.62	2.51	2.42	2.36	2.30	2.25
	.025	5.72	4.32	3.72	3.38	3.15	2.99	2.87	2.78	2.70	2.64
	.01	7.82	5.61	4.72	4.22	3.90	3.67	3.50	3.36	3.26	3.17
	.005	9.55	6.66	5.52	4.89	4.49	4.20	3.99	3.83	3.69	3.59
	.001	14.03	9.34	7.55	6.59	5.98	5.55	5.23	4.99	4.80	4.64

(continued)

df₁											
12	**15**	**20**	**24**	**30**	**40**	**60**	**120**	**240**	**inf.**	**α**	**df_2**
1.40	1.38	1.37	1.36	1.35	1.34	1.33	1.32	1.31	1.30	.25	**19**
1.91	1.86	1.81	1.79	1.76	1.73	1.70	1.67	1.65	1.63	.10	
2.31	2.23	2.16	2.11	2.07	2.03	1.98	1.93	1.90	1.88	.05	
2.72	2.62	2.51	2.45	2.39	2.33	2.27	2.20	2.17	2.13	.025	
3.30	3.15	3.00	2.92	2.84	2.76	2.67	2.58	2.54	2.49	.01	
3.76	3.59	3.40	3.31	3.21	3.11	3.00	2.89	2.83	2.78	.005	
4.97	4.70	4.43	4.29	4.14	3.99	3.84	3.68	3.60	3.51	.001	
1.39	1.37	1.36	1.35	1.34	1.33	1.32	1.31	1.30	1.29	.25	**20**
1.89	1.84	1.79	1.77	1.74	1.71	1.68	1.64	1.63	1.61	.10	
2.28	2.20	2.12	2.08	2.04	1.99	1.95	1.90	1.87	1.84	.05	
2.68	2.57	2.46	2.41	2.35	2.29	2.22	2.16	2.12	2.09	.025	
3.23	3.09	2.94	2.86	2.78	2.69	2.61	2.52	2.47	2.42	.01	
3.68	3.50	3.32	3.22	3.12	3.02	2.92	2.81	2.75	2.69	.005	
4.82	4.56	4.29	4.15	4.00	3.86	3.70	3.54	3.46	3.38	.001	
1.38	1.37	1.35	1.34	1.33	1.32	1.31	1.30	1.29	1.28	.25	**21**
1.87	1.83	1.78	1.75	1.72	1.69	1.66	1.62	1.60	1.59	.10	
2.25	2.18	2.10	2.05	2.01	1.96	1.92	1.87	1.84	1.81	.05	
2.64	2.53	2.42	2.37	2.31	2.25	2.18	2.11	2.08	2.04	.025	
3.17	3.03	2.88	2.80	2.72	2.64	2.55	2.46	2.41	2.36	.01	
3.60	3.43	3.24	3.15	3.05	2.95	2.84	2.73	2.67	2.61	.005	
4.70	4.44	4.17	4.03	3.88	3.74	3.58	3.42	3.34	3.26	.001	
1.37	1.36	1.34	1.33	1.32	1.31	1.30	1.29	1.28	1.28	.25	**22**
1.86	1.81	1.76	1.73	1.70	1.67	1.64	1.60	1.59	1.57	.10	
2.23	2.15	2.07	2.03	1.98	1.94	1.89	1.84	1.81	1.78	.05	
2.60	2.50	2.39	2.33	2.27	2.21	2.14	2.08	2.04	2.00	.025	
3.12	2.98	2.83	2.75	2.67	2.58	2.50	2.40	2.35	2.31	.01	
3.54	3.36	3.18	3.08	2.98	2.88	2.77	2.66	2.60	2.55	.005	
4.58	4.33	4.06	3.92	3.78	3.63	3.48	3.32	3.23	3.15	.001	
1.37	1.35	1.34	1.33	1.32	1.31	1.30	1.28	1.28	1.27	.25	**23**
1.84	1.80	1.74	1.72	1.69	1.66	1.62	1.59	1.57	1.55	.10	
2.20	2.13	2.05	2.01	1.96	1.91	1.86	1.81	1.79	1.76	.05	
2.57	2.47	2.36	2.30	2.24	2.18	2.11	2.04	2.01	1.97	.025	
3.07	2.93	2.78	2.70	2.62	2.54	2.45	2.35	2.31	2.26	.01	
3.47	3.30	3.12	3.02	2.92	2.82	2.71	2.60	2.54	2.48	.005	
4.48	4.23	3.96	3.82	3.68	3.53	3.38	3.22	3.14	3.05	.001	
1.36	1.35	1.33	1.32	1.31	1.30	1.29	1.28	1.27	1.26	.25	**24**
1.83	1.78	1.73	1.70	1.67	1.64	1.61	1.57	1.55	1.53	.10	
2.18	2.11	2.03	1.98	1.94	1.89	1.84	1.79	1.76	1.73	.05	
2.54	2.44	2.33	2.27	2.21	2.15	2.08	2.01	1.97	1.94	.025	
3.03	2.89	2.74	2.66	2.58	2.49	2.40	2.31	2.26	2.21	.01	
3.42	3.25	3.06	2.97	2.87	2.77	2.66	2.55	2.49	2.43	.005	
4.39	4.14	3.87	3.74	3.59	3.45	3.29	3.14	3.05	2.97	.001	

Percentage points of the F distribution (df$_2$ between 25 and 30)

df$_2$	α	1	2	3	4	5	6	7	8	9	10
						df$_1$					
25	.25	1.39	1.47	1.46	1.44	1.42	1.41	1.40	1.39	1.38	1.37
	.10	2.92	2.53	2.32	2.18	2.09	2.02	1.97	1.93	1.89	1.87
	.05	4.24	3.39	2.99	2.76	2.60	2.49	2.40	2.34	2.28	2.24
	.025	5.69	4.29	3.69	3.35	3.13	2.97	2.85	2.75	2.68	2.61
	.01	7.77	5.57	4.68	4.18	3.85	3.63	3.46	3.32	3.22	3.13
	.005	9.48	6.60	5.46	4.84	4.43	4.15	3.94	3.78	3.64	3.54
	.001	13.88	9.22	7.45	6.49	5.89	5.46	5.15	4.91	4.71	4.56
26	.25	1.38	1.46	1.45	1.44	1.42	1.41	1.39	1.38	1.37	1.37
	.10	2.91	2.52	2.31	2.17	2.08	2.01	1.96	1.92	1.88	1.86
	.05	4.23	3.37	2.98	2.74	2.59	2.47	2.39	2.32	2.27	2.22
	.025	5.66	4.27	3.67	3.33	3.10	2.94	2.82	2.73	2.65	2.59
	.01	7.72	5.53	4.64	4.14	3.82	3.59	3.42	3.29	3.18	3.09
	.005	9.41	6.54	5.41	4.79	4.38	4.10	3.89	3.73	3.60	3.49
	.001	13.74	9.12	7.36	6.41	5.80	5.38	5.07	4.83	4.64	4.48
27	.25	1.38	1.46	1.45	1.43	1.42	1.40	1.39	1.38	1.37	1.36
	.10	2.90	2.51	2.30	2.17	2.07	2.00	1.95	1.91	1.87	1.85
	.05	4.21	3.35	2.96	2.73	2.57	2.46	2.37	2.31	2.25	2.20
	.025	5.63	4.24	3.65	3.31	3.08	2.92	2.80	2.71	2.63	2.57
	.01	7.68	5.49	4.60	4.11	3.78	3.56	3.39	3.26	3.15	3.06
	.005	9.34	6.49	5.36	4.74	4.34	4.06	3.85	3.69	3.56	3.45
	.001	13.61	9.02	7.27	6.33	5.73	5.31	5.00	4.76	4.57	4.41
28	.25	1.38	1.46	1.45	1.43	1.41	1.40	1.39	1.38	1.37	1.36
	.10	2.89	2.50	2.29	2.16	2.06	2.00	1.94	1.90	1.87	1.84
	.05	4.20	3.34	2.95	2.71	2.56	2.45	2.36	2.29	2.24	2.19
	.025	5.61	4.22	3.63	3.29	3.06	2.90	2.78	2.69	2.61	2.55
	.01	7.64	5.45	4.57	4.07	3.75	3.53	3.36	3.23	3.12	3.03
	.005	9.28	6.44	5.32	4.70	4.30	4.02	3.81	3.65	3.52	3.41
	.001	13.50	8.93	7.19	6.25	5.66	5.24	4.93	4.69	4.50	4.35
29	.25	1.38	1.45	1.45	1.43	1.41	1.40	1.38	1.37	1.36	1.35
	.10	2.89	2.50	2.28	2.15	2.06	1.99	1.93	1.89	1.86	1.83
	.05	4.18	3.33	2.93	2.70	2.55	2.43	2.35	2.28	2.22	2.18
	.025	5.59	4.20	3.61	3.27	3.04	2.88	2.76	2.67	2.59	2.53
	.01	7.60	5.42	4.54	4.04	3.73	3.50	3.33	3.20	3.09	3.00
	.005	9.23	6.40	5.28	4.66	4.26	3.98	3.77	3.61	3.48	3.38
	.001	13.39	8.85	7.12	6.19	5.59	5.18	4.87	4.64	4.45	4.29
30	.25	1.38	1.45	1.44	1.42	1.41	1.39	1.38	1.37	1.36	1.35
	.10	2.88	2.49	2.28	2.14	2.05	1.98	1.93	1.88	1.85	1.82
	.05	4.17	3.32	2.92	2.69	2.53	2.42	2.33	2.27	2.21	2.16
	.025	5.57	4.18	3.59	3.25	3.03	2.87	2.75	2.65	2.57	2.51
	.01	7.56	5.39	4.51	4.02	3.70	3.47	3.30	3.17	3.07	2.98
	.005	9.18	6.35	5.24	4.62	4.23	3.95	3.74	3.58	3.45	3.34
	.001	13.29	8.77	7.05	6.12	5.53	5.12	4.82	4.58	4.39	4.24

(*continued*)

12	15	20	24	30	40	60	120	240	inf.	α	df₂
										df₁	
1.36	1.34	1.33	1.32	1.31	1.29	1.28	1.27	1.26	1.25	.25	**25**
1.82	1.77	1.72	1.69	1.66	1.63	1.59	1.56	1.54	1.52	.10	
2.16	2.09	2.01	1.96	1.92	1.87	1.82	1.77	1.74	1.71	.05	
2.51	2.41	2.30	2.24	2.18	2.12	2.05	1.98	1.94	1.91	.025	
2.99	2.85	2.70	2.62	2.54	2.45	2.36	2.27	2.22	2.17	.01	
3.37	3.20	3.01	2.92	2.82	2.72	2.61	2.50	2.44	2.38	.005	
4.31	4.06	3.79	3.66	3.52	3.37	3.22	3.06	2.98	2.89	.001	
1.35	1.34	1.32	1.31	1.30	1.29	1.28	1.26	1.26	1.25	.25	**26**
1.81	1.76	1.71	1.68	1.65	1.61	1.58	1.54	1.52	1.50	.10	
2.15	2.07	1.99	1.95	1.90	1.85	1.80	1.75	1.72	1.69	.05	
2.49	2.39	2.28	2.22	2.16	2.09	2.03	1.95	1.92	1.88	.025	
2.96	2.81	2.66	2.58	2.50	2.42	2.33	2.23	2.18	2.13	.01	
3.33	3.15	2.97	2.87	2.77	2.67	2.56	2.45	2.39	2.33	.005	
4.24	3.99	3.72	3.59	3.44	3.30	3.15	2.99	2.90	2.82	.001	
1.35	1.33	1.32	1.31	1.30	1.28	1.27	1.26	1.25	1.24	.25	**27**
1.80	1.75	1.70	1.67	1.64	1.60	1.57	1.53	1.51	1.49	.10	
2.13	2.06	1.97	1.93	1.88	1.84	1.79	1.73	1.70	1.67	.05	
2.47	2.36	2.25	2.19	2.13	2.07	2.00	1.93	1.89	1.85	.025	
2.93	2.78	2.63	2.55	2.47	2.38	2.29	2.20	2.15	2.10	.01	
3.28	3.11	2.93	2.83	2.73	2.63	2.52	2.41	2.35	2.29	.005	
4.17	3.92	3.66	3.52	3.38	3.23	3.08	2.92	2.84	2.75	.001	
1.34	1.33	1.31	1.30	1.29	1.28	1.27	1.25	1.24	1.24	.25	**28**
1.79	1.74	1.69	1.66	1.63	1.59	1.56	1.52	1.50	1.48	.10	
2.12	2.04	1.96	1.91	1.87	1.82	1.77	1.71	1.68	1.65	.05	
2.45	2.34	2.23	2.17	2.11	2.05	1.98	1.91	1.87	1.83	.025	
2.90	2.75	2.60	2.52	2.44	2.35	2.26	2.17	2.12	2.06	.01	
3.25	3.07	2.89	2.79	2.69	2.59	2.48	2.37	2.31	2.25	.005	
4.11	3.86	3.60	3.46	3.32	3.18	3.02	2.86	2.78	2.69	.001	
1.34	1.32	1.31	1.30	1.29	1.27	1.26	1.25	1.24	1.23	.25	**29**
1.78	1.73	1.68	1.65	1.62	1.58	1.55	1.51	1.49	1.47	.10	
2.10	2.03	1.94	1.90	1.85	1.81	1.75	1.70	1.67	1.64	.05	
2.43	2.32	2.21	2.15	2.09	2.03	1.96	1.89	1.85	1.81	.025	
2.87	2.73	2.57	2.49	2.41	2.33	2.23	2.14	2.09	2.03	.01	
3.21	3.04	2.86	2.76	2.66	2.56	2.45	2.33	2.27	2.21	.005	
4.05	3.80	3.54	3.41	3.27	3.12	2.97	2.81	2.73	2.64	.001	
1.34	1.32	1.30	1.29	1.28	1.27	1.26	1.24	1.23	1.23	.25	**30**
1.77	1.72	1.67	1.64	1.61	1.57	1.54	1.50	1.48	1.46	.10	
2.09	2.01	1.93	1.89	1.84	1.79	1.74	1.68	1.65	1.62	.05	
2.41	2.31	2.20	2.14	2.07	2.01	1.94	1.87	1.83	1.79	.025	
2.84	2.70	2.55	2.47	2.39	2.30	2.21	2.11	2.06	2.01	.01	
3.18	3.01	2.82	2.73	2.63	2.52	2.42	2.30	2.24	2.18	.005	
4.00	3.75	3.49	3.36	3.22	3.07	2.92	2.76	2.68	2.59	.001	

Percentage points of the F distribution (df$_2$ at least 40)

df$_2$	α	1	2	3	4	5	6	7	8	9	10
40	.25	1.36	1.44	1.42	1.40	1.39	1.37	1.36	1.35	1.34	1.33
	.10	2.84	2.44	2.23	2.09	2.00	1.93	1.87	1.83	1.79	1.76
	.05	4.08	3.23	2.84	2.61	2.45	2.34	2.25	2.18	2.12	2.08
	.025	5.42	4.05	3.46	3.13	2.90	2.74	2.62	2.53	2.45	2.39
	.01	7.31	5.18	4.31	3.83	3.51	3.29	3.12	2.99	2.89	2.80
	.005	8.83	6.07	4.98	4.37	3.99	3.71	3.51	3.35	3.22	3.12
	.001	12.61	8.25	6.59	5.70	5.13	4.73	4.44	4.21	4.02	3.87
60	.25	1.35	1.42	1.41	1.38	1.37	1.35	1.33	1.32	1.31	1.30
	.10	2.79	2.39	2.18	2.04	1.95	1.87	1.82	1.77	1.74	1.71
	.05	4.00	3.15	2.76	2.53	2.37	2.25	2.17	2.10	2.04	1.99
	.025	5.29	3.93	3.34	3.01	2.79	2.63	2.51	2.41	2.33	2.27
	.01	7.08	4.98	4.13	3.65	3.34	3.12	2.95	2.82	2.72	2.63
	.005	8.49	5.79	4.73	4.14	3.76	3.49	3.29	3.13	3.01	2.90
	.001	11.97	7.77	6.17	5.31	4.76	4.37	4.09	3.86	3.69	3.54
90	.25	1.34	1.41	1.39	1.37	1.35	1.33	1.32	1.31	1.30	1.29
	.10	2.76	2.36	2.15	2.01	1.91	1.84	1.78	1.74	1.70	1.67
	.05	3.95	3.10	2.71	2.47	2.32	2.20	2.11	2.04	1.99	1.94
	.025	5.20	3.84	3.26	2.93	2.71	2.55	2.43	2.34	2.26	2.19
	.01	6.93	4.85	4.01	3.53	3.23	3.01	2.84	2.72	2.61	2.52
	.005	8.28	5.62	4.57	3.99	3.62	3.35	3.15	3.00	2.87	2.77
	.001	11.57	7.47	5.91	5.06	4.53	4.15	3.87	3.65	3.48	3.34
120	.25	1.34	1.40	1.39	1.37	1.35	1.33	1.31	1.30	1.29	1.28
	.10	2.75	2.35	2.13	1.99	1.90	1.82	1.77	1.72	1.68	1.65
	.05	3.92	3.07	2.68	2.45	2.29	2.18	2.09	2.02	1.96	1.91
	.025	5.15	3.80	3.23	2.89	2.67	2.52	2.39	2.30	2.22	2.16
	.01	6.85	4.79	3.95	3.48	3.17	2.96	2.79	2.66	2.56	2.47
	.005	8.18	5.54	4.50	3.92	3.55	3.28	3.09	2.93	2.81	2.71
	.001	11.38	7.32	5.78	4.95	4.42	4.04	3.77	3.55	3.38	3.24
240	.25	1.33	1.39	1.38	1.36	1.34	1.32	1.30	1.29	1.27	1.27
	.10	2.73	2.32	2.10	1.97	1.87	1.80	1.74	1.70	1.65	1.63
	.05	3.88	3.03	2.64	2.41	2.25	2.14	2.04	1.98	1.92	1.87
	.025	5.09	3.75	3.17	2.84	2.62	2.46	2.34	2.25	2.17	2.10
	.01	6.74	4.69	3.86	3.40	3.09	2.88	2.71	2.59	2.48	2.40
	.005	8.03	5.42	4.38	3.82	3.45	3.19	2.99	2.84	2.71	2.61
	.001	11.10	7.11	5.60	4.78	4.25	3.89	3.62	3.41	3.24	3.09
inf.	.25	1.32	1.39	1.37	1.35	1.33	1.31	1.29	1.28	1.27	1.25
	.10	2.71	2.30	2.08	1.94	1.85	1.77	1.72	1.67	1.63	1.60
	.05	3.84	3.00	2.60	2.37	2.21	2.10	2.01	1.94	1.88	1.83
	.025	5.02	3.69	3.12	2.79	2.57	2.41	2.29	2.19	2.11	2.05
	.01	6.63	4.61	3.78	3.32	3.02	2.80	2.64	2.51	2.41	2.32
	.005	7.88	5.30	4.28	3.72	3.35	3.09	2.90	2.74	2.62	2.52
	.001	10.83	6.91	5.42	4.62	4.10	3.74	3.47	3.27	3.10	2.96

(continued)

				df_1								
12	15	20	24	30	40	60	120	240	inf.	α	df_2
1.31	1.30	1.28	1.26	1.25	1.24	1.22	1.21	1.20	1.19	.25	**40**
1.71	1.66	1.61	1.57	1.54	1.51	1.47	1.42	1.40	1.38	.10	
2.00	1.92	1.84	1.79	1.74	1.69	1.64	1.58	1.54	1.51	.05	
2.29	2.18	2.07	2.01	1.94	1.88	1.80	1.72	1.68	1.64	.025	
2.66	2.52	2.37	2.29	2.20	2.11	2.02	1.92	1.86	1.80	.01	
2.95	2.78	2.60	2.50	2.40	2.30	2.18	2.06	2.00	1.93	.005	
3.64	3.40	3.14	3.01	2.87	2.73	2.57	2.41	2.32	2.23	.001	
1.29	1.27	1.25	1.24	1.22	1.21	1.19	1.17	1.16	1.15	.25	**60**
1.66	1.60	1.54	1.51	1.48	1.44	1.40	1.35	1.32	1.29	.10	
1.92	1.84	1.75	1.70	1.65	1.59	1.53	1.47	1.43	1.39	.05	
2.17	2.06	1.94	1.88	1.82	1.74	1.67	1.58	1.53	1.48	.025	
2.50	2.35	2.20	2.12	2.03	1.94	1.84	1.73	1.67	1.60	.01	
2.74	2.57	2.39	2.29	2.19	2.08	1.96	1.83	1.76	1.69	.005	
3.32	3.08	2.83	2.69	2.55	2.41	2.25	2.08	1.99	1.89	.001	
1.27	1.25	1.23	1.22	1.20	1.19	1.17	1.15	1.13	1.12	.25	**90**
1.62	1.56	1.50	1.47	1.43	1.39	1.35	1.29	1.26	1.23	.10	
1.86	1.78	1.69	1.64	1.59	1.53	1.46	1.39	1.35	1.30	.05	
2.09	1.98	1.86	1.80	1.73	1.66	1.58	1.48	1.43	1.37	.025	
2.39	2.24	2.09	2.00	1.92	1.82	1.72	1.60	1.53	1.46	.01	
2.61	2.44	2.25	2.15	2.05	1.94	1.82	1.68	1.61	1.52	.005	
3.11	2.88	2.63	2.50	2.36	2.21	2.05	1.87	1.77	1.66	.001	
1.26	1.24	1.22	1.21	1.19	1.18	1.16	1.13	1.12	1.10	.25	**120**
1.60	1.55	1.48	1.45	1.41	1.37	1.32	1.26	1.23	1.19	10	
1.83	1.75	1.66	1.61	1.55	1.50	1.43	1.35	1.31	1.25	.05	
2.05	1.94	1.82	1.76	1.69	1.61	1.53	1.43	1.38	1.31	.025	
2.34	2.19	2.03	1.95	1.86	1.76	1.66	1.53	1.46	1.38	.01	
2.54	2.37	2.19	2.09	1.98	1.87	1.75	1.61	1.52	1.43	.005	
3.02	2.78	2.53	2.40	2.26	2.11	1.95	1.77	1.66	1.54	.001	
1.25	1.23	1.21	1.19	1.18	1.16	1.14	1.11	1.09	1.07	.25	**240**
1.57	1.52	1.45	1.42	1.38	1.33	1.28	1.22	1.18	1.13	10	
1.79	1.71	1.61	1.56	1.51	1.44	1.37	1.29	1.24	1.17	.05	
2.00	1.89	1.77	1.70	1.63	1.55	1.46	1.35	1.29	1.21	.025	
2.26	2.11	1.96	1.87	1.78	1.68	1.57	1.43	1.35	1.25	.01	
2.45	2.28	2.09	1.99	1.89	1.77	1.64	1.49	1.40	1.28	.005	
2.88	2.65	2.40	2.26	2.12	1.97	1.80	1.61	1.49	1.35	.001	
1.24	1.22	1.19	1.18	1.16	1.14	1.12	1.08	1.06	1.00	.25	**inf.**
1.55	1.49	1.42	1.38	1.34	1.30	1.24	1.17	1.12	1.00	10	
1.75	1.67	1.57	1.52	1.46	1.39	1.32	1.22	1.15	1.00	.05	
1.94	1.83	1.71	1.64	1.57	1.48	1.39	1.27	1.19	1.00	.025	
2.18	2.04	1.88	1.79	1.70	1.59	1.47	1.32	1.22	1.00	.01	
2.36	2.19	2.00	1.90	1.79	1.67	1.53	1.36	1.25	1.00	.005	
2.74	2.51	2.27	2.13	1.99	1.84	1.66	1.45	1.31	1.00	.001	

Source: Computed by P. J. Hildebrand.

Answers to Selected Exercise Problems

CHAPTER 1

1.1 (c) Bimodal, skewed to the right

(d)

```
0 | 256
1 | 2455689
2 | 023578
3 | 023566789
4 | 001233445567789
5 | 0167899
6 | 0235579
7 | 158
8 | 024
9 | 06
```

1.3 (b) [20, 25)   (c) It is close to symmetric.   (d) (ii)

1.5 (c) 170–190

1.7 (a) 8   (b) 3.65

1.9 $\bar{x} = 20.05$, $s = 5.40$

1.11 (a) 68   (b) 76.5   (c) 81   (d) 83.5   (e) 87   (f) 90

1.13 (a) 81   (b) $Q_1 = 77$, $Q_3 = 85$ (c) 81.32   (d) 6.8   (e) 94%   (f) sample range=30, IQR=8
 (g) 92   (i) There are no outliers.

1.15 (a) skewed to the left   (b) The median is greater than the mean.   (c) 87

1.17

(a)   8 | 4.2 6.9 7.7 7.9 8.1 8.8

              9 | 0.2 1.3 1.4 1.7 3.5 3.9 4.3 5.3

(b)   84 | 2

             85 |

             86 | 9

             87 | 79

             88 | 18

             89 |

             90 | 2

             91 | 347

             92 |

             93 | 59

             94 | 3

             95 | 3

(c) The second plot is more efficient.

1.19

(a)   0 | 7

              1 | 026689

              2 | 0234799

              3 | 2479

              4 | 38

(b) $\bar{x} = 25.25$, $s = 11.15$ (c) median= 23.5, $Q_1 = 17$, $Q_3 = 33$ (e) There are no outliers.

1.21

(a) median= 24, $Q_1 = 22$, $Q_3 = 26$ (b) $\bar{x} = 24.09$, $s = 2.22$ (c) 69%

(d) sample range= 10, IQR= 4   (e) 27

1.23

(a) >score=c(69,84,52,93,81,74,89,85,88,63,87,64,67,72,74,55,82,91,68,77)   >hist(score)

(b) >stem(score)   (c) >mean(score)   >sd(score)   (d) > quantile(score, c(0.5, 0.25, 0.75))

(e) > quantile(score, 0.8)   (f) > IQR(score)   (g) > boxplot(score)

1.25 (a) −0.363   (b) −0.652   (c) −0.26

1.27 (a) > men=c(22880,28548,43056,55380,56888,55016,52260)

           > women=c(21008,26416,37084,45604,44252,45188,41600)

           >genders=c("Men","Women")

           > boxplot(men, women, names=genders, main="Annual Salary")

     (b) >plot(men, women, main="Annual Salary")

     (c) >cov(men, women)

     (d) >cor(men, women)

1.29 (b) 0.87

CHAPTER 2

2.1
(a) $E \cup F = \{A\spadesuit, 2\spadesuit, 3\spadesuit, \ldots, 10\spadesuit, J\spadesuit, J\heartsuit, J\diamondsuit, J\clubsuit, Q\spadesuit, Q\heartsuit, Q\diamondsuit, Q\clubsuit, K\spadesuit, K\heartsuit, K\diamondsuit, K\clubsuit\}$
(b) $E \cap F = \{J\spadesuit, Q\spadesuit, K\spadesuit\}$
(c) $F^c = \{A\spadesuit, A\heartsuit, A\diamondsuit, A\clubsuit, 2\spadesuit, 2\heartsuit, 2\diamondsuit, 2\clubsuit, \ldots, 10\spadesuit, 10\heartsuit, 10\diamondsuit, 10\clubsuit\}$
(d) E and F are not disjoint because $E \cap F$ is not empty.
(e) F and G are disjoint.
2.3 $A \cup B = \{1, 3, 5, 6, 7, 9\}$, $A \cap B = \{3, 9\}$
2.5 (a) 56   (b) 210   (c) 504   (d) 840
2.7 216
2.9 20
2.11 2,568,960
2.13 (a) 720   (b) 120
2.15 (a) 11/42   (b) 9
2.17 0.96
2.19 0.82
2.21 (a) 0.54   (b) 0.26
2.23 0.387
2.25 (a) 0.85   (b) 0.15   (c) 0.5
2.27 0.5
2.29 2/3
2.31 (a) 0.3   (b) 0.8   (c) 0.43
2.33 (a) 1/3   (b) 0.3   (c) 2/3   (d) 0.2   (e) 0.6
2.35 0.261
2.37 (a) 0.36   (b) 0.44   (c) 0.818
2.39 (a) 4/13   (b) independent
2.41 (a) 0.18   (b) 0.2   (c) not independent
2.43 (a) 0.2   (b) 0.8   (c) 0   (d) 0.6
2.45 (a) 0.88   (b) They cannot be mutually exclusive.
2.47 (a) 0.260   (b) They cannot be mutually exclusive.
2.49 0.58
2.51 (a) 0.31   (b) 0.54   (c) not independent
2.53 0.9216
2.55 (a) 0.6   (b) not enough information   (c) 0.3   (d) 1/4   (e) 0.3047   (f) 1/12   (g) not enough information   (h) 0.8
2.57 (a) 0.735   (b) 0.321

2.59 (a) 0.0073   (b) 0.6981
2.61 (a) 0.052   (b) 0.462
2.63 (a) 0.34   (b) 0.993

CHAPTER 3

3.1

x	$P(X=x)$
1	1/4
2	1/2
3	1/4

3.3

(a)

x	2	3	4	5	6	7	8
$P(X=x)$	1/16	1/8	3/16	1/4	3/16	1/8	1/16

(b)

$$F(x) = \begin{cases} 0 & \text{if } x < 2 \\ 1/16 & \text{if } 2 \le x < 3 \\ 3/16 & \text{if } 3 \le x < 4 \\ 3/8 & \text{if } 4 \le x < 5 \\ 5/8 & \text{if } 5 \le x < 6 \\ 13/16 & \text{if } 6 \le x < 7 \\ 15/16 & \text{if } 7 \le x < 8 \\ 1 & \text{if } x \ge 8 \end{cases}$$

(c) 3/4

3.5 (a) 7   (b) 28

3.7 (a) 0.2   (b) 0.5   (c) 1.9   (d) 1.29

3.9

(a)

x	−2	−1	0	1	2
$f(x)$	1/8	1/4	1/4	1/4	1/8

(b) $E(X) = 0$, $\sigma = 1.2247$ (c) 1   (d) 3/4

3.11 (a) 9/16   (b) 5/8   (c) 5   (d) 2.5

3.13 0.5786

3.15 (a) 0.0065   (b) 0.2447   (c) 0.9435   (d) 0.0210   (e) 0.4044   (f) 0.9435

3.17 15/32

3.19 (a) 0.8491   (b) 0.8315

3.21 (a)

$$F(x) = \begin{cases} 0, & x < 0 \\ \dfrac{16}{25}, & 0 \le x < 1 \\ \dfrac{24}{25}, & 1 \le x < 2 \\ 1, & x \ge 2 \end{cases}$$

(b) 8/25   (c) 2/5   (d) 8/25

3.23 1/216

3.25 (a) 0.9185   (b) 0.5   (c) 0.0064

3.27 3

3.29 (a) 0.3223   (b) 0.9737

3.31 (a) 0.257   (b) 0.3586

3.33 0.367

3.35 (a) 0.2924   (b) 0.2816

3.37 (a) 0.1954   (b) 0.2381   (c) 4

3.39 (a) 0.234   (b) 0.385

3.41 (a) 0.224   (b) 0.950

3.43 (a) 0.2149   (b) 0.6512

3.45 (a) 0.6065   (b) 0.3033   (c) 0.0902   (d) 0.0898

3.47 0.875

3.49 He should stop after two attempts. The assumption is that each theft is independent and the probability that the thefts are solved does not change.

3.51 (a) 0.073   (b) $E(X) = 4$, $Var(X) = 10/3$ (c) 0.0804   (d) 6

3.53 (a) >dgeom(2, 0.4)   (b) >pgeom(2, 0.4)

3.55 (a) $\frac{b^2 r}{(r+b)^3}$ (b) $\frac{b^3}{(r+b)^3}$

3.57 0.96

3.59 0.0889

3.61 (a) 0.0983   (b) 0.9421

CHAPTER 4

4.1 (a) 0.0922   (b) 0.3456   (c) 1,288 hours

4.3 (a) 0.6988   (b) 0.3835   (c) 0.0907

4.5 (a) 1/4 (b) 0   (c) 2   (d) 5/8   (e) 15/32   (f) $\sqrt{2}$

4.7 (a) 1.53

(b)

$$F(x) = \begin{cases} 0, & x \leq 0 \\ \dfrac{x^3}{3}, & 0 < x < 1 \\ \dfrac{x^2}{2} - \dfrac{1}{6}, & 1 \leq x < \sqrt{21}/3 \\ 1, & x \geq \sqrt{21}/3 \end{cases}$$

(c) 55/96 (d) 1.15

4.9 (a) 1

(b)

$$F(x) = \begin{cases} 0, & x \leq 1 \\ (x-1)/5, & 1 < x < 6 \\ 1, & x \geq 6 \end{cases}$$

(c) 3/5   (d) 2/5   (e) 2/5   (f) 5

4.11 (a) $F(x) = 1 - \frac{2}{x}$, $x \geq 2$ (b) 4   (c) 1/3   (d) 2/3   (e) 1/6

4.13 (a) $f(x) = 2e^{-2x}, x \geq 0$ (b) 0.6321   (c) 0.0025   (d) 0.3466
(e) Exponential distribution with mean 0.5

4.15 (a) $f(x) = 2/x^2, x \geq 2$ (b) 8/3   (c) 1/3   (d) 4/15

4.17 (a) 1/4
(b)

$$F(x) = \begin{cases} 0, & x < 2 \\ \dfrac{x^2 - 2x}{8}, & 2 \leq x \leq 4 \\ 1, & x > 4 \end{cases}$$

(c) 19/6   (d) 5/8   (e) $1 + \sqrt{3}$

4.19 (a)

$$F(x) = \begin{cases} 0, & x \leq 0 \\ x^2/4, & 0 < x < 2 \\ 1, & x \geq 2 \end{cases}$$

(b) $P(0.5 < X < 1.5) = 1/2, P(X > 1.5) = 7/16$ (c) $\sqrt{2}$ (d) 4/3   (e) 2/9

4.21 (a) 1/8

(b)

$$F(x) = \begin{cases} 0, & x \leq 0 \\ \dfrac{x^2}{16}, & 0 < x < 4 \\ 1, & x \geq 4 \end{cases}$$

(c) 1/2   (d) 63/64   (e) 3/4   (f) 8/3   (g) 8/9

4.23 (a) 1/2
(b)

$$F(x) = \begin{cases} 0, & x \leq 2 \\ \dfrac{(x-2)^2}{4}, & 2 < x < 4 \\ 1, & x \geq 4 \end{cases}$$

(c) 3/16   (d) $E(X) = 10/3, Var(X) = 2/9$ (e) 3.897

4.25

(a) 4

(b)

$$F(x) = 1 - \frac{4}{x}, \ x \geq 4$$

(c) 8

4.27 (a) 0.9370   (b) 0.0668   (c) 0.0934   (d) 0.8638   (e) 0.6990   (f) 0   (g) 0.1479

4.29 (a) 0.28   (b) −0.62   (c) 1.17   (d) −1.07   (e) 1.28

4.31 (a) 0.3853   (b) −0.4125

4.33 (a) 0.7257   (b) 95.25

4.35 (a) 0.0912   (b) 119.2

4.37 (a) >qnorm(0.3, 527, 120)   (b) >1-pnorm(700, 527, 120)

    (c) >pnorm(600, 527, 120)-pnorm(400, 527, 120)   (d) >pnorm(650, 527, 120)

4.39 (a) 0.1587   (b) 110   (c) 0.0912

4.41 (a) $\mu = 1, \ \sigma^2 = 16$ (b) 0.3413

4.43 (a) 32nd percentile   (b) 0.8185

4.45 79

4.47 0.0624

4.49 (a) 0.8018   (b) 0.0567   (c) 0.8330   (d) 0.2266   (e) 0.4669

4.51 (a) 0.1957   (b) 5 feet 7 inches   (c) 0.0997

4.53 0.1977

4.55 (a) 0.0688   (b) 0.0564

4.57 (a) 0.3085   (b) 0.0459

4.59 (a) 0.3297   (b) 0.2699   (c) $E(X) = 5, Var(X) = 25$

4.61 (a) $f(x) = 4x^3$ when $0 < x < 1$

(b)

$$F(x) = \begin{cases} 0, & x \leq 0 \\ x^4, & 0 < x < 1 \\ 1, & x \geq 1 \end{cases}$$

(c) 0.3125   (d) $E(X) = 4/5, Var(X) = 2/75$ (e) 0.8409

4.63 (a) $f(x) = 12(x - 2x^2 + x^3)$ when $0 < x < 1$

(b)

$$F(x) = \begin{cases} 0, & x \leq 0 \\ 3x^4 - 8x^3 + 6x^2, & 0 < x < 1 \\ 1, & x \geq 1 \end{cases}$$

(c) 2/5   (d) 0.8192   (e) 0.2094

4.65 (a) 1 (b) 0.75 (c) 3 (d) NaN (error)

CHAPTER 5

5.1 (a) 0.08   (b) 0.28   (c) 0.30   (d) 0.40

(e)

$$f_1(x_1) = \begin{cases} 0.42, & x_1 = 5 \\ 0.30, & x_1 = 10 \\ 0.28, & x_1 = 20 \\ 0 & \text{otherwise} \end{cases} \qquad f_2(x_2) = \begin{cases} 0.37, & x_2 = 0 \\ 0.23, & x_2 = 10 \\ 0.40, & x_2 = 20 \\ 0, & \text{otherwise} \end{cases}$$

(f) 0.58   (g) 0.60

5.3 (a) $f_1(x) = 1/3$, $x = 1, 2, 3$; $f_2(y) = 1/4$, $y = 1, 2, 3, 4$

5.5

$$\text{(a) } f_1(x_1) = \begin{cases} 0.22, & x_1 = 0 \\ 0.36, & x_1 = 1 \\ 0.42, & x_1 = 2 \\ 0, & \text{otherwise} \end{cases} \qquad \text{(b) } f_2(x_2) = \begin{cases} 0.3, & x_2 = 0 \\ 0.4, & x_2 = 1 \\ 0.3, & x_2 = 2 \\ 0, & \text{otherwise} \end{cases}$$

(c) $f_1(0|0) = 0.2$, $f_1(1|0) = 0.3$, $f_1(2|0) = 0.5$, $f_1(0|1) = 0.2$, $f_1(1|1) = 0.425$, $f_1(2|1) = 0.375$, $f_1(0|2) = 4/15$, $f_1(1|2) = 1/3$, $f_1(2|2) = 0.4$

(d) $f_2(0|0) = 3/11$, $f_2(1|0) = f_2(2|0) = 4/11$, $f_2(0|1) = 0.25$, $f_2(1|1) = 17/36$, $f_2(2|1) = 5/18$, $f_2(0|2) = f_2(1|2) = 5/14$, $f_2(2|2) = 2/7$ (e) 0.58 (f) 0.7 (g) 0.4 (h) 0.23

5.7

(a)

x	Event	$P(X = x)$
2	{(1,1)}	1/36
3	{(1,2), (2,1)}	1/18
4	{(1,3), (2,2), (3,1)}	1/12
5	{(1,4), (2,3), (3,2), (4,1)}	1/9
6	{(1,5), (2,4), (3,3), (4,2), (5,1)}	5/36
7	{(1,6), (2,5), (3,4), (4,3), (5,2), (6,1)}	1/6
8	{(2,6), (3,5), (4,4), (5,3), (6,2)}	5/36
9	{(3,6), (4,5), (5,4), (6,3)}	1/9
10	{(4,6), (5,5), (6,4)}	1/12
11	{(5,6), (6,5)}	1/18
12	{(6,6)}	1/36

(b)

$$F(x) = \begin{cases} 0 & \text{if } x < 2 \\ 1/36 & \text{if } 2 \le x < 3 \\ 1/12 & \text{if } 3 \le x < 4 \\ 1/6 & \text{if } 4 \le x < 5 \\ 5/18 & \text{if } 5 \le x < 6 \\ 5/12 & \text{if } 6 \le x < 7 \\ 7/12 & \text{if } 7 \le x < 8 \\ 13/18 & \text{if } 8 \le x < 9 \\ 5/6 & \text{if } 9 \le x < 10 \\ 11/12 & \text{if } 10 \le x < 11 \\ 35/36 & \text{if } 11 \le x < 12 \\ 1 & \text{if } x \ge 12 \end{cases}$$

(c) 1/3

5.9 (a) $f(x, y) = 6e^{-2x-3y}$, $x > 0$, $y > 0$ (b) independent
(c) $f_1(x|y) = 2e^{-2x}$, $x > 0$ (d) $e^{-9} - e^{-13}$ (e) X has exponential distribution with mean 1/2 and Y has exponential distribution with mean 1/3.   (f) $Var(X) = 1/4$, $Var(Y) = 1/9$
(g) 0

5.11 (a) 1

(b)

$$f_1(x) = \frac{2(x+1)}{3}, 0 < x < 1; f_2(y) = \frac{4y+1}{3}, 0 < x < 1$$

(c)

$$f_1(x|y) = \frac{2(x+1)}{4y+1}, 0 < x < 1, 0 < y < 1$$

(d) 5/12

5.13

(a)

$$f_1(x) = \frac{12x^2 + 6x}{7}, \ 0 < x < 1 \qquad f_2(y) = \frac{2y+4}{21}, 0 < y < 3$$

(b)

$$F_1(x) = \begin{cases} 0 & x \le 0 \\ \dfrac{4x^2 + 3x^2}{7}, & 0 < x < 1 \\ 1, & x \ge 1 \end{cases} \qquad F_2(y) = \begin{cases} 0 & y \le 0 \\ \dfrac{y^2 + 4y}{21}, & 0 < y < 3 \\ 1, & y \ge 3 \end{cases}$$

(c) 4/7   (d) 17/84   (e) 5/21   (f) not independent

(g)

$$f_2(y|x) = \frac{2(3x+y)}{9(2x+1)}, 0 < y < 3, \ 0 < x < 1$$

(h) 1/3

5.15 (a) 0.4

(b)

$$f_1(x) = \begin{cases} 0.3, & x = 1 \\ 0.45, & x = 2 \\ 0.25, & x = 3 \end{cases} \qquad f_2(y) = \begin{cases} 0.6, & y = 1 \\ 0.4, & y = 2 \end{cases}$$

(c)

$$f_1(x|y=2) = \begin{cases} 0.5, & x=1 \\ 0.375, & x=2 \\ 0.125, & x=3 \end{cases}$$

(d) not independent   (e) $E(X) = 1.95$, $E(Y) = 1.4$ (f) $Var(X) = 0.5475$, $Var(Y) = 0.24$
(g) -0.3586

5.17 (a) 0.6

(b)

$$f_1(x) = \begin{cases} 0.3 & x=1 \\ 0.25, & x=2 \\ 0.45, & x=3 \end{cases} \qquad f_2(y) = \begin{cases} 0.35 & y=1 \\ 0.3, & y=2 \\ 0.35, & y=3 \end{cases}$$

(c) 1/6   (d) not independent   (e) $E(X) = 2.15$, $E(Y) = 2$
(f) $Var(X) = 0.7275$, $Var(X) = 0.7$ (g) -0.14

5.19 (a) $f_1(x) = 3x^2$, $0 < x < 1$; $f_2(y) = 6y(1-y)$, $0 < y < 1$
(b) $f_1(x|y) = 1/(1-y)$, $0 < y < x < 1$ (c) 3/4   (d) $E(X) = 3/4$, $E(Y) = 1/2$
(e) $Var(X) = 3/80$, $Var(Y) = 1/20$ (f) 1/40   (g) 0.5774

5.21 (a) $f_1(x) = 3(x^2 + 1)/4$, $0 < x < 1$; $f_2(y) = (9y^2 + 1)/4$, $0 < y < 1$

(b)

$$F_1(x) = \begin{cases} 0 & x \le 0 \\ \dfrac{x^3 + 3x^2}{4}, & 0 < x < 1 \\ 1, & x \ge 1 \end{cases} \qquad F_2(y) = \begin{cases} 0 & y \le 0 \\ \dfrac{3y^3 + y}{4}, & 0 < y < 1 \\ 1, & y \ge 1 \end{cases}$$

(c)

$$f_1(x|y) = \frac{3(x^2 + 3y^2)}{9y^2 + 1}, 0 < x < 1, 0 < y < 1$$

$$f_2(y|x) = \frac{x^2 + 3y^2}{x^2 + 1}, 0 < x < 1, 0 < y < 1$$

(d) $E(X) = 9/16$, $E(Y) = 11/16$ (e) $Var(X) = 107/1280$, $Var(Y) = 233/3840$
(f) $-3/256$ (g) 1/15   (h) 7/10

5.23 (a) 3/4   (b) 6

5.25 $E(U) = 9, Var(U) = 106$

5.27 (a) $3\mu_1 - 2\mu_2$ (b) $13\sigma^2$ (c) $Y \sim N(3\mu_1 - 2\mu_2, 13\sigma^2)$

5.29 (a) 0.923   (b) 0.9213   (c) 0.6421   (d) 0.9586

CHAPTER 6

6.1 6

6.3 0.8664

6.5 (a) 0.8664   (b) 0.5638

6.7 1.5

6.9 0.8413

6.11 0.0234

6.13 0.9772

6.15 (a) $N(70, 1.875^2)$ (b) 0.0038

6.17 0.3385

6.19 (a) 1/3   (b) 1/18   (c) 0.1344

6.21 (a) 5/12   (b) 0.0708   (c) 0.9342

6.23 (a) 0.2858   (b) 0.0346

6.25 (a) 5.0753   (b) >3*qchisq(0.95, 14)/14

6.27 (a) >pf(1.7298, 7, 11)   (b) >pf(1.7298, 7, 11)-pf(0.3726, 7, 11)

CHAPTER 7

7.1 (a) 9.667   (b) 2.16   (c) 0.882

7.3 (a)   (c)   (b)   (d)

7.5 (a)   (b)   (b)

7.7 $H_0: \mu \geq 400$ versus $H_1: \mu < 400$

7.9 (a) $X \sim \text{Bin}(5, p)$ (b) $H_0: p = 0.5$ versus $H_1: p > 0.5$ (c) 0.0312   (d) 0.5905

7.11 (a) $H_0: p \leq 0.3$ versus $H_1: p > 0.3$ (b) type I error probability = 0.0173

(c) type I error probability = 0.05   (d) type II error probability = 0.7869   (e) 0.05

7.13 (a) $H_0: \mu \leq 200,000$ versus $H_1: \mu > 200,000$ (b) 0.0228   (c) 0.7475

CHAPTER 8

8.1 (a) 0.0066   (b) 0.0066   (c) 0.0132

8.3 (a) 0.1   (b) 0.1   (c) 0.2

8.5 $n = 68$

8.7 $n = 189$

8.9 (a) The sample size needs to be 1/4 of the original one.   (b) The sample size needs to be quadrupled.

8.11 (a) $\beta(62) = 0.5319$ (b) $n = 60$

8.13 (a) The required sample size is 34. The sample size of 40 is enough.   (b) Reject H_0
     (c) $1 - \beta(8) = 0.995$

8.15 (a) (i) For $H_0: \mu = \mu_0$ versus $H_1: \mu > \mu_0$, the p value decreases as n increases.
(ii) For $H_0: \mu = \mu_0$ versus $H_1: \mu < \mu_0$, the p value decreases as n increases.
(iii) For $H_0: \mu = \mu_0$ versus $H_1: \mu \neq \mu_0$, the p value decreases as n increases.
(b) (i) For $H_0: \mu = \mu_0$ versus $H_1: \mu > \mu_0$, $\beta(\mu')$ decreases as n increases.
(ii) For $H_0: \mu = \mu_0$ versus $H_1: \mu < \mu_0$, $\beta(\mu')$ decreases as n increases.
(iii) For $H_0: \mu = \mu_0$ versus $H_1: \mu \neq \mu_0$, $\beta(\mu')$ decreases as n increases.

8.17 (a) $(68554, 72046)$ (b) $(68835, 71765)$

8.19 (a) $(0.65, 1.95)$ (b) The error is significantly different from zero at level $\alpha = 0.01$.

8.21 (a) Reject H_0 (b) $(219.3, 240.7)$ (c) Do not reject H_0 (d) The decision made in part (c) is correct.

8.23 (a) >x=c(29.4, 24.2, 25.6, 23.6, 23.0, 22.4, 27.4, 27.8) >t.test(x)
     (b) >t.test(x, mu=25, alt="greater")

8.25 Do not reject H_0

8.27 (a) $(4.45, 4.75)$ (b) Do not reject H_0. p-value $= 0.1754$ (c) Reject H_0. p-value $= 0.0877$

8.29 (a) (1709.8, 2243.6)   (b) Do not reject H_0.

8.31 >prop.test(42, 500, correct=F)

8.33 (a) 97 (b) 62

8.35 (a) $(0.201, 0.359)$ (b) 216

8.37 (a) 6147   (b) $(0.0641, 0.1209)$

8.39 (a) Do not reject H_0 (b) 0.3178

8.41 (a) $H_0 : p = 0.5, H_1 : p > 0.5$ (b) Reject H_0 (c) 0.0245

8.43 > prop.test(802, 1180, p=0.7, correct=F)

8.45 (a) Reject H_0 (b) p-value $= 0.066$. Since $0.066 < \alpha = 0.1$, we have strong evidence that the rate is different from 0.5.

8.47 $(59.5, 112.8)$

8.49 (a) $(26.7, 98.3)$ (b) $(28.31, 67.81)$

8.51 (a) Do not reject H_0 (b) $(74.9, 170.1)$

8.53 (a) Reject H_0 (b) Reject H_0 (c) 0.0034

CHAPTER 9

9.1 (a) 3.1875   (b) The test statistic is $Z = (\bar{X} - \bar{Y})/1.785$. We evaluate this using the data and compare it with the z critical value.   (c) $z \leq -1.645$

9.3 There is sufficient evidence that mean drying time of the paints made by company A is shorter. p-value $= 0.007$

9.5 $(-19.9, 41.9)$

9.7 (a) The data support the researcher's expectation.   (b) Do not reject H_0.   (c) Type I error

9.9 Do not reject H_0

9.11 (a) $(1.04, 18,76)$ (b) Reject H_0 (c) Do not reject H_0

9.13 Reject H_0

9.15 Reject H_0

9.17 (a) $(-1.76, 7.56)$ (b) Do not reject H_0

9.19 (a) > x=c(17,22,20,14,15,21,24)   > y=c(15,19,18,13,15,19,23) >t.test(x, y, paired=T)
(b) >t.test(x, y, alt="greater", paired=T)

9.21 Reject H_0

9.23 Reject H_0

9.25 Reject H_0

9.27 $(-0.053, -0.005)$

9.29 (a) $(-0.252, -0.116)$ (b) Reject H_0 (c) p-value≈ 0

9.31 Reject H_0

9.33 (a) Do not reject H_0 (b) $(-0.95, 16.55)$ (c) Reject H_0 (d) Do not reject H_0

9.35 (a) Do not reject H_0 (b) $(0.225, 4.182)$ (c) The assumption of equal variance seems justified.

9.37 Reject H_0. p-value$= 0.0093$

Index